한국의 도시재생
Urban Regeneration in Korea

도시를 살리는 다섯 가지 해법

한국의 도시재생
도시를 살리는 다섯 가지 해법

한국도시설계학회

Urban Regeneration in Korea
Five Ways to Revitalize Old Cities

도서출판 대가

추천의 글

한국도시설계학회에서는 최근 우리 일상에서 삶의 질을 높여온 도시설계 사례들을 모아 배경과 의미를 기록한 '한국 도시설계의 사례'를 출간하게 되었습니다.

도시설계 분야 전문가 13분이 우리 도시 환경을 개선하고 시민과 주민의 행복한 일상에 기여한 사례 중 전국에서 20개 프로젝트를 발굴하였습니다. 이후 선정된 프로젝트들의 진행 배경, 계획 수립과 추진 과정, 물리적 환경의 설계 특성, 그리고 조성 후 평가 등의 내용을 고찰하며 그 의미와 가치를 찾아 기록했습니다.

이러한 노력은 앞으로 우리 도시들이 어떠한 문제를 갖고 성장(쇠퇴)하고 있으며, 이러한 문제들은 어떠한 사회적 가치와 공공의 노력으로 대처하고, 이를 통해 우리의 생활환경은 어떻게 더 개선될 수 있는가의 질문에 관한 경험을 공유하고 지혜를 찾아가는 과정이 될 것입니다.

우리 학회는 이러한 노력을 지속적으로 해나가면서 시민의 삶 개선에 기여할 수 있도록 노력하고자 합니다. 이 책이 이러한 노력과 공론화의 좋은 시작점이기를 기대하며, 앞으로 한국 도시설계에 따뜻한 관심과 응원을 부탁드립니다.

한국도시설계학회 회장

이제선

Recommendation Letter

The Urban Design Institute of Korea has recently published 'Urban Design Cases in Korea', which records the background and meaning of contemporary urban design cases that have significantly improved the quality of urban life in Korea.

Thirteen urban design professional and scholars both in the academia and field of urban design have discovered 20 projects across the country that have improved our urban environment and contributed to the daily life of people. In the meantime, the background of these urban design projects, planning and implementation process, design characteristics of the physical environment, and post-construction evaluation were discussed to identify and record their meaning and value.

These efforts are made to share experience and find wisdom on the question of with what problems our cities have grown (or declined) with in the future, what social values and public efforts to deal with these problems, and how our living environment can be further improved through this.

I hope that this book will be a good starting point for and valuable addition to these efforts and public debate, and I would like to ask for your warm interest and support for urban design in Korea.

Lee, Jea-sun

President, Urban Design Institute of Korea

서문

　이 책은 최근 한국에서 진행된 대표적인 도시설계 사례들을 모아, 물리적인 도시환경을 개선하려는 도시설계의 가치와 실행과정의 이슈들을 이해하고 공론화하기 위해 출간되었습니다. 이러한 노력은 응용학문으로서 도시설계가 이론을 넘어 공공과 민간을 포함한 다수의 이해집단의 협력작업에서 보다 넓은 참여를 통해 합리적인 문제해결과 그 해답을 찾아나갈 수 있다는 믿음에서 비롯됩니다.

　집필진은 2000년대 전후부터 현재까지 한국 도시들에서 추진되어온 20개의 대표적인 도시설계 사례들을 추출하고, 그 사례들의 목적과 당위성, 참여자와 계획수립 과정, 도시설계적 특성과 가치 등을 기록했습니다. 특히 이 책은 영문으로도 함께 준비되어, 대외적으로 한국 도시설계를 공유하여 객관적인 평가의 기회를 마련했다는 점에서 큰 의미를 두고 있습니다.

　이 책은 도시설계 사례의 기본 특성과 추진 주체에 따라 5개의 챕터로 분류하였다. 챕터 1에서는 동네의 변화와 헤리티지의 보존 노력, 챕터 2는 공공주도의 재생과 재개발사업, 챕터 3은 대규모 유휴지와 수변공간의 재생 노력, 챕터 4는 도시의 대중교통체계와 교통환승거점의 조성, 챕터 5는 일상환경에서의 생활문화거점 조성의 노력으로서 도시설계를 다루었습니다.

　이 책의 제한된 지면상 더 많은 사례들을 다루지 못해 아쉬우며, 앞으로도 다양하고 좋은 사례들을 소개할 수 있기를 기대합니다.

Introduction

The aim of the publication are two folds: publicize the issues and value of urban design in Korea and worldwide; improve our understanding of urban design process by discussing urban design cases in Korea.

These efforts stem from the firm belief that urban design, as an applied study, often goes beyond theory, and can find out answers through a wider participatory process and rational problem solving in the cooperative work of multiple interest groups of both public and private.

The authors identifies 20 representative urban design cases that have improved the quality of urban environment in Korean cities from around the 2000s to the present, and recorded the purpose and background of the projects, the participants and the planning process, and the characteristics of urban design. In particular, this book is also prepared in English, and it is of great significance in that it provides an opportunity to share Korean urban design with readers and professionals overseas.

This book is divided into five parts based on the characteristics of urban design cases: Chapter 1 addresses the issue of neighborhood change and heritage preservation efforts; Chapter 2 examines public sector-led urban regeneration and redevelopment projects and Chapter 3 focusing on efforts to regenerate large-scale factories' brownfields and riverfront areas; Chapter 4 deals with urban public transportation systems and transit hubs and urban design efforts to create living and cultural hubs in everyday environments in Chapter 5.

It is our hope that more urban design experiences and lessons in Korea find you in the next publication in the near future.

목차

Contents

Chapter 1

동네의 변화와 로컬 헤리티지

Neighborhood Change and Local Heritage

도시 쇠퇴와 커뮤니티의 해체는 인구감소, 산업구조의 변화, 주거환경의 노후화, 도시 외곽 개발로 인한 구도심의 쇠퇴 등 다양한 이유에 기인한다. 지방도시의 쇠퇴와 수도권으로의 집중 현상은 1990년대부터 이미 시작되었으며, 쇠퇴하는 지역을 경제적, 사회적, 물리적, 환경적으로 활성화하기 위한 노력이 2000년대부터 본격화되기 시작하였다. 특히 부산은 2000년대 중후반부터 마을 미술 프로젝트를 시작하였으며, 이는 감천문화마을을 만들어내는 시발점이 되었다. 이 장에서 소개되는 네 가지 사례들은 지역의 역사와 가치를 보전하면서 사회적, 경제적으로 마을을 살리기 위한 지난 10여 년간 진행된 노력의 결과이다.

Urban decline and community disintegration are due to various reasons, such as population decline, changes in the industrial structure, deterioration of the residential environment, and the decline of the old downtown due to the development of the outskirts of the city. The decline of local cities and concentration in the metropolitan area already started in the 1990s, and efforts to revitalize the declining areas economically, socially, physically and environmentally began in earnest from the 2000s. In particular, Busan started a village art project in the mid to late 2000s, and this became the starting point for creating the Gamcheon Culture Village. The four cases introduced in this chapter are the result.

충주 원도심 청년의 D.I.Y. 실험

임시점유와 사회실험을 적용한 충주시 도시재생 일반사업 활성화계획

최순섭 [한국교통대학교 건축학부 교수]

들어가며

이 사업은 국토교통부 주관하에 2016년에 선정된 충주시 도시재생 일반사업 근린재생사업(중심시가지형)으로 2016년부터 2020년까지 진행되었다. 대상지는 충주시 원도심인 성내충인동(젊음의 거리의 성서동과 관아골인 성내동)이며 총 200여 억원이 투입되었다. 충주시 주관부서는 도시재생과이며 건축공간연구원에서 도시재생지원기구 역할을 담당했다. 활성화계획에서는 (구)성내동 우체국 유휴시설을 리모델링한 문화창업재생센터, 유휴지 활용 청년창업플랫폼 2곳, 빈 점포 리모델링 20개소가 주요사업으로 제안되었다. 이 글은 2015년에서 2017년까지 충주 원도심에서 활동하던 청년들의 D.I.Y. 작업과 창업과정을 활성화계획에 적용한 내용을 중심으로 서술되었다. 단, 2017년 말 청년플랫폼사업이 백지화된 이후 청년들은 독립적으로 여행자 플랫폼인 BTLM1960을 충주 목행동에 구축하여 운영하고 있다.

도시재생 일반사업 진행과정 : 명확한 역할 구분

2013년 말 '도시재생특별법'이 제정 및 시행된 후, 국토교통부가 주관하여 2014년 13개

D.I.Y. Experiment in the Old Town Area, Chungju

Revitalization Plan of Chungju-si Urban Regeneration General Project with Social Experiment and Temporary Occupation

Soonsub Choi (Professor, Korea National University of Transportation, Dept of Architecture)

Introduction

This project is a Neighborhood Type (Urban Center District Type) Chungju-si Urban Regeneration General Project that was selected in 2016 by the Ministry of Land, Infrastructure and Transport, and conducted from 2016 to 2020. The target site was Seongnae Chungin-dong (Seongseo-dong including the Street of Youth and Seongnae-dong called Gwanagol) in the original city center of Chungju, and approximately 20 billion won was invested. The department in charge of Chungju-si was the Urban Regeneration Division, and AURI was involved as the urban regeneration assistant agency. The revitalization plan proposed the Cultural Start-up Regeneration Center remodeling the unused facilities of the former Seongnae-dong post office, two Youth Start-up Platforms using idle public land, and 20 Empty Store Remodeling Projects as major projects. This article focuses on the contribution of the D.I.Y. works of local young people, who were desperately working in the original downtown Chungju from 2015 to 2017, to the Revitalization Plan. However, after the Youth Start-up Platform Project was suddenly scrapped at the end of 2017, young people left the organization,

소의 선도지역 대상지 지정을 시작으로 2016년에는 일반사업 23개소, 2017년부터 도시재생 뉴딜사업으로 25개소, 2018년 99개소, 2019년 98개소, 2020년 70개소, 2021년 52개소가 선정되어 전국적으로 430여 개가 넘는 도시재생사업을 진행하였다. 국무조정실 해양수산부의 어촌뉴딜, 농림축산식품부의 농촌재생뉴딜과 관계부처 합동으로 진행하는 생활형 SOC 복합화사업 등 지역 활성화를 목표하는 '재생' 관련 사업을 포함하면 지난 10여 년간 '도시재생'은 주요 도시정책의 관리 개념과 방향이었다.

특히, 도시재생 뉴딜사업부터 이전 일반사업의 활성화계획 수립과는 선정과정에서 차별화된다. 즉, 전략계획 수립 후 활성화계획을 제안하여 여러 차례 현장 및 관문심사를 거쳐 최종 선정되던 방식 대신 사업계획서 제출 후 선정으로 간소화되었고, 선정 후 본격적인 활성화계획을 수립하게 된다. 또한 세부사업 유형과 예시, 가이드라인을 국토교통부에서 사전 제공하여 효율적 계획수립이 가능해졌다.

충주시는 2013년 일반사업에 지원했으나 탈락하였고 이후 2016년 재도전을 위해서 2015년부터 전략계획과 활성화계획 수립 자문단을 운영하였다. 자문단은 충주시청 관계자, 지역의 도시·건축 전공 교수들로 구성되었다. 특히, 전략계획과 활성화계획을 수립할 용역업체 선정을 위해 통계 데이터 분석 기반 전문 도시 및 건축계획사무소와 지역활성화사업 전문회사의 협업이 권장되었다. 선정된 두 업체의 컨소시엄이 조사와 활동을 토대로 계획을 제안하면 자문단 컨설팅을 거쳐 수차례 수정 보완하는 방식으로 2016년 도시재생 일반사업이 준비되었다.

총괄 코디네이터의 역할 : 지역자원 조사 및 발굴

2015년 도시재생 일반사업을 지원할 당시 충주에는 도시재생 관련 센터 조직은 전무했다. 지원센터를 구성하려면 조례를 개정하고 직원을 채용해야 하나 선정이 확정되지 않은 이유로 처리가 계속 늦춰졌다. 그러나 도시재생 가이드라인에 의해 센터 설립은 의무였으며, 관문심

and established BTLM 1960 (traveler's platform) in Mokhaeng-dong, Chungju.

The Process of Urban Regeneration General Project : Classifying Clear Roles

After the 'Urban Regeneration Special Act' was enacted and implemented at the end of 2013 and 13 Lead Projects in 2014 were initiated, 23 General Projects in 2016, 25 Urban Regeneration New Deal Projects in 2017, 99 locations in 2019, 98 locations in 2019, 70 locations in 2020, and 52 locations in 2021 were selected, which means that more than 430 Urban Regeneration Projects across the country are underway, while being supervised by the Ministry of Land, Infrastructure and Transport. Moreover, given the 'Fishing Village New Deal Project' of the Ministry of Oceans and Fisheries, the 'Rural Regeneration New Deal Project' of the Ministry of Agriculture Affairs, Food and Rural, and the 'SOC Complex Project' jointly conducted by the related ministries, 'Urban Regeneration' has become a main concept and stream of continuous urban policy and management

The main difference in the process of establishing and selecting a revitalization plan between the previous General Regeneration Project and the Urban Regeneration New Deal Project is that the process of the revitalization plan being proposed after the strategic plan was established, and selected through several on-site and gate examinations, was simplified to the submission of only a business proposal plan, and then the establishment of a full revitalization plan after selection. In addition, the Ministry of Land, Infrastructure and Transport provided the types, examples, and guidelines of sub-projects, which could be referred to during the planning to provide

사과정에서도 센터 유무가 중요한 평가항목이었다. 따라서 행정절차를 기다리기 보다는 총괄 코디네이터가 센터 역할을 우선 담당하였다.

무엇보다도 회의와 교육공간이 필요했는데, 예산이 책정되지 못한 상황에서 활성화계획에 포함된 유휴 우체국 부지의 식당건물(문화창업재생센터 사업부지)이 확보되었다. 회의실 및 사무실을 위해 우체국에서 쓰지 않던 의자와 책상을 기부 받았으며, 도시재생사업 책자와 타 지역의 도시재생 자료를 열람할 수 있는 아카이브 공간을 위해 배식공간이 활용되었다. 간판 은 다른 장소의 도시재생 대학 및 활동에서도 사용되도록 저렴한 스탠드형으로 제작되었다. 필요 기자재도 기부 받아 예산집행과는 별도로 임시 센터 공간이 최종 구축되었다.

둘째, 지역 연구·조사 및 활동이 진행되었다. 충주지역 자산을 기록하는 일이 대학 수업과 연계하면서 광범위한 조사가 이루어졌다. 특히, 사업대상지인 성내동(관아골)은 일제시대 본 전통이라 불리던 지역으로 사라진 역사자산이 많았다. 이를 기록하기 위하여 건축학과 학생 들이 참여한 인터뷰 기록인 '관아골 스토리북'이 출간되었다. 또한 활성화계획 수립에 필요한 빈집 조사가 시급하여 'WISET' 사업과 연계하여 대학 연구실과 충주지역 고등학교 학생들이 참여하여 성내·충인동 공가와 공점포 조사 및 기록작업이 진행되었다.

셋째, 온라인 소통 채널이 운영되었다. 활성화계획 수립 및 주민협의체 회의 내용을 공유하 기 위해 총괄 코디네이터는 페이스북과 블로그를 직접 운영했다. 특히, 네이버 밴드는 용역업 체, 주민, 행정 등 이해관계자들과 실시간 소통 채널로 사용되었다.

그림 1. 충주시 도시재생사업 현장지원센터 / Chungju-si urban regeneration support center

more efficient management of those sub-projects.

The Urban Regeneration General Project of Chungju-si was firstly prepared and submitted in 2013, but after not being selected, the consultant advisory group revised the strategic plan and revitalization plan during 2015 to resubmit it in 2016. This advisory group consisted of several city officials and local urban and architectural professors. To select a project service company to proceed with the strategic plan and activation plan, the collaboration method between companies specializing in statistics and data-based urban and architectural planning, and in regional activation projects, was strongly recommended through the task orders and guidelines. After that, two plans were prepared in 2016 through the process of revising and supplementing them after frequent consultative meetings with the advisory group.

The Project Coordinator's Role:
Conducting Research, Documentations, and Activities for
Local Assets

When submitting the Chunju-si Revitalization Plan in 2015, the City of Chungju did not have any assistance organizations supporting urban regeneration projects. To establish the assistance agency center, employees could be hired only after revision of the ordinance, but it had been delayed continuously because of the conventional reason, 'not to be done at all if not confirmed'. However, according to the urban regeneration guidelines, it was mandatory to establish an assistance agency center, so during gate screening, the presence or absence of the center was also an important factor. Therefore, the project coordinator had to do something on his own initiative,

그림 2. 관아골 스토리북, 건축자산 조사 연계 수업, 성내·충인동 빈집 조사 활동 및 지도
Gwana-gol story book, class linked to architectural asset survey, and empty houses surveying and mapping activity

활성화계획의 주체 : 실행력 있는 주체의 발굴

충주시 도시재생사업에서 집중적으로 공략한 부분은 세부사업의 '실행주체 찾기'였다. 사업 후 지역에서 다양한 활동이 지속되기 위해서 단순한 관여자가 아닌 실행주체의 주도적 참여가 중요했다. 다행히도 지역조사 과정에서 활성화계획에 반영할 수 있는 다양한 잠재적 실행자들이 발굴되었다.

무엇보다도 충주에는 실행력이 강한 의식있는 청년조직이 존재하고 있었다. 관아골에서 푸드바이크 마켓을 운영하던 '너나들이(대표 김인혁)'가 대표적이었다. 푸드바이크는 야간에 운영하는 이동식 음식점이었다. 기존 일반노점상과의 차이는 유동인구가 많은 기존 상권에 의존하기 보다는, 외진 곳에서 독특한 아이템과 콘텐츠로 소비자를 모객한다는 점이었다. SNS를 통해 어느 장소에 어떤 이벤트와 음식 등을 팔 것인지 사전에 공지하고 실시간 그 모습을 공유하면서, 충주의 젊은 수요층과 여행객들을 성공적으로 끌어들였다. 특히, 방문객들은 야외에 돗자리를 깔거나 벤치나 계단에 앉아 음식과 음악을 자유롭게 즐길 수 있었다. 물론 지

instead of waiting for administrative procedures that might take a few months or longer.

First, physical space, like meeting and education rooms, was urgently needed. However, in a situation where the budget was insufficient, the only option was to rent the abandoned restaurant building on the site of the under-used post office, as the site of the Cultural Business Start-up Regeneration Center Project. Unused chairs and desks donated from the post office were arranged. In addition, the food distribution table was decorated as an archive space, so that visitors could access brochures and materials related to other urban regeneration projects. The signage of this temporary center was manufactured as an inexpensive stand, so that it could be moved and used for educational activities anywhere. Eventually, other equipment, like printers and computers, were also donated and installed to set up the temporary center space, regardless of budget execution.

Second, some local research and activities were immediately required. As Seongnae-dong (Gwana-gol) had been called 'Bonjuntong' during the Japanese colonial Period, there were once many historical assets, but most of them had disappeared. So, the project coordinator had to record the assets of the Chungju region, and conduct necessary investigations in connection with university classes. In addition, because it was urgent to identify vacant houses to establish the revitalization plan, their investigation and record in Seongnae Chungin-dong were conducted by university lab members and high school students in Chungju as the 'WISET' project, which was managed by the Korea Women in Science and Technology Support Center.

그림 3. 충주 너나들이 협동조합의 푸드바이크 마켓(2015) / Food bike market hosted by Neonadeuli(2015)

역 상인회의 승인과 협조 하에 푸드바이크는 운영되었고, 술과 음료는 주변 가게에서 구입하도록 하여 지역상권과의 갈등을 최소화였다.

이들은 '옥상도시' 프로젝트도 진행했다. 창업자금이 여의치 않은 청년들에게 옥상은 매력적인 유휴공간이었고, 건물주 동의를 얻어 버려진 옥상을 무상으로 빌릴 수 있었다. 영화상

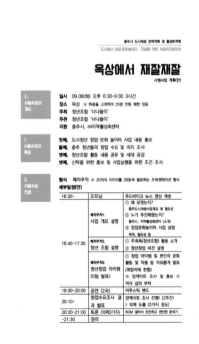

그림 4. 옥상도시 프로젝트 / Rooftop city project

Third, online communication channels were inevitable. The coordinator directly operated SNS such as Facebook and blogs to share the process and result of the frequent meetings, and managed Naver Band as a real-time communication channel with stakeholders, who included planning service companies, residents, and administration.

The implementation body in the Revitalization Plan : Excavating Participants as Players

The Chungju-si Urban Regeneration Project was mostly focused on finding the implementation groups of each sub-project. Considering the continuity of voluntary activities after the end of the project, a strong implementation entity was supposed to be formed as a group of active players, not passive participants. During the regional survey, the coordinator could fortunately meet various potential implementers to be referred to the establishment of the revitalization plan.

First, the youth organization had strong executive power in Chungju. 'Neonadeuli' (leader: Kim In-hyuk), which regularly ran the 'Food Bike Market' in Gwanagol, was a representative example. Food Bike was a kind of mobile restaurant that opened only at night. The main difference between street vendors was that it attracted and brought new consumers with unique items and contents in an under-used area, rather than relying on busy streets. The market drew young consumers and travelers by sharing information on what events and food would be provided in advance through SNS. Visitors could enjoy food and music while lying on mats outdoors or sitting on benches

영, 전시, 토크 콘서트가 가능한 아지트와 같은 루프탑 공간을 D.I.Y.로 직접 만들었다. 이와 함께 지현동에서는 공가를 직접 개조하여 4분의 1의 비용으로 카페를 만들기도 했다. 이처럼 충주 원도심에서는 청년조직의 절실한 창업사례들이 만들어지고 있었다.

이 외에도 본인 가게를 수백 만원에 창업한 경험으로 창업 아이템 선정부터 인테리어 공사까지 함께 진행하면서 브랜드와 운영 노하우를 공유한 '루트 66' 청년 사장님도 있었고, DJ, 밴드, 연극 등 다양한 예술문화 주체들이 함께 활동하고 있었다. 이들이 도시재생사업의 실행주체로 참여하여 창업과 활동을 지속하고, 이를 통해 새로운 예비 주체들이 다시 유입되는 과정과 시스템을 만드는 것이 충주의 도시재생 활성화계획의 핵심이었다.

활성화계획으로 적용 : 청년조합 활동의 동기와 의도

지역 청년들의 활동 근거나 동기를 종합하여 다음과 같이 압축할 수 있었다.

첫째, 확장된 수요 창출이 절실했다. 충주에서 지역 일자리는 한정되고 축소되어 지속적으로 사람들이 타지로 떠나고 있었다. 남아 있거나 남아 있을 수 있는 일자리를 만들기 위해서는 소비수요를 확장해야 했다. 푸드바이크가 대표사례였다. 인접한 성서동 젊음의 거리에 집중된 일반 F&B 콘텐츠와 차별화된 전략으로 관아골에서 새로운 수요층을 만들었다. 조용하고 자유로운 분위기, 가성비 있는 음식과 다양한 선택이 가능하다는 점에서 방문객들의 호응을 이끌어냈다.

충주 성서동 게스트하우스 1호점은 외국인 유학생 중심 여행자 수요를 만들기 위한 것이었다. 외국인 유학생들에게 한국의 로컬 문화를 가성비 있게 경험하게 할 장소로서, 관아 투어, 조정 체험 등 충주시 운영 무료체험 프로그램과 연계하여 패키지 여행상품을 기획했다. 또한 자체적으로 티셔츠 만들기, 옥탑 콘서트, 바비큐 파티를 운영하면서 3만원이 넘지 않는 숙박비 포함 여행 패키지로 유학생 방문객을 유치했다.

지현동에 만든 바이크 플랫폼도 여행객 유치를 위한 작업이었다. 충주 여행객들의 불편한

and stairs while relaxing. Of course, the Food Bike Market was run under the approval and cooperation of the nearby local merchants' association, and alcohol and beverages were recommended to be purchased at stores nearby to minimize conflicts and create synergies with local commercial operators.

'Neonadeuli' also carried out the 'Rooftop City' project. Many unused rooftops were attractive spaces for the young founder with poor start-up funds. The abandoned rooftops could be rented free with the consent of the building owner. They built one attractive rooftop space like a hideout, where movie screenings, exhibitions, and talk concerts could be held through the D.I.Y. process for low cost. In addition, diverse desperate start-ups were taking place in the old town center, such as the collaborative work of changing the empty old house in Jihyeon-dong into a café, which cost a quarter.

In addition, another young founder of 'Root 66' shared the restaurant brand and operation with new start-ups by selecting items and conducting construction together, based on the experience of starting his own business with only millions of won, a very small budget. Also, the project coordinator could meet various artists, such as DJ, music bands, and dramatic actors and craft people They were eventually recruited as actual implementers of the Urban Regeneration Project, and the revitalization plan was ultimately established to create a system in which new potential candidates flowed into this region over time, and were inspired and led by these strong players' activities.

그림 5. 충주 너나들이 1호 게스트하우스와 여행 프로그램 / 1st guesthouse operated by Neonadeuli

이동문제를 해결하기 위해 충주시 지원을 받아 지현동 동사무소 공터에 자전거 대여소가 설치되었다. 한정된 예산에서 자전거를 구입하는 데 대부분의 비용을 지출하게 되면서, 청년들은 컨테이너를 직접 가져와 D.I.Y.로 공간을 만들었다. 이후 너나들이에서는 홍보를 위해 '골목장'이라는 행사를 어떠한 지원 없이 자체적으로 진행하였다.

둘째, 지역 청년들에게는 조직이 필요했다. 예를 들어, 푸드바이크 마켓은 조합원 각자 능력을 모아 협업하여 만들고 운영하는 것이었다. 팀으로서 외부업체를 쓰지 않고 조리, 목공, 공사, 홍보, 이벤트 운영 등을 품앗이로 진행하여 비용을 최소화하였다. 조직은 청년들에게 많은 것을 실행할 역량과 기회의 원천이었다. 따라서 청년조합에서 기획한 옥상도시 콘서트 및 영화제, 계단 영화제와 같은 행사는 조직화와 확장을 위한 의미있는 활동이었다.

특히, 참여형 공공사업에서 청년들에게 조직이 필요한 이유는 또 있었다. 좁은 지역사회에서 기존 상인회, 유관단체 등 토착조직의 눈치를 봐야 하는 상황이 많았다. 정작 필요한 장소

Application to the Revitalization Plan :
Learning from Youth Group Activities

The lessons from the field and reality learned from the youth's activities were as follows: First, they expanded demand. Local jobs were becoming limited and reduced, so local people were constantly leaving to secure jobs. So, the young generation desperately tried to expand consumption demands to create jobs, which could allow them to remain in their hometown. Food Bike Market was the representative example. New visit demand had been created in Gwanagol with differentiated strategy compared to the general F&B contents, which were concentrated on the adjacent streets of youths in Seongseo-dong. The advantages provided to visitors were the quiet and free atmosphere, cost-effective food, and various choices they could make.

The first guest house opened in Seongseo-dong was also intended to create a new group of visitors, that of foreign students across the country. It was a place where foreign students could experience real Korean local culture in a cost-effective manner. The guesthouse, linked to free experience programs run by the City of Chungju, such as the Gwana tour and boat racing, provided reasonable travel packages. Along with the connected programs, they could make their own T-shirts, and attend rooftop concerts, and barbecue parties, which program did not exceed 30,000 won in all travel costs, including accommodation.

The Bike Platform built in Jihyeon-dong was another project to attract new travelers. To solve the most inconvenient mobility issues for local travelers, it was installed in the empty place at Jihyeon Community Center with support from the City of Chungju. Because most of the limited budget was spent on purchasing bicycles, a used container was brought and remodeled via D.I.Y. construction into bicycle storage and a small

그림 6. 바이크 플랫폼과 골목장 / Bike platform and Golmok Jang

에서의 활동이 텃새와 견제 때문에 힘든 경우도 있었다. 그렇기 때문에 지역의 실행력 있는 청년들은 공공사업에 크게 관심이 두지 않았다. 결국, 새로운 조직화와 이들에게 힘을 실어주는 과정이 지역 쇄신에 필요하였다.

셋째, 가짜 실행자들의 감별이 필요했다. 청년조합 조합원으로서 주요 기준은 '절실함'과 '생존력'이었다. 특히, 'D.I.Y. Spirit'과 'Team Spirit'을 가지고 있는지가 중요했다. 예를 들어, 푸드바이크의 뼈대는 조합에서 함께 만들지만 결코 완성시켜주지는 않았다. 운영자가 나머지를 D.I.Y.로 완성해야 했는데 이것은 감별과정이었다. 힘든 작업을 싫어하는 사람은 버티지 못해 그만두었고, 완성한 사람들은 조합원이 될 수 있었다. '창업의 환상'이 아닌 '전쟁 같은 장사판'에서 지속적으로 살아남는 팀 만들기 과정이었다.

결국 청년들의 활동과 의도를 반영하고, 쇠퇴 지역상권을 쇄신하고 재생하기 위한 방법과 과정이 활성화계획에 적용되었다. 물론 기존 지역 주체들과의 상생과 공생방법도 중요한 주제였다.

start-up office. After the completion, a few events called 'Golmok Jang' (local alley market) were also held on their own, without any support and promotion from the public sector.

Second, young people in the region were eager to build their organization in specific ways. For example, the Food Bike was made and operated in collaboration with the sum of each member's ability. Without outsourcing, all of the processes (cooking, woodworking, construction, promotion, and event hosting) were carried out together as one team to minimize costs. They believed that the more their organization grew, the more diverse things they could do by themselves. Therefore, Rooftop City Concerts, Film Festivals, and Stair Film Festivals were the sort of activities to find and recruit new members.

There were also other reasons why young people required their organizations in public participatory projects, including the Urban Regeneration Project. In a narrow community of a local area, it was difficult to engage in new activities to change the contents of the area, because existing organizations that including merchant associations were very territorial. This was also the reason why they had not previously been interested in many public projects.

Third, it was necessary for them to differentiate between fake implementers and real executors. Above all, the main criterion to select members of the youth association was the will to be desperate and survive. In other words, whether they had 'D.I.Y. Spirit' and 'Team Spirit' was the most important criterion for evaluation. For example, the framework of Food Bike was made together, but never completed. The new operator should finish the rest of it by himself through D.I.Y., which was a kind of gateway to distinguish between a real executor and a fake founder. Those who completed it by themselves could become the union member with strong will, while those who were

활성화계획 내용 : 지역경제 생태계의 순차적 변화

충주시 도시재생사업 활성화계획의 목표는 첫째, 산발적으로 추진되고 있는 대상지 내 사업을 도시재생사업으로 통합할 것, 둘째는 충주시를 대표하는 청년 중심의 문화 창업 허브를 구축하는 것이었다. 성내·충인동에서는 2015년 관아골문화클러스터 사업, 청년몰 조성사업이 진행 중이었고, 반기문 꿈자람길 조성사업, 충주천 생태하천복원사업, 보행환경개선사업, 관아골 경관조성사업, 활력 있는 전통시장 육성사업, 예성로 전선 지중화사업이 이미 완료된 상태였다. 따라서 타 부서 사업과 연계하고 집중된 활성화사업으로 만드는 것이 첫 번째 목표였다. 그리고 사업 수요층을 명확히 하였는데, 어린이 동반 가족과 젊은 소비층, 여행객들이었다. 이들을 유치·수용하기 위한 상권 콘텐츠 변화를 위해서 청년창업 인큐베이터 공간과 프로그램을 구축·운영하고 이를 통해 성장한 창업자들이 빈 점포를 순차적으로 채우는 것이 두 번째 목표였다.

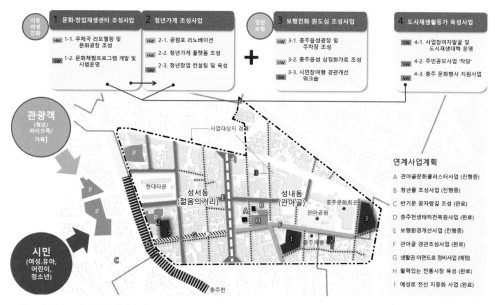

그림 7. 충주시 도시재생 일반사업 활성화계획과 연계사업
Chungju-si urban regeneration revitalization plan and linked projects

reluctant to do hard physical labor would not stand it, and would run away. This was also the procedure to create one spirit team to survive in the 'war-like market field', not the 'fantasy start-up field'.

In the end, the process made certain that they were able to devise new methods and processes to regenerate commercial districts in hopeless declining areas, so their participation in Chungju-si Urban Regeneration Project was above all that. Of course, it was also required to find ways to live together with the existing local groups.

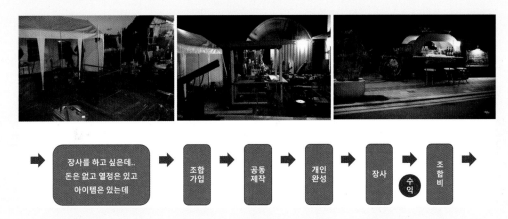

그림 8. 감별과정으로서 푸드바이크 제작과정 / The manufacturing process of food bike for distinguishment

Contents of the Revitalization Plan :
Sequentially Changing the Local Economic System

The Revitalization Plan's first goal of the Chungju-si Urban Regeneration Project was to integrate sporadic projects managed by another department into it, while the second was to establish a youth-centered cultural & start-up fosterage hub place. In Seongnae Chungin-dong, 'the Culture Cluster Project' and 'Youth Mall Construction

이에 따라 크게 네 가지의 핵심전략사업이 제안되었다. 각각은 물리적 계획사업과 프로그램 운영사업으로 나누어 제안되었는데 우선, 문화창업재생센터 조성사업은 유휴 우체국 건물과 부지에 창업과 문화 콘텐츠의 육성 및 거점공간을 만드는 것이었다. 두 번째는 문화창업재생센터에서 육성되고 실험한 후, 생존한 청년들에게 빈 점포 20곳에 창업을 지원하는 청년가게 조성사업(빈 점포 리노베이션 사업)이었다. 단, 우체국 건물 매입과 리모델링이 늦춰질 것을 예상하여 문화창업재생센터의 기능을 대신할 임시 점유시설을 성내동과 성서동 두 곳에 마련하는 청년가게(청년창업) 플랫폼사업을 추가하였다. 지역상권의 생태계를 바꿀 사업들이 먼저 진행된 후 증가할 방문객들을 위한 보행환경과 주차장 이용을 개선하는 사업이 세 번째 사업이었다. 더불어 전체 사업기간 동안 지역의 운영·관리에 필요한 주체를 육성하는 도시재생대학, 주민 공모사업과 주민참여 축제 및 행사 지원사업 등이 포함된 역량강화사업인 도시재생 활동가 육성사업도 제안되었다.

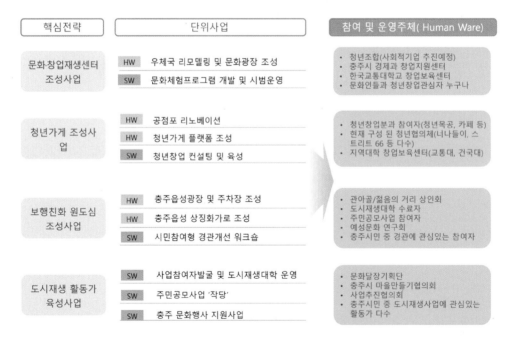

그림 9. 충주시 도시재생 일반사업 세부사업과 참여·운영주체
Chungju-si urban regeneration sub-projects and participating & operating groups

Project' were in progress in 2015, and the 'Banimun Dream Growing-street Project', 'Natural Stream Restoration Project of Chungjucheon', 'Improvement Project for Pedestrian Environment', 'Gwanagol Landscape Maintenance Project', and 'Electric line Underground Project of Yesung-ro Road' were already completed. So, the first goal was to make them organized and focused on 'regional revitalization' in connection with the diverse projects being undertaken by other administrative departments. This also clarified the newly extended consumers, who included families with children, young people, and local travelers. Therefore, the second goal was to establish and operate youth start-up incubator spaces and programs to change the commercial district's contents to attract and accommodate new visitors, and to fill empty stores with the incubated contents sequentially.

Accordingly, four major core strategic projects were proposed. Each was planned with both physical planning projects and program operation projects. First, the project to build the Cultural Business Start-up Regeneration Center at the unused post office buildings and sites was intended to foster a variety of start-ups and cultural players. The second was the creation of youth start-up stores (Empty Store Renovation Project), which supported 20 stores for young founders who eventually survived after being fostered and tested at the Cultural Business Start-up Regeneration Center. However, considering some delay in the purchase and remodeling of the post office buildings, the Youth Start-up Platform Creation Project was required and added later to temporarily provide the same function as the Cultural Business Start-up Regeneration Center in two areas, Seongnae-dong, and Seongseo-dong. The third project was to improve the environment for pedestrians and the parking lots for visitors after completing the prior two projects, which would change the local commercial ecosystem. In addition, the last project to foster local activists was also proposed, so that sustainable operation would

각 세부사업의 운영관리 주체는 명확하였다. 이미 지역에 존재하는 콘텐츠와 주체를 연계하고 활동을 확장할 방법으로 활성화계획이 수립되었기 때문이었다. 예를 들어, 푸드바이크, 옥상도시, 골목장 등의 이전 청년조직 활동을 참고하여 문화창업재생센터나 청년가게 플랫폼 조성사업, 빈 점포 리노베이션 사업이 도출되었다. 특히, 단순히 수동적으로 모집하는 관행적 사업과정을 지양하고 실행주체를 직접 섭외하고 감별하며 육성하는 청년창업플랫폼 조성사업과 같은 사회실험 프로젝트가 제안되었다.

활성화계획 수립 완료 후 진행(2016-2017) : 실행으로써의 주민참여

충주시 도시재생사업이 2016년 10월 관문심사를 통과하여 사업대상지로 선정된 후 도시재생대학이 운영되기 시작했다. 이 과정을 통해 '실행주체'를 발굴하고 여러 팀을 구성하여 사업의 실행주체로 안착하고자 하는 의도였다. 따라서 5개 분과(문화분과, 청년창업분과, 상인분과, 도시재생분과)로 나누었으며, 이후 목표가 확실한 맞춤형 교육과 활동으로 연결되었다. 특히, 특정 단체나 조직이 사업의결기구인 협의체를 독점하지 않도록 각 분과의 추천인으로 주민협의체가 구성되도록 규정을 만들었다.

각 분과에서 소규모 팀을 조직하여 제안한 사업계획서는 주민공모사업으로 연결되었으며, 그 첫 번째 활동이 '문화달장' 기획과 실행이었다. 이는 문화분과와 청년창업분과가 주축이 된 주민이 실행하는 소규모 축제였다. 그동안 지역행사에서 주민참여가 저조하고 똑같은 콘텐츠이며 외주업체가 독식하여 지역경제에 기여하지 못하는 문제에 대해 모두가 공감하여 제안되었다. 그러나 첫 행사인 만큼 예상치 못한 상황도 있었는데, 부족한 예산으로 행정상 집행의 제한사항이 많았다. 메인 이벤트의 위치를 정하는 과정에서는 상인회들의 이해관계가 엇갈렸고 청년조합의 푸드바이크도 제대로 운영을 못하는 등 잠재된 갈등이 드러나기도 하였다.

예비창업인을 육성할 임시 점유시설인 청년창업플랫폼 운영자로는 청년창업분과와 문화분과 구성원에서 팀을 이룬 청년협동조합이 최종 선정되었다. 이들은 각각 카페, 여행, 예술

be maintained, even after the end of the Urban Regeneration Project, which included Urban Regeneration University management, the resident offer & implementation project, and the residents' participation festival & event.

Each sub-project had its own clear operation and management groups. This was because the Revitalization Plan was established by linking and expanding the contents and subjects that were already active in this area. The Cultural Business Start-up Regeneration Center, Youth Start-ups Platform Creation project and Empty Store Renovation project were 'derived' from existing activities, such as Food Bikes, Rooftop Cities, and Golmok Jang. In particular, the social experimental project of the Youth Start-up Creation Platform was the core to avoid the conventional way of simply recruiting, but to identify and foster the strong implementing teams.

Implementation of the Revitalization Plan (2016-2017) : Involving as Strong Players

After Chungju-si Urban Regeneration Revitalization Plan passed the gate review in October 2016, and was then finally selected, the urban regeneration educational program (Urban Regeneration University) began first. Through this program, it was intended to discover the executive player, create strong teams, and settle them as the players of the sub-projects. Therefore, it was divided into five divisions (cultural division, youth start-up division, merchant division and urban regeneration division), and provided customized education & activities based on the interests and goals of each groups In particular, the members of the resident consultative group, the voting body, were also recommended equally from each division to prevent monopolization by

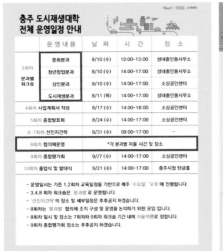
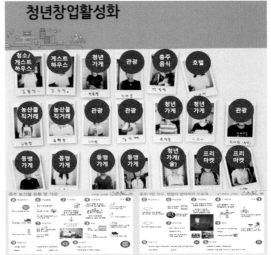

그림 10. 분과별 도시재생대학 일정과 참여자, 팀별 사업제안서 예시
Urban regeneration education schedule by division and participants, sample of team business proposal.

분야에서 '공동'창업하여 '공동'운영하고 새로운 예비창업창업자 육성 프로그램과 각종 행사를 진행할 주체였다. 공동창업이 강조된 것은 수익의 불균형으로 어느 한쪽 사업장이 들러리가 되는 것을 방지하기 위한 것이었다. 이는 F&B 소비는 활발하지만 공방이나 체험 프로그램은 수요가 적을 경우 나타나는 타 지역 청년창업 실패 사례를 참고한 결과였다. 따라서 품앗이를 통해 문제를 해결하고, 매출을 증대할 프로그램을 지속적으로 도모하는 경제활동의 공동 운명체 운영방식이 우선되었다. 이럴 경우 여행, 카페, 공방, 체험 프로그램이 하나의 운영 사업

그림 11. 문화달장 기획 및 행사 / Munhwa Daljang planning and hosting

any group.

The small teams that formed in each division started to devise business plans, continuing towards resident-offered projects step-by-step. As the result, the first project called Munhwa Daljang (Cultural Market) was proposed by the entire division together. This was the festival mainly led by the Cultural Division and the Youth Startup Division. They commonly shared the problems of local festivals not contributing to the local economy because of the same boring contents every year and outsourcing ways, which together made resident participation low in the meantime. However, unexpected situations frequently arose, as it was the first residents-led project. The budget was insufficient, and there were many limitations in its execution guideline set by the administration. In the process of determining the location of the main event, some conflicts were revealed, and there was a controversy with the existing merchant associations on the issue of whether the Food Bike Market would be run.

For another sub-project, one youth cooperative union was finally selected as the operator team of the Youth Start-up Platform on the temporary occupancy site. They were a new team from the youth start-up division and the culture division. As the co-founders in the fields of cafe, travel, and craft based on mutual benefits, they were supposed to conduct new start-up fostering programs and various events. The emphasis on joint start-ups ways was to prevent either side from playing second to others, causing an imbalance in profits. In other words, as witnessed commonly in similar projects, it referred to the fact that F&B consumption would normally make better profits than the workshop programs. As one group sharing a common destiny of economy and life, working with each other would be required to continuously devise new programs to solve problems together to increase the profit. For example, if travel services, cafe, workshops, and leisure programs are grouped together as one

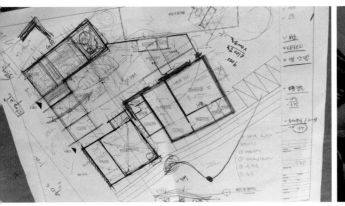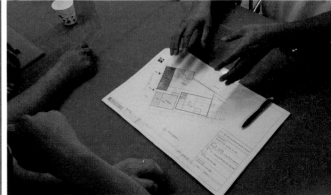

그림 12. 청년창업플랫폼 운영자 참여 설계 / Operators participation design of Youth start-up platform

체에 묶여 패키지 상품 기획과 공동 이벤트 실행이 용이해질 수 있었다.

그러나 청년창업플랫폼은 우여곡절 끝에 우체국 후면마당의 현장지원센터와 유휴 택배 오토바이 주차시설을 활용하여 1억 내외의 예산으로 조성되는 것으로 최종 변경되었다. 축소된 예산으로 인해 미리 선정된 운영자들과 지역 건축사가 같이 협의하여 공간을 계획하였다. 최소한 시설만 공공에서 제공하고 나머지는 운영자들이 D.I.Y.로 구축하는 방식으로 적은 예산으로도 공간 구축이 가능하도록 하였다. 결과적으로는 D.I.Y. 과정이 팀으로써의 결속력을 높이는 계기가 될 수 있었다.

계획 후 시공업체가 선정되어 완성될 때까지 몇 개월의 기간이 필요했다. 따라서 운영자들은 청년창업플랫폼의 콘텐츠를 사전에 선보이고 수익으로 D.I.Y. 재료를 구입하기 위해 주말 행사인 '관아골 프리덤' 마켓을 기획 및 진행하였다. 예산 지원 없는 운영자의 자발적 참여 행사였다. 특히, 빈 점포 리모델링 사업에 선정된 운영자들도 자신의 아이템을 전시 및 판매할 수 있도록 푸드바이크는 아트바이크로 개조되었다. 결국, 이 행사는 음식을 파는 푸드바이크, 홍보하고 물건을 파는 아트바이크, 게릴라 공연 등으로 구성된 사회실험 또는 임시점유 프로젝트로 성황리에 진행되었다. 그러나 2017년 10월 청년창업플랫폼사업이 전면 백지화되면서 청년협동조합팀은 흩어지고 총괄 코디네이터도 사임하게 되면서 충주 원도심 청년들의 D.I.Y. 실험은 끝나게 되었다.

operating business, it can be easy to plan new package products & programs, and to continuously implement joint events.

After many twists and turns, the Youth Start-up Platform was finally changed to utilize the abandoned parking lot of motorcycles in the rear yard of the post office, while using the small budget of around 100 million won. For that reason, it was led to plan the space together with the youth cooperative union (operators) and local architect (designers). This extraordinary process could dramatically save money during co-planning, because they reached the agreement that the minimum necessary facilities would be provided by the public, and the rest would be built together by the operators through D.I.Y. This was the only way for them to have their desired space at the fixed budget of around 100 million won, while making more opportunities to increase their solidarity as a team.

It took several more months to select the construction contractor through a public tendering procedure. To introduce the contents of the Youth Start-up Platform in advance and purchase materials for D.I.Y., they planned and implemented the 'Gwanagol Freedom' market, a weekend event, by themselves. This was the operator's voluntary event without any public support. Art Bikes converted from Food Bikes were made to provide some spaces for other operator teams selected for the Empty Store Remodeling Project to sell their items. This was a kind of social experiment or temporary occupancy project that introduced new consumption culture, with Food Bikes serving food, Art Bikes selling goods, and Guerrilla Music Band offering performances. However, as in October 2017, the Youth Start-up Platform Project was completely canceled by the administration without prior consultation, the youth union team was disappointed and dispersed, and after that, the project coordinator resigned. This meant that the youth activities and implementations to renovate the local economy were over.

맺으며

 충주시 도시재생사업은 운영주체의 발굴 및 연계 사업화전략으로 관문심사와 충청북도 심의에서 긍정적 평가를 받았다. 목표부터 전략, 세부사업이 충주의 물적, 인적, 문화자원과 연계되어 '도출'되었기 때문이었다. 특히, 지역의 실행자들을 존중하고 연계할 방법을 찾아 확장하려고 의도한 결과였다. 또한 주민참여를 '참견'과 '관여'만이 아닌 직접 '실행'하는 것으로 정의하고 실행주체들에게 기회와 권한이 부여되었다. '공동 실행'으로 공간을 만들고 축제와 행사를 진행하며 교육 프로그램을 운영한다면 상대적으로 적은 예산으로 많은 것을 할 수 있고 이 과정에서 끈끈한 조직이 만들어져 사업 후 지속된 활동이 가능하기 때문이었다. 따라서 임시점유와 사회실험 프로젝트들이 중요했고 점차 더 큰 사업으로 순차적으로 진행되도록 하는 것이 활성화계획의 핵심이었다.

 총괄 코디네이터와 청년조합(대표 김인혁)이 함께 제안한 성내동 청년몰사업(중소벤처기업부와 소상공인시장진흥공단 주관)도 같은 과정이었다. 제안서와 발표로만 청년상인을 선정하지 않고 임시 선정 후 공동주방, 작업공간, 판매 부스를 D.I.Y.로 함께 만들어 실제로 실험하는 과정을 거쳐 최종 선정되도록 했다. 소중한 창업의 기회가 보조금만 바라는 청년상인에

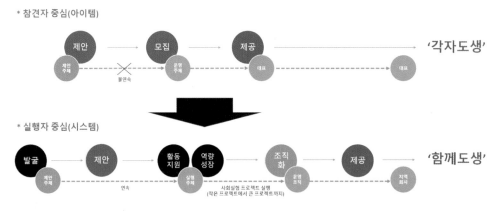

그림 13. 사회실험 적용 실행자 중심 도시재생과정 다이어그램
Urban regeneration process diagram centered on players with social experiment projects

그림 14. 청년창업플랫폼사업 부지 내 아트바이크 제작과 관아골 프리덤 행사
Art bike manufacturing process and Gwana-gol freedom event

Closing Remarks

The Chungju-si Urban Regeneration Project received many positive reviews on the strategy to discover and link together as many strong operator groups as possible at the gate review and Chungcheongbuk-do review. This response naturally resulted from the emphasis on the goals, strategies, and sub-projects that should be 'derived' in connection with the diverse physical, human, and cultural resources in Chungju. It was also the intention of the project coordinator to find and expand ways to respect and link to something existing. In addition, the resident participation was defined as the way to implement new activity directly, not to interfere, and opportunity and authority in the consultative group were given to the implementing players. This was based on the belief that running educational programs and holding festivals and events could

게 소모되는 것을 막기 위한 것이었다. 따라서 임시 프로젝트를 시행하는 과정이 강조되었다. 그러나 사업이 선정되고 추진단이 꾸려지면서 관행적 진행과정으로 바뀌었다.

도시재생사업과 청년몰사업에서 얻은 교훈은 다양한 이해관계를 가진 주민을 모아 통합된 거버넌스를 구축하는 것이 지역재생의 답은 아닐 수 있다는 것이다. 이해관계에 따라 철저히 나누고 세분화하여 원하는 사업에만 집중하도록 하는 것도 필요하다. 실행할 사람들이 그렇지 않을 사람들에 의해 위축되고 내몰림 당하기 때문이기도 하다. 단, 그것의 큰 틀과 최소의 지침을 규정하는 활성화계획은 필요하며 그 틀 안에서 목표와 전략에 맞게 실행이 되도록 해야 한다. 각 주체가 잘 하는 것에 집중하도록 세분화하고 임시점유나 사회실험 프로젝트를 통해 최소비용으로 빠르게 만든 후 서로를 이해시키는 것이 먼저일 것이다. 그 후 본격적으로 사업을 진행하여 함께 연계 운영·관리하도록 유도하는 방식이 토착세력이 강한 지역에서 유효할 수 있다. 물론, 무엇보다 지역재생사업에 임하는 공무원, 주민, 청년, 중간지원조직 등 역할이 명확해야 하며 관행적 마인드도 완전히 달라져야 한다.

be done with relatively less money if a space was created through 'joint execution', which could also make a more stronger organization, so that more activities would continue, even after the public project was over. Therefore, temporary occupancy and social experiment projects were most essential in the first stage, and the sub-projects in the Revitalization Plan were framed to proceed sequentially from smaller to bigger projects.

The Youth Start-up Mall Project (organized by the Ministry of SMEs & Startups and Agency for Traditional Market Administration), proposed by the youth association (leader: Kim In-hyuk), and selected, also included the same implementation process. Instead of selecting young merchants only with their proposals and presentations, it was intended for them to use a joint kitchen, workspace, and sales booth made through their D.I.Y. for a temporary store, and then to be selected as the final beneficiaries. This process aimed to prevent valuable start-up opportunities from being consumed by 'fake' candidates who wanted only subsidies. For that reason, the process of implementing temporary projects was emphasized. However, as the project was selected and the supporting team was formed by the public sector, they changed to all becoming conventional, as happens in most public projects.

It would never the case that as many residents as possible should be gathered for desirable governance. It seems necessary to divide and subdivide thoroughly according to their interests, so that residents can focus only on the sub-projects that they want to be involved in. This is because strong players can be intimidated and expelled by conservative obtruders. However, the revitalization plan should create large frameworks and small guidelines, so that each and every sub-project could be implemented within its goals and strategies. It should first encourage each participant to quickly focus on what they are good at with minimum cost through temporary

참고문헌

1. 충주시, 충주시 도시재생 전략계획, 2015

2. 충주시, 충주시 성내충인동 도시재생 활성화계획, 2016

3. 최순섭. 도시재생사업에서 유휴 국공유지 활용의 '현장적 한계' 연구. 한국산학기술학회논문집, 2021.8

4. 최순섭, 지역재생을 위한 'D.I.Y. Spirit' 개념과 실행방식 연구, 한국산학기술학회논문집, 2021.9

5. 최순섭, 도시재생 후진지 되지 않기, 유룩출판, 2020.2

occupancy and social experiment projects, and then make other groups understand with their realized effects. After those steps, inducing each other to operate and manage the sub-projects together in earnest may be more effective, especially in the region with strong indigenous groups. Of course, the roles of participants (public officials, residents, youths, and assistance agency groups) should be clearly defined, and each group's mindset should be dramatically changed.

References

1. Chungju-si, Chungju-si Urban Regeneration Strategic Plan, 2015
2. Chungju-si, Chungju-si Urban Regeneration Revitalization Plan, 2016
3. Soon-Sub, Choi, A Study on 'Limitation in the Field' of Urban Regeneration Project using Under-used Public Owned Land, Journal of the Korea Academia-Industrial cooperation Society, 2021.8
4. Soon-Sub, Choi, A Study on the Concept, and Implementation Method of 'D.I.Y. Spirit' for Regional Regeneration, Journal of the Korea Academia-Industrial cooperation Society, 2021.9
5. Soon-Sub, Choi, Learning from U·R Project, Youlook Press, 2020.2

경기 연천 백의리마을호텔

소멸마을의 빈집 활용 경제기반 생태계 혁신 계획

최순섭 [한국교통대학교 건축학부 교수]

들어가며

경기도의 정주환경개선사업은 군사시설 보호구역과 개발제한구역의 규제, 군부대 재편에 따른 배후 마을의 정주환경 악화에 대응한 소득창출 및 생활편의시설을 개선하는 사업이다. 연천 백의리 마을공유호텔사업은 첫 번째 시범사업으로 2018년부터 경기도청 균형발전기획 실과 연천군청 투자진행과에서 담당하여 진행하였다. 대상지는 연천군 청산면 백의2리로 총 20여 억원이 초기 투입되었다. 주요사업으로는 (구)파출소 유휴건물을 활용한 체크인 센터 리모델링, (구)옥류장 여관을 리모델링한 제1호 게스트하우스, (구)보리밭 식당부지의 시니어 창업플랫폼 사업 등이 있다. 특히 이 프로젝트는 여행자플랫폼인 충주시 목행동 BTLM 1960 에서 실험한 콘텐츠를 확장하여 계획이 수립되었다. 여행자플랫폼은 충주시 도시재생사업에 서 청년창업플랫폼 내 창업 아이템으로서 게스트하우스를 중심으로 주변 맛집과 여행코스 등 을 묶어 제공하는 오프라인 여행거점 개념이다. 여행자들이 자고, 먹고, 마시고, 즐기는 콘텐 츠를 패키지를 통해 저렴하게 예약하여 이용 가능한 공간으로 기획되었지만 끝내 실현되지 못하였다.

Yeoncheon Baekui-ri Village Share Hotel Project, Gyeonggi-do

Innovation Plan of Local Economic Ecosystem Using Vacant Houses in Extinction Village

Soonsub Choi (Professor, Korea National University of Transportation, Dept of Architecture)

Introduction

This project was to increase income and improve amenities in response to deterioration of the living environment in the border village, caused by the protected zones of military facilities, restricted development zones, and restructuring of military units. The Yeoncheon Baekui-ri Village Share Hotel was the first pilot project starting in 2018, and it has been in charge of the Director of Balanced Development (Kyungi-do Provincial Government) and the Department of Investment Promotion (Yeonchen-gun). Approximately 2 billion won was initially invested in Baekui 2-ri village, Cheongsan-myeon, Yeoncheon-gun. The main projects included remodeling of the Check-in Center using the unused police station buildings, the first Guesthouse remodeling of the abandoned Okryujang Inn, and the Senior Start-up Platform Project on the unused Boribat restaurant site. This project was planned in reference to the contents experimented in BTLM 1960, a traveler's platform. The traveler's platform was the concept of an offline travel agency that combines nearby restaurants and travel courses around guesthouses. This was intended to be a start-up item of the Youth

충주시 목행동 BTLM 1960 : 마을공유 호텔 시스템의 사전 실험

충주시 도시재생사업에서 구현되지 못한 지역생태계 변화를 통한 일자리 및 수익증대 실험의 장소로 목행동이 최종 선택되었다. 여행자가 전무한 동네였으나 여행자플랫폼을 통해 충분히 매력적인 로컬 여행자원이 많이 남아 있었다.

목행동은 한국전쟁 후 1959년 미국 원조를 받아 비료공장과 사택단지가 생기면서 형성된 상권의 배후지역이었지만 1983년 비료공장이 문을 닫으면서 급속히 쇠락한 동네였다. 그러나 로컬 여행자들이 좋아할 만한 1960년대 건물들과 가게들이 그대로 남아 있었다. 이 중 60년대에 건축되어 여인숙으로 사용되다 버려진 한옥이 여행자플랫폼 장소로 선택되었다. 한옥은 관광진흥법에 의해 내국인 여행자를 수용할 수 있으며 음식 판매가 가능한 이점이 컸다. 또한 한옥의 방들은 도미토리 형식으로 전환하면 많은 여행객들을 수용할 수 있었고 마당에서는 다양한 이벤트가 가능할 수 있었다.

여행자플랫폼 창업을 위한 세 가지 원칙은 다음과 같았다.

첫째, 최대한 낮은 비용의 시공방식이 고려되었다. 부족한 창업예산과 임대기간이 끝난 후 철수할 수도 있는 최악의 상황을 대비한 것이다. 둘째, 로컬 여행지로서 확장된 수요를 만들기 위한 가성비 높은 프로그램이 필요하였다. 이는 도미토리 게스트하우스에 숙박할 여행객 수요에 맞춘 것이다. 마지막으로 주변 가게와 레저 사업장의 적극적 연계가 필요했다. 협의된 주변 가게(동맹가게)를 이용하고 충주의 액티비티 사업장과 연계한 패키지를 여행객들이 제공받도록 하였다. 1인 운영으로 재료 저장, 음식 조리의 노력을 덜고 여행객들에게는 다양한 선택이 가능하도록 한 이유였다.

이러한 원칙에서 업사이클링 방식의 D.I.Y.로 리모델링이 진행되었다. 가구는 직접 제작되거나 고물상에서 수집되어 재가공되었다. 특히 60년대 모습의 마을에서 수집된 빈티지 물건들을 전시할 공간이 있는 여행자플랫폼의 이름을 BTLM 1960(Back To Local Modern 1960)으로 정하였다. 결국, 전기공사를 제외한 모든 공정이 D.I.Y.로 진행된 결과 업체 시공 기준 비용 대비 10분의 1의 비용으로 완성되었다.

Start-up Platform in the Chungju Urban Regeneration Project, where through the package, local travelers could book for sleeping, eating, drinking, and enjoying being there at a low price.

Chungju Mokhaeng-dong BTLM 1960 : Prior Experiment of the System of the Village Share Hotel

After many field trips with Kim In-hyuk, leader of Neonadeuli, Mokhaeng-dong in Chungju was finally chosen to experiment the possibility of increasing jobs and profits through the change of the local economic ecosystem, which failed to realize in the Chungju-si Urban Regeneration Project. This village was the neighborhood that few local travelers ever visited, even though it had sufficient and attractive resources to draw them.

Mokhaeng-dong was once the village at the rear of the abundant commercial district that flourished from the effect of the Chungju Fertilizer Factory and Company Housing with U.S. aid (UNKRA) after the Korean War in 1959. However, after the factory closed in 1983, it rapidly decline, though most of the old buildings and shops built in the 1960s remained the same, which was expected to fascinate local travelers. Among the vacant buildings, one small Hanok abandoned after being used as an inn in the 60s drew our attention. Hanok had some advantages of accommodating foreign and Korean local travelers while selling food and beverage under the approval of the Tourism Promotion Act. In addition, by switching to dormitory rooms, its room structure could accommodate as many tourists as possible, while the wide yard could be used for various events outside.

그림 1. 충주시 목행동 50-60년대 건물들 / 50-60s buildings in Mokhaeng-dong

　한편, 주변 짚라인, 요트, 수상스키 사업장과 협의한 후 한옥 게스트하우스 숙박과 결합된 할인 패키지 상품이 기획되었다. 특히 숙박고객들에게 60년대 지폐를 활용해 만든 로컬페이 (지역화폐)가 제공되었다. 로컬페이로 여행객들은 동맹가게와 게스트하우스에서 현금처럼 쓸 수 있으며 각 동맹가게에 수집된 로컬페이는 BTLM1960에서 현금으로 환전되었다. 또한 온라인 플랫폼을 이용한 음식 주문방식의 오프라인 버전 플랫폼(Platform)과 펍(pub)의 합성어인 '플랫펍(PLATPUB)' 방식을 기획하였는데, 마당에서 여행자들이 목행동 정육점, 편의

When initiating a traveler's platform project, three principles were agreed in advance:

First, it had to be constructed as cheaply as possible. Insufficient start-up budget, as well as the worst situation of having to leave at the end of the lease period, were the main reasons. Second, cost-effective contents for visitors had to be considered to meet more expanded demands from local travelers, which made us choose the dormitory type of guesthouse. Lastly, it was necessary to actively link neighborhood stores and businesses. So, after the negotiation with nearby stores (alliance stores) and leisure businesses, a variety of travel packages were planned. This also aimed to reduce the effort of one-person operation, such as the storage and cooking of food, allowing travelers to make various choices.

Based on those principles, remodeling was carried out in the upcycling method through D.I.Y. For example, every furniture item was made by the project, or reprocessed after being collected from junk shops. In particular, a small but notable space to display vintage items of the 60's collected in these villages was created on the main façade to symbolize the birth of the local village and this Hanok. This traveler's platform was also named (Back To Local Modern 1960 (BTLM 1960). Because of the D.I.Y. construction processes, tt was finally completed at 1/10th of the normal construction cost, through only for electrical work.

Meanwhile, travel package programs were planned with the nearby zip line shop, yacht, and water-skiing business to provide a discount rate to travelers staying in the guesthouse. In particular when they checked in, travelers were issued Local Pay (local currency) using banknotes of the 60s. They could then use Local Pay like cash in alliance stores and guesthouses, and Local Pay collected in the stores was converted at BTLM1960 into real cash. In addition, travelers were allowed to cook with

그림 2. BTLM1960에서 D.I.Y. 및 업사이클링 시공과정 / Up-cycling & D.I.Y. process in the construction of BTLM1960

점, 농수산물시장 등에서 직접 구매한 재료로 요리하거나 동네 가게의 배달음식을 시켜 먹을 수 있도록 하였다. 이는 저렴하면서도 질 높은 로컬 음식을 여행자들이 경험할 수 있는 장소와 시스템이었다.

BTLM1960으로 인한 파급효과는 바로 나타났다. 오픈 후 3개월 동안 300명이 넘는 여행객들이 찾아왔으며 주변 식당과 액티비티 사업장 매출도 동반하여 증가했다. 특히, 동맹 가게들은 로컬페이를 받고 다시 현금화하는 과정에서 새로 창출된 수요를 명확히 체감하였다. BTLM1960과 동맹가게를 맺지 못한 곳에서는 지속적으로 협조 요청을 하였으며 여행자들의 주문이 빈번하지 않은 배달 음식점들은 음식 질의 개선방법을 자발적으로 찾으려 노력하였다. 결국, 이전 도시재생사업에서 구현하지 못한 청년창업플랫폼을 축소해 만든 BTLM1960를 통해, 의도했던 새로운 수요층 유입에 기반한 새로운 지역상권 생태계의 변화 가능성이 확인되었다. 그리고 이는 로컬 자원을 활용한 새로운 경제기반 생태계 구축을 위한 '마을공유호텔'의 사전 실험 프로젝트였다.

fresh ingredients or enjoy food delivered from local stores, including butcher shops, convenience stores, and the Argo-Fishery Market in Mokhaeng-dong. The name 'PLATPUB' referred to the combination of platform (like Doordash) and pub, which might be defined as the food delivery platform of the offline version. In this system, travelers could experience cheaper but higher quality local food.

The ripple effect of BTLM1960 appeared immediately. Over 300 travelers visited during the first three months, and the sales at nearby restaurants and activity business places also increased together. In particular, village stores were able to clearly feel the effects of the newly created consumption in the process of receiving Local Pay and cashing back. There were continuous requests from other stores for alliance, and many delivery restaurants receiving few orders from travelers tried to find a way to improve their food quality on their own. As a result, BTLM1960's success was the opportunity to confirm the possibility of a new local economic ecosystem with the influx of new consumer groups, which had not previously been proved, due to the cancellation of the Youth Start-up Platform in the Chungju-si Urban Regeneration Project. So, the 'Village Share Hotel' could be the preliminary experimental project of building a new economic-based ecosystem using local assets.

The Planning of Baekui-ri Village Share Hotel : Dividing Roles as Professional Action Groups

The increase of the number of vacant houses in Baekui-ri has become a serious problem in Yeoncheon-gun, Gyeonggi-do. The results of the survey of vacant houses in June 2017 showed that the situation was more serious than expected, because

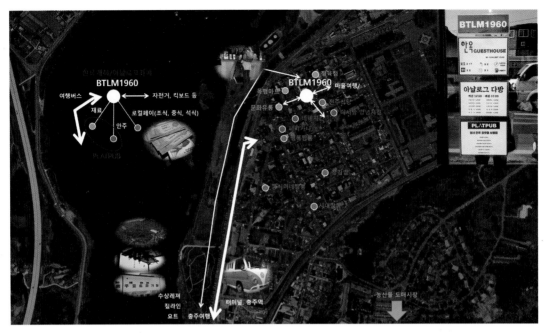

그림 3. BTLM1960의 마을 자원과 연계한 운영 시스템 / Operating system of BTLM1960 connected to village resources

그림 4. BTLM1960의 특화공간과 로컬페이 / Specialized spaces and local pay of BTLM1960

28 of 170 households (excluding military official residences) were abandoned, or approximately 16 %. Since the withdrawal of the U.S. military, expansion of the garrison area and two floods, the village's economic foundation has completely collapsed. Almost all of the merchants, the main residents in the declined commercial district, had no foundation to change their jobs. Therefore, the restoration of the economic ecosystem and jobs were the priority to improve the settlement environment in Baekui 2-ri. However, no clear alternative was evident, even after the public hearing with the representatives of the residents, senior citizens' association, youth association, 5th Military Division, and Baekui Elementary School. Eventually, the concept of 'Village Share Hotel' was proposed to innovate the village's economic ecosystem through on-site visits and meetings with expert advisory groups in May, August, and September of the same year. The plan was finally selected as the targeted village for 'the Improvement Project for Settlement Environment Using Empty Houses in the Border Area' in 2018, and secured a budget of 2 billion won.

After the selection, the devising of a master plan as well as the running of an educational program for residents began at the end of 2018. The participating groups were the two administrations, residents, related organizations, planning service companies, and the project coordinator. First, as a precedent pilot project, Gyeonggi-do and Yeoncheon-gun strongly recommended the master plan, building plan, and implementation design, and resident participation programs to be conducted each by a professional firm to provide high-quality performance with efficiency. So, the planning service company joined as a consortium between the architects' office and local business companies, so that the regional survey, master plan, architectural design, operation management plan, and resident education program were able to proceed at the same time. Most importantly, the project period was shortened because of the

백의리 마을공유호텔의 기획 및 계획 : 전문 실행팀으로서 역할분담

군부대 개편과 위수지역 확대에 따른 배후 마을에서 빈집 증가는 경기도 연천군의 심각한 문제로 대두되었다. 이에 따라 2017년 6월 연천군 백의2리 빈집 현황을 조사한 결과 군 관사를 제외한 주택 170세대 중 폐가는 28세대로 16%에 달했다. 미군 철수, 이수지역 확대와 두 차례 홍수로 인해 마을의 경제적 기반이 무너진 원인이 컸다. 무너진 상권을 대체할 기반이 없었다. 따라서 백의2리의 정주환경 개선을 위해서는 경제생태계 복원과 일자리 창출이 급선무였다. 마을 이장님과 노인회, 청년회 등 주민대표와 5사단 군부대, 백의초등학교를 통해 의견이 수렴되었으나 뚜렷한 대안을 찾지 못하였다. 따라서 같은 해 5월, 8월과 9월에 전문가 자문 그룹의 현장방문 및 회의를 통해 마을의 경제생태계 혁신을 위한 빈집 활용 '마을공유호텔' 프로젝트가 제안되었고, 2018년 '접경지역 빈집 활용 정주여건 개선 공모사업'에 선정되어 도비와 군비를 합쳐 20억의 예산을 확보하게 되었다.

선정 후 2018년 말까지 마스터플랜 수립과 주민역량강화 교육이 시작되었다. 참여주체는 행정, 주민, 유관기관, 계획수립 용역업체와 총괄 코디네이터였다. 우선 경기도와 연천군은 시범사업으로 효율적이고 우수한 시설과 공간이 제공되도록 마스터플랜, 건축계획, 실시설계와 주민교육 및 활동을 함께 이행할 용역을 통합하여 발주하였다. 이에 따라 선정된 용역업체는 건축사사무소와 지역재생사업 전문회사의 컨소시엄이었고 지역조사, 마스터플랜, 건축설계, 운영관리 계획, 주민교육 등이 동시에 진행되었다. 무엇보다 이 프로젝트에서는 적극적으로 부지매입 협의를 진행한 결과 시간을 단축할 수 있었다. 한편, 주민과 유관기관은 최초 기획 단계에서는 소극적이었지만, 교육 프로그램부터는 적극적으로 참여하기 시작했다. 특히, 백의2리 정착 2세대분들의 참여가 적극적으로 이루어지면서 운영주체인 협동조합이 조직되었다. 2019년 종합계획 수립 완료 후 2020년 체크인 센터, 2021년 게스트하우스 1호점과 시니어창업플랫폼이 완공되었다.

close cooperation, for example, the planning company was quickly reviewing many alternatives, while the administration actively progressed the site purchase. Though in the initial planning stage, residents and related organizations were passive, as the project proceeded, their participation in education gradually increased. Eventually, the cooperative union consisting of the second generation of Baekui 2-ri settlement was founded as the operating group of the Village Share Hotel. After the completion of the master plan in 2019, the Check-in Center was first built in 2020, and then the first guesthouse and the Senior Start-up Platform were completed in 2021, in order.

The Role of the Project Coordinator : Creating a Village Scenario through On-site Survey

As after the Korean War, the U.S. military unit, Camp St. Barbara, was stationed across Baekui Bridge, Baekui 2-ri as a village to the rear was born and grew. Shops for soldiers were lined up around the village's inner road, which met Cheongchang-ro vertically connecting to Baekui Bridge, and entertainment streets such as Gijichon (military camp side town) flourished along the hidden inner road. It was said that the first site of Baekui-ri village was bare land. However, one businessman who was close to the military unit obtained permission for land development, and sold the rights to the land with no ownership. As a result of dividing the lot on the bare land using nails and thread, a grid-shaped structure of the village was formed. The village, where the U.S. and Korean troops were once the main customers in the 60's, was rich enough in dollars for a rampant loan business. An interviewee also said that the village was so densely occupied that no empty place could be found for new buildings. The fact that

총괄 코디네이터의 역할 : 마을조사를 통한 스토리텔링 구상

 한국전쟁 후 백의교 건너편에 세인트 바바라 미군부대(Camp St. Barbara)가 주둔하면서 상업 배후 마을로서 백의2리가 형성되었다. 백의교와 연결된 청창로와 수직으로 만나는 마을 안 길 주변에는 상회와 가게들이 즐비했고 숨겨진 안쪽 길에는 술집과 기지촌으로 형성된 유흥거리가 번창했다. 원래 한국전쟁 후 백의리 마을의 대지들은 나대지였다. 그런데 군부대와 친분이 있던 업자가 허가를 얻어 주인이 없던 토지를 분할하여 팔았다. 나대지 위 필지 분할은 현장에서 못과 실을 이용하여 구획한 결과 격자형 마을구조를 형성하게 되었다. 미군과 한국군이 주 고객층이었던 마을에는 달러가 풍족하게 돌았고 대부업도 성행했다. 마을은 건물 지을 땅이 없을 정도로 밀집도가 높아졌고 서울에서 백의리 마을 버스터미널까지 직통버스가 다닐 정도로 유동인구가 많았다.

 1973년 미군이 철수한 후 한국군이 주둔하여 상권은 어느정도 유지되었지만 이수지역 확

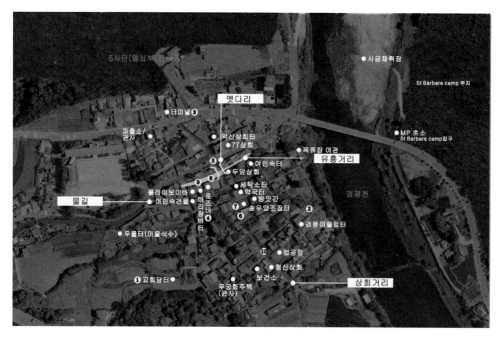

그림 5. 백의2리 60년대 장소 지도 / Place map of Baekui-ri in the 60s

direct buses traveled from Seoul to Baekui-ri Village Bus Terminal testified to its glory days too.

After the withdrawal of the U.S. military unit in 1973, although the commercial district was maintained to some extent, the economic base collapsed due to the expansion of the garrison area, and to two floods in 1996 and 1999. More and more residents left their home in debt, and the remaining residents lived in temporary cargo containers and tents. The number of soldiers visiting the village dramatically decreased. As they moved to Jeongok for a night leave or vacation, Baekui-ri remained a desolate village, the same as it is now. The old and abandoned signs of arcades, video stores, laundries, karaoke rooms, pubs, and stationery stores only showed the traces of the once-crowded neighborhood.

Ironically, the collapsed village has been preserved in the exact same mood of the 50s and 60s. In particular, the signboards engraved on the walls, unique crafted interiors, American-style bars, and vintage commercial buildings convey the feeling of a film set that young people in their 20s and 30s had not experienced before. This village is also the record of people's lives struggling to survive after the Korean War. For example, the ruins of inns, where up to hundreds of Yang Saek-si (Yang Gongju) stayed and earned money as military wives in the narrow rooms right behind the bar, testify to it.

In the end, it was agreed that Baekui 2-ri Village offered much potential to create a local retro atmosphere while preserving the sad history, which would likely be as a kind of local travel core that could deliver the story of 'hee-ro-ae-rak' (human feelings). In addition, there were abundant places for various local trips around Baekui-ri, which include the Columnar Joints Roads, Gold Harvesting Grounds, Jaein Falls, Hantang River Dam, Water Culture Center, Jeongok Pre-history Museum, Yeongang Gallery

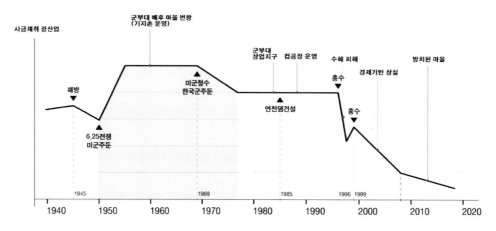

그림 6. 백의2리 마을 희로애락 里生 그래프 / 'Village life' graph of Baekui-ri

대와 1996년과 1999년 두 차례의 수해로 인해 경제기반은 회복될 수 없을 정도로 무너졌다.
무너진 집을 그대로 두고 빚을 지고 떠나는 주민들이 늘었고 남아 있던 주민들은 컨테이너와
천막으로 버티며 살았다. 폐허가 된 마을에는 장병의 발길이 줄었고 인근 전곡읍으로 외박과
휴가를 가기 시작하면서 지금과 같은 황량한 마을로 남게 되었다. 오락실, 비디오가게, 세탁
소, 노래방, 호프집, 문방구 등 간판과 빈 점포들은 북적이던 마을 모습을 유추할 수 있는 흔적
들이다.

한순간에 무너진 마을 상황으로 인해 아이러니하게도 50, 60년대 상회거리 모습이 보전
되었다. 특히, 지붕 벽에 양각으로 새긴 간판과 인테리어, 미국식 술집과 상회건물은 현재의
20-30대가 경험 못한 역사적 현장의 느낌을 그대로 전달하고 있었다. 또한 마을 전체가 한국
전쟁 후 생존을 위해 몸부림쳤던 삶의 현장기록이기도 하였다. 예를 들어, 미군을 상대했던
양색시(양공주)들이 한때 몇 백명이 머무르기도 했는데, 술집 바로 뒤편에 폐허로 남은 여인
숙 건물의 좁은 방들이 이를 증명하고 있었다.

백의2리 마을은 지역 자체가 레트로한 분위기를 형성하면서 슬픈 삶의 역사도 간직하고 있
음에 따라 '희로애락'의 스토리텔링 체험이 가능한 여행지로서 테마가 제안되었다. 또한 백의
리 주변에 주상절리 길, 사금채취장, 재인폭포, 한탄강 댐, 물 문화관, 전곡선사박물관, 민통선

within the Civilian Control Line, Cranes Habitat, and Steam Locomotive Water Tower. On the other hand, there were not enough places for families or young people to stay. For the same reason, right after their DMZ tour, most foreign travelers skipped and left this region. Therefore, if Baekui 2-ri becomes the center of local travel as the Village Share Hotel, it could be expected to be able to revive the ecosystem of the local economy by creating new demands, just as BTLM1960 did before.

그림 7. 백의2리 마을의 빈집들 / Empty houses in Baekui-ri

내 연강갤러리, 두루미 서식지, 증기기관차 급수탑 등 여러 로컬 여행자원들이 풍부했다. 그러나 상대적으로 가족이나 젊은 층이 숙박할 공간이 미비하였다. DMZ 관광 후 외국인들도 그대로 이 지역을 빠져나가는 상황이기도 했다. 따라서 백의2리가 마을공유호텔로 로컬 여행의 거점이 된다면 BTLM1960과 같이 새로운 수요를 창출하여 지역경제 생태계를 변화시킬 수 있을 것으로 기획과정에서 합의가 되었다.

마을공유호텔 마스터플랜 : 희로애락 콘텐츠와 공간구상

백의리 마을공유호텔의 방향 설정 후, 우선 유휴 건물정보가 기록된 빈집 카드가 제작되었다. 이를 통해 철거 또는 활용 가능한 빈집을 구분하고 리모델링 주안점이 기록되었다. 빈집 카드를 통해 추후 백의리에 창업하는 사람들에게 제공되어 건물의 숨겨진 가치를 전달하여 마을 테마를 유지하기 위한 목적이기도 했다.

이후 BTLM1960 프로젝트 경험과 백의리 조사를 바탕으로 콘텐츠와 필요공간들이 구상되었다. 특히, 여행자들이 느끼는 즐거움(喜)과 기쁨(樂)의 장소만들기 시나리오가 제안되었다.

"여행객들은 호텔 로비에서 체크인을 한다. 기다리는 동안 카페에서 커피를 마시며 마을과 여행 프로그램 정보를 얻는다. 키를 받고 난 후 배정받은 방으로 이동하여 짐을 푼다. 자전거나 킥보드를 대여하여 이동할 수 있다. 짐을 푼 후 여행 패키지로서 예약한 액티비티 프로그램을 이용하거나 주변 여행지들을 방문한다. 물론 마을에서도 갤러리, 식당, 바, 기념품 가게 등을 이용할 수 있다."

이에 따라 마을 전체에 구축할 장소와 시설이 도출되었다. 먼저 호텔 로비, 카페(여행정보 공유), 자전거·킥보드 대여장소를 통합하여 체크인 센터가 제안되었다. 인지가 잘되는 장소인 도로변 유휴 파출소와 관사부지가 선택되었다. 체크인 센터에는 이벤트를 진행할 마당과 루프탑이 계획되었고 관사에는 마을공유호텔을 운영할 마을기업(협동조합) 사무실이 제안되었다.

The Master Plan of the Village Share Hotel :
Devising the Contents and Spaces of 'Hee-ro-ae-rak'

After deciding the theme of Baekui-ri Village Share Hotel, the empty house record cards were first made. Through this procedure, whether each vacant building was demolished or remained was decided, and then the main characteristics to be considered during the remodeling of the surviving buildings were suggested. This was also intended to provide the value of the building to those who wanted to start new businesses in Baekui-ri later, so that they could maintain harmony with the village and landscape.

Based on the BTLM1960 project experience and frequent field surveys, the main usages and necessary spaces began to be conceived. In particular, the space scenario was created in terms of the traveler's feelings, such as joy and pleasure.

"Travelers check in at the hotel lobby. They can drink coffee at the cafe and gain information on village and travel programs while waiting. After receiving their key, they go to their rooms and unpack. They can rent bicycles or electric kickboards to explore, attend leisure program places booked in advance by travel packages, or visit nearby travel destinations. Of course, there are also many places to enjoy day and night in the village, such as galleries, restaurants, bars, and souvenir shops."

According to this scenario, necessary places and facilities for the hotel were derived. First of all, a Check-in Center was needed to accommodate the hotel lobby, cafe (travel bureau), and personal mobility rental shop. Also, it should be a place that was easy to recognize from the gate of Baekui-2 ri, so the abandoned police substation and the official residence along the road were finally selected for it. Additionally, a versatile community yard in the site was also proposed with rooftop space to hold a variety of

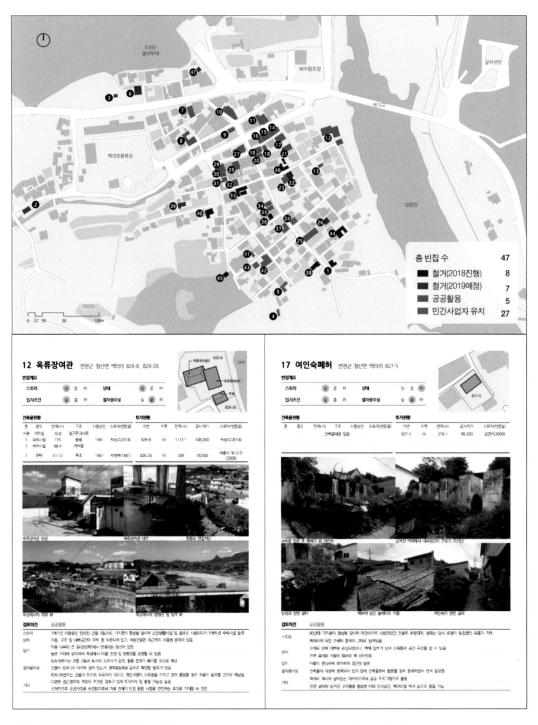

그림 8. 백의2리 빈집 현황 및 빈집 카드 / The distribution and record cards of empty houses in Baekui-ri

events, and an abandoned residence was planned for the office of the cooperative union operating the entire Village Share Hotel.

The second project was the first guesthouse. Like BTLM1960, it was intended as the core place where visitors could enjoy delivery food and barbecue party while staying, so the surrounding scenery should be outstanding to satisfy local travelers using SNS. Therefore, the abandoned inn (Okryujang) located on the cliff of Yeongpyeong Stream was considered the first option. After long negotiation with the owner, it was finally purchased. In particular, a foot bath where travelers could relieve fatigue while looking at the amazing landscape was designed, as well as versatile rooftop spaces.

The last place was the Senior Start-up Platform for multicultural programs. This included multi entertainment space like the iPad book café, as well as a snack bar served by village union members for soldiers and travelers. In addition, a terraced small auditorium was placed at the 2nd story to hold diverse events for them. Those functions and spaces were derived from a survey of the soldiers (potential consumers) of the 5th Military Division unit in front of the village.

Due to the limited budget, only three major facilities were first completed, but additional projects were also proposed in the master plan. The memorial parks using the ruins of inns to preserve and display the lives of Yang Saek-si as the theme places of sadness and anger, another community park, and a gallery project utilizing vacant buildings and sites were proposed as well. With those place-making projects, educational programs were also organized to support the operation of the leisure business, including Yeoncheon travel guide tours, alluvial mining, and kayaking & biking.

그림 9. 백의리 마을공유호텔 전체 조감도 / Birdeye view of Baekui-ri village share hotel

두 번째 공간은 게스트하우스 1호점이었다. BTLM1960과 같이 숙박기능과 함께 배달음식을 먹거나 바비큐 파티가 가능한 공간이면서, SNS를 즐기는 여행자들을 위해 주변 경관이 우수해야 했다. 따라서 영평천 절벽 위에 위치한 옛 옥류장 여인숙 유휴 건물을 매입하여 객실과 루프탑 공간이 계획되었다. 천을 조망하면서 피로를 풀 수 있는 족욕장도 로비에 제안되었다.

마지막 공간은 멀티 문화소비공간으로 시니어창업플랫폼이었다. 주민들이 여행객들과 외출 군인들에게 필요한 서비스를 제공하는 공간으로서 아이패드 북카페 겸 분식점이 되는 멀티방 개념의 공간이다. 또한 주민과 군인들이 사용할 다목적 공간으로서 계단식 소강당도 제안되었는데, 이러한 공간과 용도들은 마을 전면에 위치한 5사단 열쇠부대 장병들을 상대로 설문조사를 진행한 결과를 적용한 것이었다.

한정된 예산으로 3개의 주요시설을 우선 완성한 후 추가 진행할 사업도 구상되었다. 마을에서 슬픔(哀)과 노여움(怒)의 테마 장소로서 양색시들의 삶을 기록하고 전시할 옛 여인숙 폐허를 활용한 메모리얼 파크, 유휴건물과 부지를 정비한 마을공원과 갤러리 조성사업이 제안

체크인 센터	시니어창업플랫폼	멀티공간		게스트하우스
호텔로비/카페	(Ⓐ 구 파출소)	(Ⓑ 구 보리밭 식당)	(ⓒ 구 옥류장)
2020년 완공	2021년 완공	2021년 완공

그림 10. 백의리 마을공유호텔 이용 행태도(마스터플랜) 및 주요시설
Activities plan (master plan) and major facilities of Baekui-ri village share hotel

The Operation and Management Plan of the Village Share Hotel : Creating the 'NET'-Working System

The experience from BTLM1960 confirmed that the building as 'a system to connect

되었다. 또한 연천 여행가이드, 사금채취 체험, 카약 및 MTB 등 액티비티 프로그램 운영 지원 사업도 마스터플랜에 포함되었다.

마을공유호텔의 운영관리계획 : 'NET' 연계 체계의 구축

BTLM1960을 통해 '아이템보다 시스템' 또는 '아이템의 연결 시스템'을 기획하고 구축하는 것이 지역재생에서 더 중요하다는 것이 확인되었다. 특히 로컬페이는 마을 전체가 상생과 공생 시너지를 위한 시스템으로 묶는 주요 매체가 될 수 있었다.

따라서 백의리 마을공유호텔에서도 로컬페이를 중심으로 여행 패키지가 예약되고 공간이 이용되도록 하였다. 예를 들어, 백의리에서 숙박과 액티비티 프로그램(모빌리티 대여, 사금채취 및 카약 체험), 마을가게(기념품 가게, 음식점)에 사용가능한 로컬페이 포함 패키지 상품이 기획되었다. 게스트하우스나 카페만 수익이 나고 나머지 사업장들이 들러리가 되는 타 지역 실패사례를 참고한 것이다. 마을 여행객이 증가하면 마을 각 사업장에도 자연스럽게 수익이 돌아가는 공생의 연계(NET)구조를 만든 것이다. 단, 패키지 이용 시 가장 큰 할인과 혜택을 받고 각 사업장들은 여행객들이 만족할 가성비 있는 콘텐츠가 갖춰지도록 하였다.

한편, 백의2리가 연천군 여행자플랫폼 역할을 하도록 로컬 여행버스 및 여행가이드 프로그램도 제안되었다. 어느정도 마을이 활성화되었을 때 진행하는 후속사업이었다. 60년대 백의리 터미널에 정차했던 우리나라 최초의 버스인 하동환 버스나 군부대 트럭을 이용한 로컬 여행버스로서 연천역과 전곡역에서 여행객들을 픽업하거나 시티투어 버스와 같이 연천의 관광지를 둘러볼 수 있는 프로그램이다. 이와 함께 백의2리 마을을 장병 우대 마을로 지정되도록 하였다. 마을 가게들과 협의 후 장병들에게는 가격할인 또는 특별 서비스를 제공하여 군부대 마을의 부정적 통념을 쇄신하여 군인들을 다시 수요층으로 전환시키려는 의도였다.

items' was more important for regional regeneration. Local Pay could be the major medium with which the entire village would be managed in one system to create win-win synergies.

Therefore, Baekui-ri Village Share Hotel was also intended to provide Local Pay purchased automatically when travelers reserved travel packages, which currency could be used in the local shops and restaurants. So, travelers could not help but use it for accommodation, activity programs (personal mobility rental, alluvial mining, kayaking, etc.), and village stores (souvenir shops, restaurants). This strategy was referred to many failures in other regions where only guesthouses and cafes were profitable, while the remainder of the businesses became ancillary. So, the consumption through Local Pay could create the linked motivation (NET) for a symbiosis in which profits were naturally returned or distributed to each business in the village. To further activate this system, the biggest discount would be provided while all shops and restaurants should maintain cost-effective contents, which would satisfy young local travelers.

Meanwhile, a local travel tour bus and travel guide program were also proposed, so that Baekui 2-ri would become the traveler's platform in Yeoncheon-gun. This was a follow-up project when the village's economy would to some extent be activated again. 'Hadonghwan Bus' was recommended to be restored, because it was the first bus manufactured by Koreans with drums the U.S army discarded, which used to stop at Baekui-ri Terminal in the 1960s. Like a city tour bus, it would pick up tourists at Yeoncheon Station, and take them to their travel destination. In addition, Baekui 2-ri was guided to become a preferential village for soldiers. The intention was to erase the negative images of the village living off soldiers. So, it was highly recommended for merchants to offer some benefits, such as discounts and special free items, only to

"여행 패키지를 통한 지역경제 재생"

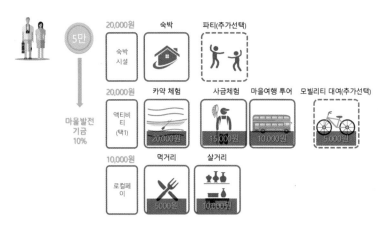

"지역상생구조 마을사업 운영"

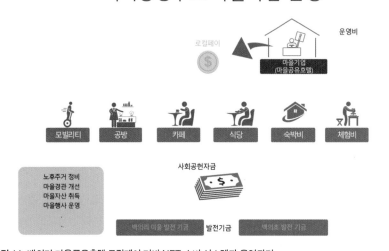

그림 11. 백의리 마을공유호텔 로컬페이 기반 NET-소비 시스템과 운영관리
NET-consumption system and operation management based on local pay of Baekui-ri village share hotel

마을 전체가 여행 콘텐츠로 연계된 계획은 당연히 운영관리 측면에서도 하나의 시스템이 필요했다. 따라서 시설과 프로그램 운영은 백의리 마을기업이 맡는 목표로 2020년 말 백의리 정착 2세대 주민 중심으로 협동조합이 구성되었다. 운영조직은 게스트하우스, 모빌리티 및 체험, 시니어창업플랫폼을 운영관리 할 세 부서로 제안되었다. 또한 조합원으로 등록된 마을

soldiers.

The upper plan to link all places of the village with travel contents naturally needed to have a unified system in terms of operation and management. Therefore, the cooperative union, where most members were the young second-generation, was organized in 2020 to operate facilities and programs. The proposed operating organization consisted of three departments: the first team to operate and manage guesthouses, the second for the mobility and activity business, and the third to manage the Senior Start-up Platform. Villagers registered as union members could be hired as working staff. The income management was structured to be unified, so some of the profits, excluding direct and indirect expenses, were used for village development funds and scholarships for students of Baekui Elementary School. In addition, it was proposed to invest some profits in recruiting founders or cultural artists, so that attractive contents using empty houses could flow into it consistently.

Closing Remarks

First, most of the pending issues, including the planning, designing, and construction of this project, could be solved completely with close cooperation, and new ideas could be continuously reflected to the plan without hesitation. This was because each participant faithfully played their own roles as 'solvers', not 'supervisors'. For example, many problems in the budget and administrative procedures were clearly solved by the officials of Gyeonggi-do and Yeoncheon-gun through vigorous consultation with their own and other departments, and planning service companies continuously tried to find ways to apply the solutions without reducing ideas, while implementing them

주민들이 스텝으로 고용되도록 했다. 마을공유호텔의 수입 관리 창구를 일원화하였고 직·간접비 등 경비를 제외한 수익 일부를 마을의 발전기금이나 백의 초등학교의 장학금 등으로 사용되도록 제안되었다. 또한 창업자 또는 문화예술인을 모집하여 빈집 활용 콘텐츠가 지속 확대되도록 마을기업의 수익을 환원하도록 했다.

맺으며

우선, 이 사업의 기획과 계획, 시공에 이르기까지 당면한 현안들이 대부분 해결되면서 아이디어가 계획에 지속적으로 반영되는 과정이 중요했다. 이는 각 참여자들이 각자의 영역에서 '해결사'로서 역할을 충실히 했기 때문이었다. 예산과 행정절차의 문제는 경기도와 연천군에서 자체 또는 타 부서와의 협의를 통해 해결하였고, 용역업체는 아이디어를 축소하지 않고 실제 구현할 방법을 찾으면서 업무를 일정에 맞게 이행하였다. 또한 연천군은 기획, 기본설계와 실시설계를 통합하여 건축사가 수행하고 시공의 감리까지 참여하도록 하여 계획한 공간이 의도에 맞게 구현될 수 있었다. 총괄 코디네이터는 지역조사와 차별화된 스토리텔링 구성을 맡아 관행적 사업이 아닌 실제 수익과 일자리를 만들 수 있는 '비즈니스'로서 사업이 구상되도록 중심을 잡았다.

또한 경제공동체로서 마을공유호텔의 시스템 구축이 계획의 핵심이었다. 여행객들이 소비할 프로그램들이 'NET' 운영관리가 되는지, 수익배분은 이루어지는지, 동반 일자리 창출이 가능한지에 따라 마을공유호텔 여부가 결정된다. 여행가이드 프로그램을 같이 운영하거나 주변 가게에서 할인되는 정도만 연계 운영하는 게스트하우스로는 절대 마을공유호텔이 될 수 없기 때문이다. 반면 로컬페이와 같은 공생의 매체로 상호 연계되도록 구상되어야 했다. 마을 전체가 팀과 조직으로서 문제를 해결하고 새로운 기획과 실행이 지속될 수 있는 운명공동체가 되는 운영관리방안이 핵심이었다.

이런 공생과 상생의 시스템 만들기는 일반 지역재생사업에서 대부분 놓치는 부분이기도 하

at the right moment. Even in the construction process, Yeoncheon-gun allowed the original architect to participate in the supervision until the end of construction, so that the planned space could be properly made according to the original design. The project coordinator engaged in vigorous regional research, and then tried to devise a unique scenario of places, while keeping a business point of view to create actual profits and jobs, unlike other conventional projects.

The team mostly focused on making the system of the Village Share Hotel build new economic community again. This was based on the belief that whether there is a 'NET'-Working system of consumption and management, profits are distributed, and it is possible to create jobs synergically, will determine whether it is a real Village Share Hotel or not. Normal guesthouses that only operate some tour guide programs, or provide discount coupons that are used at nearby stores, can never achieve the Village Share Hotel concept. Beyond that, the system should be linked through some substantial medium, like Local Pay. The only way to maintain strong implementation is to unify the group into one economic community, which will motivate them to solve any of the regional problems that consistently arise.

The notion that a system of win–win symbiosis must be constructed during a regional project is frequently ignored. The main task should not be to arrange items only advertised beautifully, but to build the framework of change to the local ecosystem. In addition, just because it is 'a public' project', it should not be stingy with any profits, and the quality and strategy of spaces and programs should not be lower than those in the private sector. Since such spaces and projects are created and managed by the public, more attractive and interesting contents could be provided. In this respect, Baekui-ri Village Share Hotel may present an alternative to the conventional Korean-style regeneration project, especially in the extinction villages.

다. '아름답게만 보이는' 아이템들을 배치하는 사업이 아니라 지역 생태계를 바꾸는 시스템 구축사업이 바로 재생사업이어야 한다는 것이다. 또한 공공사업이라는 미명 아래 수익과 매출에 인색하여 공간과 프로그램 품질이 민간사업보다 떨어져서도 안 된다. 공공에서 진행하므로 민간에서 할 수 없는 더 매력적이고 흥미로운 공간과 프로그램을 만들 수 있다는 변화된 생각이 중요하다. 이런 점에서 백의리 마을공유호텔 프로젝트는 소멸지역에 현실적으로 적용될 수 있는 한국형 재생사업의 대안일 수 있다.

단, 이 사업에서는 아직 실행할 세부사업들과 협동조합이 마을기업으로 성장할 동기와 기회가 지속적으로 부여되어야 하는 숙제도 남았다. 또한 젊은 여행객들의 트랜드에 맞게 공간과 프로그램을 운영해야 하는 것도 필요하다. 지금까지 현장에서 어떠한 지원조직 없이 참여자들의 고된 노력으로 계획과 최소의 공간들이 수월하게 만들어졌으나 지역 경제생태계가 과연 쇄신하고 변화할지에 대해서는 이제부터 시작일 것이다.

그림 12. 백의리 마을공유호텔 협동조합과 실무자회의 / Baekui-ri village share hotel cooperative and working-level meeting

참고문헌

1. 연천군, 백의리 빈집 활용 정주여건 개선 사업 사업설명서, 2020.11
2. 연천군, 연천군 백의리 빈집 활용 정주여건 개선 종합계획, 2019.8
3. 최순섭, 도시재생 후진지 되지 않기, 유룩출판, 2020.2

However, many projects remain to proceed with, and more motivation and opportunities for the union of residents must be consistently provided to make them a competitive professional company. It will also be another important task to run fancy spaces and programs while attracting the younger end of the travel market. Even though the master plan and minimum spaces could be completed without any support of a field organization, there is still a long way to go in renovating the local economic ecosystem in the field. It is just the beginning of deciding the village destiny, 'To be or not to be...'

References

1. Yeoncheon-gun, Explanation of Project of Baekui-ri Improvement Project for Settlement Environment Using Empty Houses in the Border Area, 2020.11
2. Yeoncheon-gun, Comprehensive Plan of Baekui-ri Improvement Project for Settlement Environment Using Empty Houses in the Border Area, 2019.8
3. Soon-Sub, Choi, Learning from U·R Project, Youlook Press, 2020.2

부산 아미동 비석문화마을

서구 아미·초장 도시재생사업

우신구 (부산대학교 건축학과 교수)

들어가며

부산시 사하구 감천문화마을과 고개 하나를 사이에 두고 있는 서구 아미동에는 과거 일제 강점기 조선인 빈민들의 거주지와 함께 일본인들의 공동묘지와 화장장, 장제장 그리고 총천

그림 1. 아미동 일본인 공동묘지의 피난민 주거 (1953년경 Ingvar Svensson)
Refugees' shacks on the japanese cemetary

Ami-dong Tombstone Cultural Village Dreaming of Tomorrow, Busan

The Ami-Chojang Urban Regeneration Project in Seo-gu

Shinkoo Woo (Professor, Pusan National University, Dept of Architecture)

Introduction

In Ami-dong, Seo-gu, Busan, which is located just beside Gamcheon Culture Village in Saha-gu, there were Japanese cemeteries, crematoriums, funeral halls, Buddhist temples that included Chongcheonsa Temple, along with a village of poor Koreans during the Japanese colonial period. With the outbreak of the Korean War in 1950 and the January 4th retreat in 1951, numerous refugees from all over the country headed to Busan. Refugees flocked to the vacant lot, riverside, and roadside of Busan's urban area, and to the beaches and mountains around the city area, causing a flurry of temporary housing. Some refugees went up to the Japanese cemetery built on the Ami-dong hill, built tents and shanty houses, and began to live there. When the Korean War ended in a state of division, refugees who could not return to their hometowns were forced to continue to settle; then through the industrialization era of the 60s and 70s, more population flowed in, and residential areas expanded along the slopes. Even now, Japanese tombstones are scattered throughout the village. This is why it is nicknamed Biseok (tombstone) Village.

사라는 불교사찰이 있었다. 1950년 한국전쟁이 발발하고 1951년 1.4후퇴와 함께 전국의 수많은 피난민들이 부산으로 향했다. 부산 시내의 공터와 도로변, 하천변뿐만 아니라 시가지 주변의 바닷가와 산지에까지 피란민들이 몰려들어 임시주거가 난립하였다. 일부 피난민들은 아미동 고개 위에 조성된 일본인 공동묘지로 올라가 텐트나 판자집을 짓고 거주하기 시작했다. 한국전쟁이 끝나자 고향으로 돌아가지 못한 피난민들은 계속 정착할 수 밖에 없었고, 60년대와 70년대 산업화시대를 거치면서 오히려 인구가 더 많이 유입되어 경사지를 따라 주거지가 더 확대되었다. 지금도 마을 곳곳에는 일본인 비석들이 산재해 있다. 비석마을이라는 별칭으로 불리는 이유이다.

1980년 약 3만명으로 정점을 찍은 아미동의 인구는 원도심의 쇠퇴와 부산의 탈산업화로 인해 1985년 이후 감소하기 시작하여 2000년대에 접어들면서 급속히 줄어들었다. 1990년 24,307명, 1995년 20,516명, 2000년 15,198명, 2005년, 12,208명, 2010년 10,520명 그리고 2015년에는 만명 이하로 줄어들었다. 인구가 증가할 때 분동되었던 아미1동과 아미2동은 '인구의 감소와 함께' 1998년 다시 단일동으로 통합되었다. 부산 전체로 보면 인구감소 현상이 1990년대 초반 이후에 시작되는 것과 비교하면, 아미동의 인구감소는 좀 일찍 시작되었다고 볼 수 있다. 이러한 변화는 아미동이 부산의 원도심에 가깝기 때문에 원도심의 쇠퇴와 함께 인구도 함께 줄어들었다고 볼 수 있다.

아미동에서 다른 지역으로 빠져나가는 사람들은 주로 젊은 세대들이다. 좁고 가파른 도로로 접근해야 하는 경사 주거지는 자동차를 이용하기에 불편했고, 주차장과 같은 기반시설도 부족했다. 어린 자녀를 가진 젊은 부부는 아이가 초등학교나 중학교에 입학할 나이가 되면, 좋은 교육환경을 가진 지역으로 이사가는 경우가 많았다.

그 결과 아미동에는 다른 지역으로 떠날 수 있는 경제적 여유가 없는 노인, 장애인, 저소득 가정 들이 남게 되었다. 사람들은 떠나지만 새로 이주해 오는 사람들이 없자, 마을에는 자연히 빈집이 증가할 수밖에 없다. 2015년 부산대학교 도시건축연구실의 자체 현장조사에 따르면 전체의 11.6%가 폐공가로 추정될 정도였다.

The population of Ami-dong peaked at about 30,000 in 1980, then after 1985 began to decline due to the decline of the original city center and the de-industrialization of Busan, and in the 2000s, rapidly decreased. In 1990, 1995, 2000, 2005, 2010, and 2015, it decreased to (24,307, 20,516, 15,198, 12,208, 10,520, and to less than 10,000), respectively. Ami 1-dong and Ami 2-dong, which were divided when the population increased, were merged into a single dong again in 1998 with the decrease in population.

The population decline in Ami-dong began earlier, compared to the phenomenon of population decline that began after the early 1990s in Busan as a whole. The people who leave Ami-dong for other regions are mainly the younger generation. The sloping residential area, which had to be accessed by narrow roads, was inconvenient for car access, and lacked infrastructure such as parking lots. Young couples with young children often move to areas with good educational environments when their children are getting ready to enter elementary or middle schools.

As a result, the elderly, the disabled, and low-income families who cannot afford to leave for other regions remain in Ami-dong. When there are no new immigrants after people leave, the number of vacant houses in the village naturally increases. An on-site survey conducted by the Urban Architecture Lab of Pusan National University in 2015 estimated 11.6 % of the total housing to be empty.

To cope with the serious decline of the original downtown area, Busan Metropolitan City (BMC) started the 'Happy Village Making' project in 2010 and the 'Sanbok Road Renaissance' project in 2011 to take care of various classes living in the Sanbok Road area, to preserve the settlement environment, and to restore the village economy. (Busan Urban Regeneration Comprehensive Information Management System, http://www.burtis.or.kr/intro/introBusiness.do#, dated December 1, 2019).

그림 2. 아미동과 초장동의 지리적 위치 / The location of Ami-Chojang project

부산시에서는 원도심지역의 심각한 쇠퇴현상에 대응하기 위하여 2010년부터 '마을공동체 복원을 통한 통합형 도시재생사업'으로 행복마을 만들기사업을, 2011년부터 '(산복도로) 지역에 살고 있는 다양한 계층을 함께 포용하고 산복도로가 가지고 있는 공간·문화·경관·역사적 자산을 보존하면서, 정주환경을 개선하고 마을경제를 회복하는 새로운 방식의 종합재생사업'으로 산복도로 르네상스사업을 시작하였다.[1]

초장동에서는 2011년 행복마을 만들기사업에 선정되어 천마산로에 3층 규모의 '한마음 행복센터'가 만들어졌다. 아미동에서 눈에 띄는 가시적인 도시재생사업은 산복도로 르네상스사업이었다. 아미동 비석마을 일대는 인접한 감천문화마을과 함께 아미·감천구역에 포함되어 2012년부터 사업을 시작하였고, 주민들을 위한 아미골행복센터, 기찻집예술체험장, 아미문화학습관 등 공공시설과 주민거점시설이 지어지면서 주민들이 모이고, 활동할 수 있는 공간을 마련하였다.

쇠퇴하는 아미동과 초장동을 통합적으로 개선하려는 본격적인 도시재생사업은 서구청에

---

1   부산도시재생종합정보관리시스템 (http://www.burtis.or.kr/intro/introBusiness.do#) (2019.12.01.)

Chojang-dong was selected as a site of the Happy Village Making project in 2011, and a three-story 'Hanmaeum Happiness Center' was built on Cheonmasan-ro. The visible urban regeneration project in Ami-dong was the Sanbok Road Renaissance project. The area around Ami-dong Tombstone Village was included in the Ami-Gamcheon District, along with the adjacent Gamcheon Culture Village. The project began in 2012, and public facilities, such as the Amigol Happiness Center, the Railroad Art Experience Center, and the Ami Culture Learning Center, were built for residents and visitors to gather and work.

The Seo-gu District Office began in 2014 to prepare a comprehensive urban regeneration project to improve the declining Ami-dong and Chojang-dong areas. The project was planned as an 'urban regeneration general neighborhood type' (total project cost of 10 billion won), supported by the Ministry of Land, Infrastructure and Transport (MOLIT), for an area of 1.16 km² (0.71 km² in Ami-dong and 0.45 km² in Chojang-dong). Although the project has not yet been completed due to difficulties in purchasing the land necessary for the planned facilities, shrinking resident participation due to COVID-19, and delays in design and construction of buildings, it has been highly esteemed, in that children and other residents participated in the planning process, the unique history and characteristics of the village were well reflected in the plan, an active public–private cooperation system was organized, and the capabilities of active resident organizations that included teenagers were empowered.

서 2014년부터 준비하기 시작한 도시재생사업이었다. 부산광역시 서구 아미동과 초장동 일원 1.16㎢ (아미동 0.71㎢, 초장동 0.45㎢) 면적을 대상으로 국토교통부가 지원하는 '도시재생 일반근린형' (총 사업비 100억) 사업으로 사업기간은 2016년부터 2021년까지로 계획하였다. 사업에 필요한 토지매입의 어려움, 코로나19로 인한 주민참여의 위축, 시설조성사업을 위한 설계와 공사의 지연 등으로 아직 사업이 완료되지 못했지만, 어린이를 비롯한 주민들이 사업계획수립과정에 참여하였으며, 독특한 마을의 역사와 특성을 활용한 도시재생계획을 수립하였고, 적극적인 민관학 협치체계를 구축하였으며, 청소년을 비롯한 활발한 주민조직의 역량을 강화한 점에서 의미가 큰 사업으로 평가받고 있다.

도시재생사업의 활성화계획 수립과정과 내용

부산광역시 서구는 2015년 12월, 국토교통부 주관 '도시재생 일반지역 근린재생형' 사업으로 '내일을 꿈꾸는 비석문화마을: 아미·초장 도시재생 프로젝트(사업기간 2016-2020년, 5년)' 사업계획서를 제출하여 선정되었다. 2016년 2차에 걸친 관문심사(거버넌스 구축과 활성화계획 수립)를 거쳐 활성화계획을 확정하여 사업을 추진하고 있다. 2016년 도시재생 공모사업은 사업에 선정된 후 1년 정도의 시간을 투입하여 활성화계획을 수립하였기 때문에 여유 있게 주민들이 참여하는 사업계획을 수립할 수 있었다.

이 사업의 사업제안서와 활성화계획은 그 전부터 서구청과 협력관계를 맺고 있던 부산대학교 도시건축연구실(지도교수 우신구)에서 담당하였다. 2014년부터 사업제안서를 준비하면서 지역의 기관이나 단체들과 협의하면서 사업내용이나 절차, 주체 등을 협의하였다. 그 과정에서 서부경찰서, 부산기독교종합사회복지관(이하 부기사) 같은 공공기관이나 복지단체 등과의 협력체계를 구축해 갔다. 또한 도시재생대학을 통해 지역 내의 가능한 다양한 주민들의 희망과 요구를 수렴하였다.

Planning Process and Contents of the Activation Plan for Urban Regeneration

Seo-gu District Office of BMC submitted a project proposal plan for 'Tombstone Culture Village Dreaming Tomorrow: Ami-Chojang Urban Regeneration Project (project period: 2016–2020, 5 Years)' as a 'General Neighborhood Type' of urban regeneration supported by the MOLIT. After two stages of gate reviews (establishment of governance and activation plan) in 2016, the Activation Plan was finalized, and the project is currently being implemented. Since the 2016 Activation Plan was established by investing for about a year, it was possible to make a plan in which residents participated.

The project proposal plan and Activation Plan of this project were prepared by the Urban Architecture Lab of Pusan National University (supervising professor Woo Shin-Koo), which had previously been working in cooperation with the Seo-gu District Office. Since 2014, while preparing a project proposal plan, the Lab had consulted with local institutions or organizations to discuss the project details, procedures, and subjects of the project. In the process, a cooperative network with local institutions and welfare organizations, such as the Seobu Police Office and the Busan Christian Social Welfare Center (BCSWC), was established. In addition, through the Urban Regeneration School, the wishes and demands of various residents in the region were collected.

To establish the revitalization plan, more specific surveys were also conducted. In particular, the target area was formed spontaneously without sufficient infrastructure during the Korean War and industrialization era, so many of the houses still remain unauthorized and lack building registration. Therefore, the data of housing were not accurately constructed. In particular, in the Tombstone Village area, most houses

활성화계획의 수립을 위해 보다 구체적인 조사도 함께 실행하였다. 특히 대상지역은 한국
전쟁과 산업화시대를 거치면서 충분한 기반시설이 없는 상태에서 자생적으로 형성된 주거지
역이었기 때문에 상당수의 주거가 여전히 건축물대장이 없는 무허가 상태로 남아 있다. 따라
서 주거에 대한 데이터가 정확하게 구축되어 있지 않았다. 특히 비석마을 일대는 국공유지 위
에 지어진 주거도 많고, 과소필지에 도로를 접하지 않는 맹지도 적지 않다. 따라서 주거가 노
후화되더라도 신축을 할 수가 없고, 재산권을 행사하기도 어렵기 때문에 인구가 감소하면서
공가와 폐가가 급속히 증가하고 있었다.

계획을 수립한 부산대학교 도시건축연구실은 공가와 폐가에 대한 정확한 정보를 얻기 위해
아미동과 초장동 전역에 대해 대학원과 학부학생들이 함께 현장조사를 실시하였다. 그 결과
공가, 폐가, 나대지에 관한 실질적인 정보를 구축할 수 있었고, 이를 바탕으로 공폐가가 군집
으로 발생하는 지역을 발견하고 발생특성을 조사하였다.

그림 3. 마을 곳곳에 산재한 공폐가 / Empty and abandoned houses

분석결과 2015년 1월 기준 공폐가는 총 582개소로 '깨진 유리창 이론'이 시사하는 바와 같
이 대상지역에도 공폐가가 집중발생하고 있었다. 공폐가 매핑 지도를 지름 100~200m 정도
의 영역으로 구분한 결과 8개의 집중발생구역을 도출하였다. 그결과 공폐가는 주로 차량통행
이 불가능한 급경사 계단길이나 경사로 주변 그리고 막다른 골목길 주변에 집중적으로 발생
한다는 알 수 있었다. 뿐만 아니라 경찰서의 협조로 지역 내의 범죄 발생지 53개소를 매핑한

are built on national and public land, and there are many sections or plots that are not serviced by legal roads. Therefore, even if housing was becoming decrepit, it was impossible to construct new buildings and exercise property rights, so that as the population decreased, empty and abandoned houses were rapidly increasing.

The Urban Architecture Lab of Pusan National University, which was making the plan, conducted an on-site survey with graduate and undergraduate students throughout the Ami-dong and Chojang-dong area to obtain accurate information on empty and abandoned houses. As a result, practical information on empty and abandoned houses and vacant land was constructed. Based on this data, areas where vacant houses occur as clusters were discovered, and the characteristics of their occurrence were investigated.

According to the analysis as of January 2015, there were a total of 582 empty and abandoned houses, and as the 'broken window theory' suggests, they were

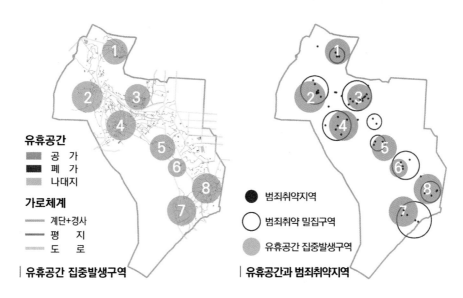

그림 4. 유휴공간과 범죄발생 현황(2015년1월 기준) / Clusters of empty houses and crimes

결과 50~200m 범위로 총 9개의 범죄 집중발생구역을 도출하였다. 8개소의 공폐가 집중발생구역과 9개소의 범죄 집중발생구역을 비교해본 결과 거의 중첩하고 있음을 알 수 있었다. 예상했던 대로 공폐가가 많은 곳에 범죄도 많이 발생하는 경향이 있었다.

그림 5. 장소통합적 도시재생사업 구상도 / Comprehensive plan for urban regeneration

이상과 같은 과정을 거쳐 확정한 아미·초장 도시재생 활성화계획은 '내일을 꿈꾸는 비석문화마을'을 비전으로 하여, '정주환경 개선, 근린경제 활력, 역사문화 보전, 주민참여 확산'의 총 4개 부문 14개 세부사업계획을 담았다.

'정주환경 개선' 부문은 평생 살고 싶은 마을을 조성하기 위해 산동네 피난민 주거지의 정주환경을 개선하는 사업으로서, 총 4개 세부사업을 계획하였다.

'근린경제 활력' 부문은 활기 가득한 마을을 조성하기 위해 쇠퇴한 고지대 근린상권의 활력

concentrated in the target area. After drawing the map into areas with a diameter of about (100 to 200) m, eight concentrated areas of occurrence were derived. The analysis showed that the empty houses mainly occur intensively around steep stairs and slopes, and dead ends where vehicles cannot pass. In addition, mapping the sites of 53 crimes that occurred in the region with the cooperation of the police office, a total of 9 crime occurrence zones were derived in the range (50–200) m. Comparison of the 8 concentrated empty house occurrence areas with the 9 concentrated crime occurrence areas showed that they almost overlapped. As expected, there was a tendency for many crimes to occur in places where there were many empty and abandoned houses.

The Activation Plan for Ami-Chojang Urban Regeneration, which was confirmed through the above process, contains 14 detailed projects in a total of four categories of 'improving settlement environment, revitalizing neighborhood economy, preserving history and culture, and expanding resident's participation'.

The 'improving settlement environment' category is to improve the environment of residential areas in shanty villages on the sloping area to create a village where residents would want to live for a lifetime, and a total of four detailed projects have been planned.

The 'revitalizing neighborhood economy' category is to restore the vitality of the declining neighborhood commercial facilities to create a lively village, and four detailed projects have been planned.

To create an attractive village, the 'preserving history and culture' category planned three detail projects to attract local tourism through the integrated use of scattered historical and cultural assets.

The 'expanding resident's participation' category is intended to support resident's participation through education and cultural program to make the village together, and

을 회복하는 사업으로서, 4개 세부사업을 계획하였다.

'역사문화 보전' 부문은 매력 넘치는 마을을 조성하기 위해 산재된 역사문화자산의 통합적 활용을 통해 지역관광 활성화사업으로서 3개 세부사업을 계획하였다.

'주민참여 확산' 부문은 함께 만드는 마을을 조성하기 위해 창작과 교육을 통한 주민참여를 확산하는 사업으로서, 3개 세부사업을 구성하였다.

각 부분별 세부사업을 간략하게 소개하면 다음과 같다.

정주환경 개선

'정주환경 개선' 부문은 산재한 노후주택과 공폐가, 좁은 차도와 위험한 골목길, 부족한 기초생활 인프라시설 등의 문제점에 대응하기 위한 사업이었다. 지역의 인적 자원과 협력하여 주민들의 주거환경을 개선하는 '마을지기 집수리사업단 운영', 공폐가와 나대지를 활용하여 주민들의 생활편의공간을 제공하고 휴식과 녹지공간을 조성하는 '마을베란다 공동이용장 조성사업'과 '우리동네 쌈지마당 만들기사업', 좁고 위험한 차도와 골목길을 CPTED기법을 활용하여 안전하게 개선하고, 안전한 버스 정류소를 제공하는 '안심골목 안전도로 조성사업' 등 4개의 세부사업을 계획하였다.

'마을베란다 공동이용장 조성사업'으로 비석마을 골목 깊숙한 곳의 나대지와 빈집을 활용하여 조성한 '골목빨래방'은 인근 주민들로부터 많은 환영을 받고 있다. 비석마을에는 집이 좁아서, 내부에 화장실이나 샤워장이 없는 집도 많다. 뿐만 아니라 부피를 많이 차지하는 세탁기를 골목에 내놓고 사용하거나 심지어 세탁기도 없는 집들도 많다. 골목이 좁고, 집 안 공간이 협소하여 큰 용량의 세탁기를 들여놓기가 어려웠다. 독거노인의 경우, 오랜 시간 세탁도 못한 이불과 옷을 계속 이용하게 된다. 이런 상황에서 좁은 골목에 오랫동안 방치된 빈집과 나대지를 활용하여 '골목빨래방'을 조성하고, 이불빨래도 할 수 있는 대형 세탁기와 건조기를 설치하여 저렴하게 이용할 수 있도록 하였다. 햇볕도 제대로 들지 않던 골목이 환해졌고, 주민들이 빨래하러 왔다가 서로 만날 수 있는 시간도 늘어났다. 물을 활용하는 빨래방의 특성을 활용하여 한켠에는 두 사람이 동시에 샤워할 수 있는 샤워실도 마련하여 어르신들이 이용할 수 있도록 하였다.

three detail projects have been planned.

A brief introduction of the detail project is as follows:

Improving the Settlement Environment.

The 'improving the settlement environment' category is to cope with problems such as old, decrepit, and empty houses, narrow streets and dangerous alleys, and insufficient basic living infrastructure facilities. This category includes the 'Operation of Housing Repair Team Project' in cooperation with local human resources, 'Neighborhood Communal Veranda Project', and 'Neighborhood Pocket Yard Project' to create facilities for living convenience, relaxation, and green space using empty houses, and the 'Safe Street and Alley Project' to improve narrow alley and road safely using CPTED.

The 'Alley Laundry Room', which was created using vacant land and empty houses deep in the alley of Tombstone Village as a project of the 'Neighborhood Communal Veranda Project', is being well received by nearby residents. Many houses in Tombstone Village are small, and lack internal toilets and/or showers. In addition, there are many houses that locate washing machines outside in the alleys, or don't even have washing machines at all. Due to the narrow alley and narrow space in the house, it is difficult to bring in a large washing machine. In the case of the elderly living alone, they continue to use blankets and clothes that they have not washed for a long time. In this situation, empty houses and vacant lands that have been left in narrow alleys for a long time were used to build an 'Alley Laundry Room', and large washing machines and dryers that can wash large blankets were installed for resident's use at inexpensive fee. The alley, where the sunlight didn't come in properly, brightened up, and the opportunities for residents to come and meet each other increased. Using the

그림 6. 공가와 나대지를 활용한 골목빨래방과 이를 이용한 미용 봉사활동 / Alley Laundry Room and volunteer hair dressers

근린경제 활력

아미동의 인구가 줄어들면서 식당이나 슈퍼, 부식가게 같은 마을의 상점은 하나둘씩 사라졌고, 젊은 세대가 유출되면서 어린이 상대의 학원이나 먹거리 가게도 폐업하였다. 치킨집이나 짜장면집이 사라지면 아이들은 즐겨먹는 간식도 제때 사 먹을 수 없게 된다. 주민들은 당장 생활에 필요한 식료품이나 편의용품을 사기 위해 마을버스를 타고 멀리 있는 시장에까지 가야 한다. 특히 신체적으로 쇠약한 고령자들은 이동성이 약하기 때문에 생활의 불편이 가중될 뿐만 아니라 필수약품도 사기 어렵기 때문에 고통받게 된다.

'근린경제 활력' 부문은 생활에 편의를 제공하는 상점들이 대부분 폐업하면서 쇠퇴한 마을의 근린상권의 활력을 회복하여 주민들에게는 생활의 편의와 함께 일자리를 제공하여 경제적 수익을 증대하는 것을 목적으로 계획하였다. 마을의 공가나 나대지와 같은 유휴공간의 소유주와 협의하여 필요로 하는 사람들과 매칭시켜 주는 '우리동네 복덕방사업', 관광객들에게 마을의 독특한 역사와 공간에 대한 체험과 뛰어난 조망을 제공하는 '비석문화마을체험 게스트하우스사업', 마당이 있는 단독주택에서 살고 싶은 은퇴자나 재택근무자들을 위한 '마당이 있는 미니주택 보급사업', 마을의 빈 상가를 리모델링하여 주민과 관광객들을 위한 상점을 운영하도록 지원하는 '근린상권 활력사업' 등 4개 세부사업이 계획되었다.

이 도시재생사업이 본격적으로 추진되던 시기에 부산지역에도 부동산 가격이 급등하면서 도시재생사업 지구에도 주택을 매입하려는 투자자들이 출입하기 시작하였다. '우리동네 복덕

characteristics of a laundry room using water, a shower room for two people to shower at the same time was also provided on one side, so that the elderly could use it.

Revitalizing the Neighborhood Economy

As the population of Ami-dong decreased, shops in villages, such as restaurants, convenience shops, and grocery shops, disappeared one-by-one, and as the younger generation left, academies and snack shops for children also closed. If a chicken restaurant or Chinese restaurant disappears, children are not able to buy and eat their favorite snacks on time. Residents must take a bus to the far market to buy the groceries and convenience items needed for their lives. In particular, the physically weak elderly suffer because of their weak mobility, and because it is difficult for them to buy essential medication.

The 'revitalizing the neighborhood economy' category was planned to increase economic profits with jobs and living conveniences by restoring the vitality of neighborhood commercial streets in declining villages, as most of the shops that provide convenience to life were closed. Four detailed projects have been planned, including the 'Neighborhood Real Estate Information Project', which matches the owners of idle properties, such as empty houses and vacant lands, in the village with prospective users; the Tombstone Culture Village Guest House Project〉, which provides tourists with opportunities to experience the unique history, space, and impressive city view; the 'Mini House Project' for retirees or telecommuters who want to live in detached houses with yards; and the 'Neighborhood Commercial Vitality Project' to help residents remodel and reopen empty shops in the village for tourists.

When this urban regeneration project was in full swing, real estate prices soared in Busan city, and investors who wanted to purchase homes began to enter the urban

그림 7. 근린상권 활성화사업 구상도 / Neighborhood Commercial Vitality Project

방사업'은 비석문화마을 주변으로 유휴공간에 대한 데이터를 구축했지만, 주민들에게 불필요한 오해를 줄 우려가 있어서 사업을 추진하지 못했다.

대상지의 가장 중요한 도로인 아미로는 과거 인구가 많은 시절에는 길 양편에 상점과 노점상이 늘어서 있었지만 현재는 슈퍼, 식당, 미용실 몇 개만 남고 거의 비어 있다. '근린상권 활력사업'에서는 아미로에 접한 빈 상가주택을 매입하여 리모델링하여 마을기업이나 사회적 경제조직에게 저렴하게 혹은 무상으로 임대하는 사업이다. 특히, 아미로에는 '김박사집'이라는 돌로 지은 주택이 공가로 남아 있었다. 1960년대에서 70년대에 걸쳐 '김박사'라는 별명으로 불린 집주인이 부인과 손수 산에서 돌을 운반해 와서 조금씩 조금씩 지은 집이었다. 독특한 이야기와 구조를 가진 주택을 어떻게 활용할 것인지에 대해 동네주민, 청소년 그리고 지역의 전문가들이 모여 2차례의 워크숍을 진행하여, 보존가치, 리모델링 방법과 향후 기능에 대해 논의하였다. 그 결과 하마터면 사라질뻔한 돌집은 리모델링을 거쳐 독특한 공간과 형태를 가진 건물로 보존될 수 있었고, 상업과 문화를 복합한 시설로 운영될 계획이다.

역사문화 보전

아미동 일대는 일제강점기의 일본인 공동묘지, 한국전쟁 시기의 피난민 주거, 산업화시대의 서민 주거지 등의 다양한 시대의 흔적들이 중첩되면서 형성되어 독특한 공간적 구조와 주거들이 남아 있다. 마을에는 여전히 일본인 비석들이 동네 곳곳에 산재해 있어서 과거의 어두

regeneration project area. The 'Neighborhood Real Estate Information Project' had built up data on idle spaces around the Tombstone Culture Village, but refrained from promoting the project due to concerns that it would be used by speculators, rather than by well-intentioned users.

Ami-ro, the most important road in the target site, had shops and street vendors lined up on both sides of the street in the past when there was a large population, but now only several shops, restaurants, and beauty salons remain. The 'Neighborhood Commercial Vitality Project' is a project that purchases and remodels empty shops and houses adjacent to Ami-ro, and rents them cheaply or free of charge to village residents or socio-economic organizations. In particular, in Ami-ro, a stone house called Dr. Kim's house had remained empty for a long time. From the 1960s to the 1970s, the landlord, nicknamed 'Dr. Kim', brought stones from the mountain with his wife and little-by-little built a stone house. Local residents, teenagers, and local experts gathered to hold two workshops on how to use the stone house with unique stories and structures to discuss preservation values, remodeling methods, and future functions. As a result, the stone house in danger of being demolished could be preserved as a building with a unique space and shape after remodeling, and there are plans to operate it as a cultural and commercial facility.

Preserving History and Culture

The Ami-dong area is formed by overlapping traces of various historical periods, such as the Japanese colonial era's Japanese cemetery, refugee shanty housing during the Korean War, and residential areas for the working class in the industrial era, leaving unique spatial structures and housing. Japanese tombstones are still scattered throughout the village, so the dark history of the past can be imagined.

운 역사를 짐작할 수 있다.

'역사문화 보전' 부문은 아미동의 독특한 역사와 생활문화자원을 활용하여 마을의 역사성과 정체성을 강화하고, 이를 통해 우리나라와 부산의 근대역사를 배우고 체험하는 기회를 방문객들에게 제공하려는 목적으로 계획되었다. 일제강점기 불교사찰인 총천사를 거쳐 일본인 묘지로 가는 주진입로였던 아미로의 역사성을 강화하는 '아미로 100년 근대역사가로화 사업', 마을의 주요 역사적 흔적을 연계하는 답사 루트이면서 동시에 관광객들로 인한 주민들의 사생활 침해를 최소화하려는 '아미·초장 탐방로 구축 및 홍보사업', 좁은 골목길과 주요 조망점 주변에 어지럽게 설치된 전선과 통신선을 정비하여 조망을 확보하려는 '하늘경관 정리사업' 등 3개 세부사업이 계획되었다.

2014년 아미로 중간 곡각지의 도로확장공사 과정에서 일본인 묘지 하부구조와 그 위에 한

그림 8. 2014년 도로확장공사 중 발견된 비석주택 / Tombstone House found in 2014

The 'Preserving History and Culture' section was designed to strengthen the history and identity of the village by utilizing Ami-dong's unique history and living cultural resources, and to provide visitors with opportunities to learn and experience the modern history of Korea and Busan. Three detailed projects were planned, including the '100-year-old Ami-ro Modern History Road Project', which strengthens the historicity of Ami-ro, the main access road to the cemetery via the Buddhist temple during the Japanese colonial era; the 'Ami-Chojang Historical Trail and Promotion Project', to minimize the invasion of residents' privacy by tourists while providing a tour route connecting historical sites of the village; and the 'Sky Landscape Clean-up Project', which aims to secure views by overhauling the dizzyingly installed wires and communication lines around narrow alleys and major viewing points.

In 2014, road expansion work in the middle of Ami-ro found a Japanese cemetery substructure and a refugee shack built on it during the Korean War. It is difficult to say that this residence, which is called Tombstone House,[1] had cultural value in itself, but it is regarded as a valuable asset that contains memories of the Japanese colonial era, the Korean War, and the subsequent industrialization era. As part of the '100-year-old Ami-ro Modern History Road Project', the Activation Plan planned the 'Refugee Life Museum and Information Center' to preserve Tombstone House, old barbershops, and vacant houses around them, restore the living situation during the refuge, and show visitors to the village. Currently, the houses around Tombstone House are being purchased as urban regeneration projects, and the interiors are being created as museums that reproduce the life of the time. It is planned to restore Tombstone House

---

1 Marking the 70th anniversary of liberation in 2015 and the 65th anniversary of the Korean War, the BMC promoted the listing of historical and cultural resources of Busan, the temporary capital during the Korean War, as a World Heritage Site. As Ami-dong, a representative refugee area, was included as one of its heritages, professional survey and research were conducted. Based on newspapers found inside the wall of Tombstone House, it was recognized as a shack dating to the Korean War, and was announced as the first registered cultural asset in Busan in January 2022.

국전쟁 시기에 지은 피난민 판잣집이 발견되었다. '비석주택'[2]이라고 불리게 된 이 주거는 그 자체로 문화재적인 가치가 있다고 보기 어려웠지만, 일제강점기와 한국전쟁기 그리고 그 이후 산업화시대의 기억을 동시에 담고 있는 소중한 자료라고 생각하였다. 활성화계획에서는 '아미로 100년 근대역사가로화사업'의 일환으로 비석주택과 그 주변의 오래된 이발소와 빈 주거들을 그대로 보존하고 피란 당시의 생활을 복원하여 마을을 찾는 탐방객들에게 보여주는 '피난생활박물관 및 정보이용원 조성사업'이 계획되었다. 현재, 비석주택 주변 주거들은 도시재생사업으로 매입하여 내부에는 당시의 생활모습을 재현한 박물관을 조성하였고, 비석주택은 부산시 등록문화재 지정을 계기로 복원작업을 시행할 계획이다.

주민참여 확산

주민들이 떠나는 마을에서 살고싶은 마을로 바꾸기 위해서는 단순히 물리적 환경만을 개선하는 사업만으로는 부족하다. 주민들 스스로 도시재생사업에 참여하면서 다양한 프로그램을 통해 마을공동체를 활성화하여, 궁극적으로는 도시재생사업으로 조성되는 게스트하우스나 근린상점을 운영할 수 있는 주민들의 역량을 강화해야 한다. 따라서 '주민참여 확산' 부문에서는 도시재생대학이나 마을축제 등을 통해 주민들을 교육하고 공동체활동을 지원하는 '주민역량강화사업', 마을에 대한 애착이 약한 청소년들의 문화, 예술, 창작활동 등을 지원하는 프로그램과 공간을 제공하는 '아미·초장 하자마을', 저소득 가정의 어린이, 청소년, 노약자, 취약계층의 질병을 예방하고 초기에 진료하는 지역 전담의사를 지정 운영하는 '우리동네 건강주치의 사업' 등의 세부사업을 계획하였다.

계획된 세부사업 중 '우리동네 건강주치의 사업'은 산동네의 좁고 노후한 주택에 거주하는 저소득가정의 어린이들이 척추측만증이나 치과 관련 질병에 취약한 특성이 있지만, 우리나라의 건강보험제도가 잘 되어 있고, 치과교정 등은 과다한 비용이 발생하는 등의 문제가 있어 기대만큼의 결실을 얻지는 못했다.

---

2 부산시는 2015년 광복 70주년, 한국전쟁 65주년을 맞아 한국전쟁 당시 임시수도였던 부산의 역사문화자원을 '피란 수도 부산 유산'으로 세계유산으로 등재를 추진하였다. 대표적인 피난민 주거지였던 아미동도 그 유산 중에 하나로 포함됨에 따라 전문적인 조사와 연구가 진행되었고, 벽체 내부의 신문지를 바탕으로 한국전쟁 당시의 판잣집임이 인정되어 2022년 1월에는 최초의 부산시 등록문화재로 고시되었다.

그림 9. 비석주택과 주변 건물을 활용한 피난생활박물관 계획 / Refugee Life Museum

in the wake of the designation of registered cultural asset in Busan.

Expanding Resident Participation

To change from a village where residents are leaving to a village where they want to live, simply improving the physical environment is not enough. Residents themselves should participate in urban regeneration projects to revitalize the village community through various programs, and this will ultimately strengthen their ability to operate guest houses or neighborhood shops created by urban regeneration projects. Therefore, in the 'Expanding Resident's Participation' section, three detailed projects were planned of the 'Resident Empowerment Project', to educate residents and support community activities through urban regeneration school and village festivals; 'Ami-Chojang Let's-do Village', which provides space and programs to support culture,

그림 10. 어린이부터 할머니까지 마을주민들이 참여해서 진행하는 아미동 마을축제

　'주민참여 확산' 부문에서 가장 크게 결실을 거둔 사업은 '아미·초장 하자마을'이었다. 마을의 유휴공간을 활용하여 '청소년 창작스페이스'를 제공하는 물리적 사업에 앞서 도시재생사업 초기부터 청소년들의 활동을 지원하는 '아미·초장 청소년 마을학교' 프로그램을 운영하였다. 특히 아미동에서 오랜 시간 동안 저소득 어린이들을 위한 다양한 프로그램을 운영해 오고 있던 부기사(부산기독교종합사회복지관)는 중학교에 진학한 이후에는 지원 프로그램이 없어서 아쉬움을 가지고 있었다. 도시재생지원센터에서는 부기사와 함께 지역의 중·고등학교 학생들을 모아 자체적인 모임의 만들도록 지원하고, 마을을 기반으로 어떤 일을 하고 싶은지를 스스로 찾도록 하였다. 청소년들은 모임을 이름을 청소년기획단 YOLO라고 정하고, 가정 형편상 다니지 못한 여행을 다니면서 세상을 널리 보기 위한 목적으로 여행학교라는 프로그램을 제안하였다. 2017년 청소년기획단 YOLO는 여름방학을 이용하여 제주를 목적지로 정하고 4·3사건과 관련된 장소를 중심으로 스스로 루트를 선정하고 대중교통만을 이용해서 첫 여행을 다녀 왔다. 스스로 기획하고 실행한 여행에서 많은 배움을 얻은 학생들은 겨울방학에는 강원도 철원의 DMZ 일대를 다녀왔고, 2018년에는 독립운동가들의 역사를 찾아 러시아 블라디보스토크까지 여행하였다.

　여행학교와 함께 성장한 청소년기획단 YOLO는 마을에도 관심을 가지게 되어 마을축제, 역사탐방, 생태여행, 제로웨이스트, 슬로라이프 등 다양한 활동을 진행하고 있다. 1기 기획단이 대학에 진학하면서 다음 세대로 계속 이어져 현재 5기가 활동하고 있으며, 대학에 진학한 선배들도 마을을 위한 다양한 활동을 계획하고 실행하고 있다. 이러한 성과를 바탕으로 2018년

art, and the creative activities of teenagers who have a weak sense of belonging to the village; and the 'Neighborhood Health Doctor Project', which designates neighborhood doctors to prevent and treat diseases of low-income families.

Among the planned detail projects, the 'Neighborhood Health Doctor Project' has not paid off as much as expected, since Korea's health insurance system is well established while orthodontics require a large amount of money, although there is some tendency for the children from low-income families living in narrow and old houses to be vulnerable to scoliosis or dental-related diseases.

The project that paid off the most in the 'Expanding Resident's Participation' section was the 'Ami-Chojang Let's-do Village'. Prior to the physical project of building 'Creation Space for Teenager' by utilizing the idle space in the village, the 'Ami-Chojang Villlage School for Teenagers' program was implemented to support youth activities from the beginning of the urban regeneration project. In particular, the BCSWC, who had been running various programs for children from low-income families for a long time in Ami-dong, was disappointed that after entering middle school, there was no support program. The Urban Regeneration Support Center, along with BCSWC, gathered local middle and high school students to help them form their own organization, and get them to find out what they want to do based on the village. Teenagers named the organization 'Youth Planning Group YOLO', and derived a program called Travel School for the purpose of seeing the world by traveling, which their families could not otherwise afford due to family circumstances. In 2017, the Youth Planning Group YOLO used summer vacation to undertake their first travel to Jeju-do, selecting routes on their own, focusing on places related to the Jeju April 3 incident, and using only public transportation. Students who learned a lot on their own planned and implemented the first travel to the DMZ area in Cheolwon, Gangwon-do during the winter vacation, and

그림 11. 2018 도시재생한마당에서 대상을 수상한 청소년기획단 YOLO
Youth Planning Group YOLO awarded grand prize

도시재생한마당에서 주민참여 경진대회 대상을 수상하기도 하였다.

아미·초장 도시재생 활성화계획의 특징

　아미·초장 도시재생 활성화계획의 가장 큰 특징은 4개 분야 14개 세부사업이 서로 연계하면서 진행하도록 계획되었다는 점이다. 예를 들면 '마을지기 집수리사업단 운영' 사업은 노후한 주거의 집수리를 실행하는 사업이지만, 그 이외에도 공가나 폐가를 활용하여 주민들을 위한 시설을 조성하는 '마을베란다 공동이용장 조성사업', '비석문화마을 체험 게스트하우스 사업'에서 추진하는 집수리 등도 지원하였다. 공폐가와 나대지 등 유휴공간에 대한 정보를 구축하는 '우리동네 복덕방사업'은 이들 유휴공간을 필요로 하는 '마당이 있는 미니주택 보급사업'이나 '근린상권 활력사업'을 위한 공간정보를 제공하도록 하였다.

in 2018, they travelled to Vladivostok, Russia, in search of the history of independence activists.

The Youth Planning Group YOLO that has grown up with the Travel School has become interested in the villages, and is conducting various activities, such as village festivals, history tours, ecological trips, and zero-waist and slow-life programs. As the first members of YOLO went to college, the project continued to the next generation, and the fifth generation is currently active; and seniors who went to college are also planning and implementing various activities for the village. Based on these achievements, the group won the grand prize i the residents' participation contest in the 2018 National Urban Regeneration Festival.

Characteristics of the Activation Plan for the Ami-Chojang Urban Regeneration Project

The most distinct feature of the Ami-Chojang Urban Regeneration Project is that 14 detailed projects in four categories were planned to be linked with each other, and so support one another. For example, the 'Operation of Housing Repair Team Project' is a project to repair old houses, but it also supports the 'Neighborhood Communal Veranda Project' and the 'Tombstone Culture Village Guest House Project' to create facilities for residents and visitors using empty or abandoned houses. The 'Neighborhood Real Estate Information Project', which builds information on idle spaces, such as empty houses and vacant land, provides information for the 'Mini House Project' or 'Neighborhood Commercial Vitality Project' that requires these idle spaces.

In particular, the 'Resident Empowerment Project' is planned to provide education

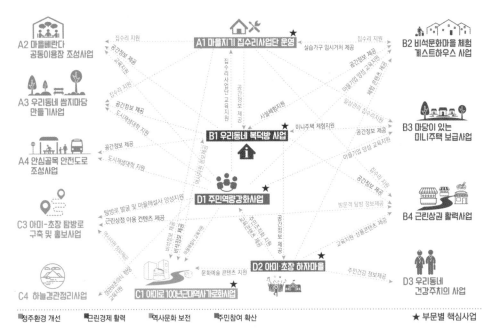

그림 12. 세부사업 연계방안 / Relationship of detail projects

　특히 '주민역량 강화사업'은 주민들이 도시재생에 대한 이해를 넓히는 교육만을 제공하는 것이 아니라 세부사업에 주민들이 참여할 수 있는 실질적인 역량을 강화하는 교육을 제공하도록 계획하였다. 예컨대, 2017년 '아미·초장 마을해설사 심화과정 및 2차 아카이빙 과정' 프로그램으로 아미동 주민들과 서구 지역 내의 구민 14명을 마을해설사로 양성하였으며, 그 중 11명이 해설사로 활동하고 있다. 이렇게 교육받은 해설사들은 방문객이 많이 찾는 주말 동안 비석주택 주변에 배치되어 마을의 유래와 비석주택을 포함한 마을 곳곳의 비석흔적과 역사를 설명하였다. 아미·초장 도시재생사업이 전국적으로 우수사례로 알려지면서 부산, 경남지역뿐만 아니라 전국의 도시재생사업을 시행하고 있는 지자체에서 공무원, 지자체 의원 그리고 주민들이 도시재생 선진사례 견학을 위해 찾아오고 있다. 그 결과 2018년 한해 동안 마을해설사 활동건수도 연 337건으로 크게 증가하였다.[3]

3　부산광역시 서구, 『아미·초장 도시재생 평가·모니터링 2018』, 부산광역시 서구, 2019, p.39

that strengthens residents' practical capabilities to participate in detailed projects, as well as broaden their understanding of urban regeneration. For example, in 2017, the Ami-Chojang Village Guide Advanced Course and 2nd Archiving Course programs trained 14 residents in the village and Seo-gu District, of which 11 are active as village guides. These educated guides were stationed around Tombstone House during the weekend when many tourists visited, explaining the origin of the village, the historical traces and tombstones throughout the village, including Tombstone House. As the Ami-Chojang Urban Regeneration Project is becoming known as an excellent case nationwide, local government officials, local council members, and residents are visiting as field trips from both Busan and the Gyeongnam area, and from all over the nation. As a result, the number of village guide activities in 2018 also increased significantly to 337 per year.

Governance System of the Ami-Chojang Urban Regeneration Project

Looking at the activation plans of most urban regeneration projects currently underway in Korea, a diagram showing public–private governance in which residents and administration cooperate is always suggested. Securing a structure in which residents and administration cooperate and residents participate in the process of planning and implementation of the project and present opinions, and in which their opinions are actually reflected in urban regeneration projects, will be the core of public–private governance. The problem is that when visiting urban regeneration project sites in Korea, public–private governance is often formed after establishing an

아미·초장 도시재생사업의 거버넌스

현재 우리나라에서 진행되는 대부분의 도시재생사업의 활성화계획을 보면 주민과 행정이 협력하는 민관 거버넌스를 보여주는 추진체계 다이어그램이 그려져 있다. 실제로 주민과 행정이 협의하고, 주민이 사업의 계획, 실행과정에 참여하면서 의견을 제시하고, 그 의견들이 실제로 도시재생사업에 반영되는 구조를 확보하는 것이 민관 거버넌스의 핵심일 것이다. 문제는 우리나라 도시재생사업 현장을 보면 도시재생 활성화계획을 수립하고 그 계획이 국가의 지원사업으로 선정된 이후부터 민관 거버넌스를 만드는 경우가 많다는 점이다.

민관 거버넌스는 도시재생 활성화계획을 작성하는 단계부터 지역의 다양한 주민주체, 지역의 민간조직, 공공기관들과 만나면서 도시재생사업에 관심이 있고, 사업에 참여하고자 하는 지역 주체들을 발굴하여 사업추진협의회(위원회)를 구성하는 것이 첫 출발이라고 할 수 있다. 이렇게 결성된 사업추진협의회가 작동하는 가장 좋은 방식은 정기적이고 의무적으로 회의를 개최하면서, 각 시기별로 진행되거나 착수하는 사업과 관련되는 여러 가지 안건들을 공개적으로 논의하고 결정하는 것이다.

이 사업을 위한 핵심적인 민관 거버넌스로서 '아미·초장 도시재생운영위원회'는 원활한 사업추진을 위한 행정, 주민, 지역기관, 중간지원조직의 협의체로서, 2015년 4월 '도시재생활성화 이행협약'을 통한 예비사업추진협의회로 공식발족하였다. 도시재생사업으로 선정되어 본격적으로 활성화계획을 수립하면서, 2016년 2월, 행정-전문가-주민의 공동의결기구로서 총 17명의 위원으로 조직을 재구성하였다. 행정협의회 대표 부구청장, 중간지원조직(현장지원센터)의 대표 총괄 코디네이터, 주민협의체 대표 3인이 공동의장을 맡아 매월 1회 정기회의를 꾸준히 진행하면서 사업추진과 관련된 사항들을 논의하고 결정하였다.

물론 '운영위원회'에서 주민대표들이 사업의 진행과정이나 사업의 내용 등에 대해서 불만을 제기하고, 수정을 요구함으로써 갈등이 발생하고, 사업의 추진이 느려지는 경우도 있었다. 이런 이유로 행정에서는 사업추진위원회의 개최에 대해 부정적인 생각을 가지게 되는 경우가 많고 정기적인 회의 개최를 기피하는 경우도 발생하였다. 하지만 현실적으로 모든 주민이 도

activation plan of urban regeneration, and the plan is selected as a national support project.

The first step must be to organize the project steering council (committee) from the stage of preparing an activation plan while meeting and discovering public and private organizations in the region that are interested and want to participate in urban regeneration projects. The best way for this project steering council to actually work is to hold meetings regularly and on a mandatory basis to discuss and decide openly on various agendas related to projects,

As a key public-private governance for this project, the 'Ami-Chojang Urban Regeneration Steering Committee' was officially launched in April 2015.as a preliminary council of resident, local government, local organizations, and supporting centers through the 'Agreement on Implementation of Urban Regeneration'. After being selected as an urban regeneration project by national government, the Committee was reorganized into a total of 17 members of joint administrator-expert-resident organization, while establishing a main activation plan in February 2016. The Committee with three co-chairs of the deputy mayor of Seo-gu District Office, the general coordinator of the urban regeneration supporting center, and the representative of the residents' group, held regular meetings once a month to discuss and decide on agendas related to the project.

Of course, there were cases in which the residents' representatives complained about the progress of the project or the details of the project, and demanded correction, resulting in conflicts, and slowing the implementation of the project. For this reason, the administration officials often have negative thoughts about hosting of the project steering committee, and there are cases where regular meetings are avoided. However, the reality is that residents will not be able to become active subjects

그림 13. 아미·초장 도시재생운영위원회 실무회의 모습 / Meeting of Urban Regeneration Steering Committee

시재생사업에 참여할 수 없는 상황에서 마을에서 가장 활발하게 활동하고 주민들과 네트워크가 많은 주민대표들이 도시재생사업에서 주민들의 목소리를 대변하고, 사업을 어느 정도 결정할 수 있는 권리를 가지지 못한다면 주민들은 도시재생사업의 주인이 되지 못할 것이다.

맺으며

도시재생사업은 사회적, 경제적, 물리적, 환경적 재생을 포함하는 통합적인 사업으로서 사업기간이 최소 4년 이상 소요되는 장기적인 프로젝트이다. 우리나라의 행정에서는 담당공무원이 일반적으로 2년에 한 번씩 보직을 순환한다. 도시재생사업의 계획을 수립할 당시 참여한 공무원이 도시재생사업의 실행에는 참여하지 못하는 경우가 대부분이다. 시장이나 구청장이 바뀌는 경우도 적지 않다. 이 경우 바뀐 담당자의 성향에 따라 도시재생사업의 추진방식이나, 심지어 내용이 바뀌기도 한다. 이런 경우를 대비하여 처음부터 끝까지 사업을 지원하는 도시재생 현장지원센터와 민간인 총괄 코디네이터를 임명하고 있다.

2020년에 닥쳐온 코로나19로 인해 우리나라의 많은 도시재생사업들이 큰 타격을 받았다. 사회적 거리두기가 의무화되고 여러 사람들의 모임이 제한되면서 아미-초장 도시재생사업도 주민역량 강화교육과 민관학 거버넌스 운영이 많이 위축되었고, 사업은 행정 주도로 진행되

of urban regeneration projects unless the representatives of residents have the right to represent the voices of the people affected, and decide to some extent the details of the project.

Closing Remarks

The urban regeneration project is an integrated project that includes social, economic, physical, and environmental regeneration, and is a long-term project that will take at least 4 years. In the administration system of Korea, public officials in charge generally circulate their positions once every two years. In most cases, public officials who participated at the time of the planning process for an urban regeneration project cannot participate in the implementation process of that urban regeneration project. There are many cases where the mayor or the head of the district office changes. In this case, the operation process or even contents of the urban regeneration project may change, depending on the preference of the new official in charge. To prevent these cases, urban regeneration site support centers and civilian general coordinators are appointed to support the project from beginning to end.

Many urban regeneration projects in Korea have been badly impacted by the COVID-19 pandemic that hit in 2020. As social distancing became mandatory and meetings of people were restricted, the Ami-Chojang urban regeneration project also shrank a lot in resident empowerment education and public–private–academic governance, and the project began to be led more by administration. As vacant houses were renovated and transformed into new buildings through urban regeneration projects, and new buildings began to be built, local demands to use them in a different

기 시작했다. 도시재생사업을 통해서 빈집이 리노베이션되어 새로운 건물로 바뀌고, 신축건물이 들어서기 시작하자, 원래 활성화계획과 다른 방식으로 이용하려는 지역의 요구가 나타나기 시작했다.

또 한 가지 문제는 도시재생사업 예산 100억이 적지 않은 금액이긴 하지만 아미동과 초장동의 부족한 인프라를 충족시키고, 거주환경을 일시에 개선하기에는 턱없이 부족한 금액이기도 하다. 그로 인해 도시재생사업으로 많은 부분이 개선되었지만, 전체적으로 볼 때, 아미동 일원은 여전히 생활하기 불편하고, 인구는 계속 감소하고, 빈집도 계속 증가하고 있다.

감천문화마을의 사례에서 보았듯이 하나의 사업으로 마을이 재생되기는 어렵다. 아미동 비석문화마을도 도시재생사업을 계기로 주민들과 행정과 지역의 전문가들이 함께 장기적인 비전을 가지고 다양한 사업들을 통합적으로 추진해 가는 것이 바람직하다. 도시재생사업을 통해 역량을 강화한 지역의 주민, 그리고 도시재생에 참여한 부기사와 같은 지역의 기관들, 그리고 마을에 관심을 가진 지역의 대학 사이의 협력이 앞으로도 지속되기를 기대한다.

참고문헌

1. 부산광역시 서구, 아미·초장 도시재생 활성화계획, 2016
2. 부산광역시 서구, 아미·초장 도시재생 평가·모니터링 2018, 부산대학교 산학협력단, 2019
3. 오재환 외, 산복도로 르네상스 사업 평가를 통한 부산형 도시재생 방향 설정, 부산: 부산연구원, 2021
4. 우신구, 아미동이야기 : 포개진 삶, 겹쳐진 공간, 서울: 국립민속박물관, 2020

way from the original Activation Plan began to emerge.

Another problem is that although the budget for urban regeneration projects is significant, it is not enough to meet the insufficient infrastructure in Ami-dong and Chojang-dong and improve the living environment at once. Although many parts have been improved by the urban regeneration project, overall, the Ami-dong area is still inconvenient to live in, the population continues to decrease, and the number of vacant houses continues to increase.

As seen in the case of Gamcheon Culture Village, it is difficult to regenerate the village through one single project. In the wake of the urban regeneration project, it is desirable for residents and administrative and local experts to promote various projects in an integrated manner with a long-term vision. We hope that cooperation between local residents who have strengthened their capabilities through urban regeneration projects, local institutions such as the BCSWC who participated in urban regeneration, and local universities interested in local communities, will continue in the future.

References

1. Seo-gu District Office of Busan Metropolitan City, Activation Plan for Ami-Chojang Urban Regeneration Project, 2016
2. Seo-gu District Office of Busan Metropolitan City, The 2018 Report of Evaluation and Monitoring of Ami-Chojang Urban Regeneration Project, Industry-Academic Cooperation Foundation of Pusan National University, 2019
3. Oh Jae-hwan et al., Busan-style Urban Regeneration Direction through the Evaluation of the Sanbok-road Renaissance Project, Busan: Busan Research Institute, 2021
4. Woo Shin-Koo, Ami-dong Story: Overlapping Life, Layered Space, Seoul: National Folk Museum of Korea, 2020

부산 태극도마을에서 감천문화마을로

독특한 경관을 활용한 일련의 마을재생사업

우신구 [부산대학교 건축학과 교수]

들어가며

한때 태극도 마을로 알려졌던 감천문화마을은 부산광역시 사하구 감천2동 천마산과 옥녀봉 사이의 해발 100~200m의 경사면에 위치한다. 면적은 약 0.62㎢이며, 인구는 2021년 현재 4,219세대 8,936명이며 65세 이상의 인구가 22%를 차지하는 고령화마을이다. 국제시장, 광복로, 용두산공원, 자갈치시장, 영도다리 등 전국적으로 잘 알려진 부산의 명소들이 불과 4km 내외의 가까운 거리에 자리 잡고 있지만, 2009년 이전만 해도 외부인에게 거의 알려지지 않았고 마을을 찾는 외부 방문객도 거의 없었다.

이렇게 잘 알려지지 않았던 마을이 지금은 우리나라의 어느 관광지보다 유명한 마을이 되었다. 감천문화마을을 방문하는 관광객들의 반응은 대개 비슷하다. '옹기종기 모여 있는 알록달록한 집'들이 참 예쁘다는 것이다. 모양이나 크기 그리고 색채가 비슷비슷한 조그만 집들이 급경사의 등고선을 따라 가지런히 늘어선 모습을 보고 '한국의 산토리니' 혹은 '한국의 마추픽추'라고 부르기도 한다.

불과 10여 년 사이에 부산의 한적한 산동네를 국제적인 관광지 감천문화마을로 변화시킨 것은 하나의 프로젝트가 아니라 마을미술프로젝트(2009), 콘텐츠융합형 관광협력사업(2010), 산복도로 르네상스사업(2011), 도시활력증진지역개발사업(2014), 새뜰마을사업

From Taegeukdo Village to Gamcheon Culture Village, Busan

A Series of Urban Regeneration Projects Using a Unique Townscape

Shinkoo Woo (Professor, Pusan National University, Dept of Architecture)

Introduction

Gamcheon Culture Village, once known as Taegeukdo Village, is located on a slope at (100–200) m above sea level between Cheonmasan Mountain and Oknyeobong Peak in Gamcheon 2-dong, Saha-gu, Busan Metropolitan City (BMC). As of 2021, the population was 8,936 in 4,219 households with a village area of about 0.62 km^2, and it was an aging village with 22 % of the population aged 65 or older. Busan's well-known attractions, which include Gukje Market, Gwangbok-ro, Yongdusan Park, Jagalchi Market, and Yeongdo Bridge, are located within only 4 km distance, but until 2009, the village was rarely known to outsiders, and attracted few visitors.

Such an unknown village has now become the most famous tourist destination in Korea. Tourists visiting Gamcheon Culture Village usually react similarly. They say that the colorful houses gathered together look very pretty. Small houses with similar shapes, sizes, and colors are neatly lined up along the steep contours, so they are sometimes called Santorini or Machu Picchu in Korea.

In just over a decade, a village in Busan was turned into an international tourist

(2015), 주거환경관리사업, 경관보전형 지구단위계획 등연속적으로 시행된 다양한 사업들이다. 하지만 무엇보다도 이 마을을 독특하게 만든 것은 이 마을만의 독특한 역사와 형성배경이다.

감천문화마을의 역사와 형성배경

감천문화마을의 독특한 풍경은 태극도라는 종교와 깊은 관계를 가지고 있다. 한국전쟁 당시 부산으로 피난 온 충청도와 경북지역의 태극도 교인들이 보수동 하천 주변에 판잣집을 짓고 모여 살았다. 전쟁이 끝나자 도심의 판자촌을 철거하려는 부산시와 태극도 교단은 협의를 통해 1955년 지금의 감천2동 일대로 집단으로 이주하였다. 이 지역은 주거지가 되기에는 상당히 급경사지였다. 그럼에도 불구하고 태극도 교단이 이 지역을 집단 거주지로 선택한 이유는 몇 가지로 추정된다.

그림 1. 천마산과 옥녀봉 사이의 계곡에 자리잡은 감천문화마을 / Gamcheon Cultural Village located between two mountains

destination and became Gamcheon Culture Village, because various projects, such as the Community Art Project (2009), Cooperation Project of Content & Tourism (2010), Sanbok-road Renaissance Project (2011), Urban Vitality Promotion Project (2014), Saetteul Community Project (2015), Residential Environment Management Project, and District Unit Planning for Townscape Conservation, were implemented. However, above all, what made this village unique is its unique history and background.

History and Background of Gamcheon Culture Village

The unique townscape of Gamcheon Culture Village shares a deep relationship with the religion of Taegeukdo (https://en.wikipedia.org/wiki/Gamcheon_Culture_Village), a branch of Jeungsanism. During the Korean War, believers of Taegeukdo in Chungcheong Province and Gyeongsangbuk-do Province, fled to Busan, built shacks around the river in Bosu-dong, and lived together. After the war ended, Busan City Government, which tried to demolish the shantytown in the city center, tried in 1955 to forge an agreement with Taegeukdo leaders through negotiation to move in groups to the current Gamcheon 2-dong area. The area was a fairly steep slope to become a residential area. Nevertheless, there are several reasons why the Taegeukdo leaders agreed to this area as a collective settlement.

First, the village was close to downtown, which at the time was the center of all economic activities in Busan. Areas close to Jagalchi Market and Gukje Market, which provided a lot of job opportunities, would have been advantageous for refugees without ties to Busan to make a living. Next is the religious reason, because the leaders of Taegeukdo viewed the narrow valley between two mountains, overlooking the sea

그림 2. 감천문화마을의 독특한 마을경관 / Unique townscape of Gamcheon Cultural Village

우선, 이 지역이 당시 부산의 모든 경제활동의 중심이었던 중구 일대와 가깝기 때문이다. 지역에 연고가 없는 피난민들이 생계를 유지하려면 일거리가 많은 자갈치시장, 국제시장 등과 가까운 지역이 유리했을 것이다. 다음으로는 종교적인 이유인데, 태극도 지도자들은 멀리 감천만의 바다가 내려다보이는 천마산과 옥녀봉 사이의 좁은 계곡을 하늘이 감추어 놓은 길한 땅, 즉 '천장길방(天藏吉方)'의 명당으로 보았기 때문이다(재단법인 태극도 홈페이지 http://www.tgd.or.kr/). 마지막으로 이 계곡이 주변의 시가지와 단절되어 그들만의 독립적인 종교공동체를 형성하기에 적합한 곳으로 생각하였다.

황무지에 가까운 급경사지를 집을 지을 택지로 조성하는 작업은 거의 인력에 의존하였다. 주변 산에서 자연석을 옮겨와 약 3m 높이의 석축을 쌓고 등고선을 따라 약 5~6m 폭의 좁고 편평한 택지를 조성하였다. 이 모든 작업은 중장비나 기계의 도움 없이 오로지 인력에 의지하였다. 이 계단형 부지 위에 보수동 판잣집을 해체하여 그 부재를 운반해 와서 긴 판자집으로 다시 조립하였다. 긴 판자집의 내부는 부엌 하나와 방 한 칸 정도가 들어갈 정도의 폭으로 균등하게 칸막이를 하여 각 세대에게 나누어 주었다. 문이나 창문도 제대로 없이 널빤지로 벽을

of Gamcheon Bay in the distance, as a propitious site for Cheonjanggilbang, a place hidden by heaven (Taegeukdo Foundation website, http://www.tgd.or.kr/). It might have been considered that by being cut off from the surrounding urban areas, this valley would be suitable for forming its own independent religious community.

The work of transforming a steep slope close to a wasteland into a housing site relied almost entirely on manpower. Natural stones were collected and moved from the surrounding mountains to build stone retaining walls of about 3 m high, creating a narrow and flat land of about (5–6) m width along the contour line. All of this work was performed without the help of heavy equipment or machinery. The houses of the shantytown in Bosu-dong were dismantled, transported by a truck supported by the Busan City Government, and reassembled into long wooden shacks on this stepped terraced site. Partitions were evenly placed inside to accommodate one kitchen and one room, and provided to each household. Old photographs of the early shacks of the village, which blocked the walls with wooden boards without proper doors or windows, and covered the roofs with roofing paper, reveal how economically needy they were.

Unlike sloped residential areas in other regions, Gamcheon Culture Village has houses of a certain size arranged side-by-side, which originated from the historical background of Taegeukdo Village, which moved collectively by building 'housing of similar size on similar land shape'.

Through the industrialization era in the 60s and 70s, and as the population of Busan increased rapidly, the size of Taegeukdo Village, located in Gamcheon Valley, grew larger. In the early 1990s, the population of Gamcheon 2-dong also increased to 29,000. Unlike the growth of the village, various problems occurred inside the Taegeukdo religious organization. When in 1958, the religious leader of Taegeukdo died suddenly, a conflict arose within the group over the initiative, and in 1968, it finally separated into

막고, 루핑지로 지붕을 덮은 초기 주거 사진을 보면 그들이 얼마나 경제적으로 궁핍한 상황이었는지 알 수 있다.

감천문화마을이 다른 지역의 산동네 주거지와 달리 일정한 크기의 집들이 나란히 배치되어 있는 경관은 '비슷한 땅 모양에 비슷한 크기의 주거'를 동시에 지어 집단적으로 이주한 태극도 마을의 형성배경에서 비롯된 것이다.

1960년대와 70년대 산업화시대를 거치면서 부산의 인구가 급증하자 감천계곡에 자리잡은 태극도마을의 규모는 점점 더 커졌다. 감천2동의 인구도 1990년대 초에는 2만9천명까지 증가하였다. 마을의 성장과는 달리 태극도 교단 내부에는 여러 가지 문제가 발생하였다. 1958년 태극도 교주가 갑자기 사망하자 교단 내부에는 주도권을 둘러싼 갈등이 발생하였고 끝내 1968년 신파와 구파로 분리되었다.

주민들은 차츰 종교보다는 경제를 더 우선하기 시작했다. 가족수의 증가, 자녀의 성장, 양호한 주거환경에 대한 양적·질적 요구는 초기에 지은 판잣집으로는 감당할 수 없었다. 이에 따라, 균등한 크기와 구조를 가지고 있던 주거를 각 세대의 경제적 능력에 따라 끊임없이 공간적으로 확장하고 개량하였다. 단열과 화재에 취약한 루핑지와 나무판으로 지은 판잣집을 블록벽/슬레이트 지붕으로 개량하였고, 형편에 따라 다시 벽돌벽/슬래브 지붕으로 개축하였다. 부족한 공간은 다락방을 만들거나, 형편이 나은 사람들은 2층 슬래브집을 지었다. 이러한 확장과 개량, 증축, 신축은 태극도 교단과 무관하게 각 세대의 경제적 능력에 따라 개별적으로 진행되었다.

1955년부터 2000년대까지 감천문화마을의 주거의 변화는 '최소한의 계단형 택지에 최소한의 재료를 이용하여 최소한의 공간을 가진 공동의 주거'가 제각각 다른 모습을 가진 집으로 나뉘는 과정이었다. 이 변화는 또한 초기의 획일적이었던 경관이 '알록달록한 집들이 옹기종기 모여 있는' 다양성을 가진 경관으로 변화해 온 과정이기도 하다.

two groups.

Naturally, residents began to prioritize the economy over religion. The initially built poor shacks could not meet the increase in the number of families, the growth of children, and the quantitative and qualitative demands for a good living environment. Accordingly, housing, which had an equal size and structure, was constantly expanded and improved according to the economic capability of each individual family. The roofing paper, which is vulnerable to insulation and fire, and the shacks built with wooden boards were improved to block wall with slate roof, then according to circumstances, renovated again into brick wall with slab roof. To supplement insufficient indoor space, they built attics, or people who were better off built two-storied slab houses. These improvements, additions, and expansions were carried out individually according to the economic capability of each family, regardless of the Taegeukdo organization.

From 1955 to the 2000s, the transformation of housing in Gamcheon Culture Village was a process in which 'common housing with minimal space using minimal materials on the minimum terraced site' evolved into various houses with different features. This change is also a process in which the landscape, which was uniform in the early days, has been transformed into a collective townscape with diversity, 'with colorful houses flocked together'.

Public Art Project

The unique history and change of Gamcheon Culture Village was the background of creating a unique townscape that could not be seen anywhere else in Korea, but it was also the main cause of the decline of the village. Narrow stairways and alleys built on

공공미술 프로젝트

감천문화마을의 독특한 역사와 변천은 우리나라 어디에서도 볼 수 없는 독특한 마을경관을 만들게 되는 배경이었지만, 마을이 쇠퇴하게 된 주원인이기도 했다. 급경사지에 계단형 부지에 조성한 좁은 계단과 골목은 젊은 사람들이 기피하였고, 이 마을에 오래 거주한 주민들도 대부분 노인이 되어 계단을 오르내리기가 힘겨워졌다. 급하게 이주한 주거지였기에 하수 및 오수처리시설 같은 도시기반시설이 빈약하였다. 여전히 집 안에 수세식 화장실이 없어서 공동화장실을 이용해야 하는 가정도 적지 않다.

이처럼 외부인들에게 마을은 경관적으로 아름답지만, 주민으로서 이 마을에 사는 것은 여러모로 불편하게 보였다. 열악한 위생환경, 쉽게 이용하기에는 부족한 녹지공간과 편의시설, 이동하기 불편한 좁고 가파른 계단 등으로 감천문화마을의 인구는 지속적으로 감소하였다. 그 결과 2000년대에는 공가와 폐가가 급증하였다.

인구가 감소하고 사회적 관심으로부터 멀어져 있던 태극도 마을에 변화가 생긴 것은 2009년부터이다. 당시 문화체육관광부는 주민과 함께하는 공공미술의 활성화, 생활 속 미술문화 향유여건 개선, 예술가의 창작활동기회 제공을 목적으로 하는 '마을미술 프로젝트'를 진행하였다.

당시 다대포지역의 공단에 자리잡고 있던 '아트팩토리 인 다대포'라는 민간 예술/공예조직을 중심으로 지역의 문화예술전문가들은 감천2동 태극도마을이 피난민들의 힘겨운 삶의 터전으로 시작되어 근현대사의 흔적과 기록을 그대로 간직하고 있어 역사적, 문화적 보존가치가 높은 곳으로 주목하고 있었다. 이들은 문화적 불모지였던 감천2동 태극도마을을 문화공간으로 조성하여 '살고 싶은 공간', '걷고 싶은 거리'로 만들려는 비전을 제시하면서 '꿈을 꾸는 부산의 마추픽추'(이하 '마추픽추 프로젝트')라는 프로젝트를 제안하여 선정되었다(문화체육관광부, 2010).

'아트팩토리 인 다대포'를 중심으로 한 문화예술인들은 '사람 그리고 새', '꿈꾸는 물고기', 'Good-Morning!', '희망의 노래를 담은 풍선', '가을여행', '무지개가 피어나는 마을' 등 10점

a steep slope and a terraced site have become a residential area that young people avoid. With time, most of the residents who lived long-term in this village also became elderly, and had difficulty in going up and down the stairs. Since it was a residential area that had been constructed in a hurry, urban infrastructure, such as sewage treatment facilities, was poor. Many families still have to use common toilets, because there are no flush toilets in their house.

As such, while the village looks beautiful to outsiders, living in this village as a resident is inconvenient in many ways. The population of the village continued to decline due to the poor hygienic environment that required the use of common toilets, insufficient green spaces and convenience facilities, and narrow and uncomfortable stairs. As a result, in the 2000s, empty and abandoned houses increased.

From 2009, some changes were brought into Taegeukdo village, which had been removed from social interest. At that time, the Ministry of Culture, Sports and Tourism started a Community Art Project aimed at encouraging public art with residents, improving conditions for enjoying art and culture in daily life, and providing opportunities for artists to be committed to creative activities.

Local cultural and artistic experts, participating in a private art/craft organization called the Art Factory in Dadaepo, which was located in the industrial complex of the Dadaepo area at the time, were paying attention to the history and culture of Taegeukdo Village in preserving the traces of modern times of Korea, including the difficult life of the refugee. They proposed a project called 'Dreaming of Machu Picchu of Busan' (the Machu Picchu Project) as a 'Community Art Project' granted by the Ministry of Culture, Sports and Tourism (MCST), presenting a vision to turn Taegeukdo Village in Gamcheon 2-dong into a cultural village with 'space you want to live', and with 'street you want to walk' (MCST, 2010).

그림 3. 2009년 설치된 공공미술작품 '사람 그리고 새' People and Birds

그림 4. 마을 외곽 옹벽에 설치된 '가을여행'과 '내 마음을 풍선에 담아' / 'Autumn Travel' and 'My Heart in Balloon' installed on the retaining wall outside the village

의 공공미술작품을 마을 곳곳에 설치하여 생동감을 불어넣고자 하였다. 그 중에는 마을의 공부방 학생들이 자신의 희망을 담은 그림을 그린 '내 마음을 풍선에 담아', 주민들의 희망메시지가 이루어지기를 바라는 마음을 담은 '달콤한 민들레의 속삭임' 등 주민들이 직접 참여한 작품도 4점이 포함되어 있었다.

공공미술작품을 설치하는 마추픽추 프로젝트를 성공적으로 수행함으로써 자신감이 생긴 사하구와 문화예술전문가들은 2010년에는 문화체육관광부의 또 다른 공모사업인 '콘텐츠융합형 관광협력 공모사업'에도 '미로미로 골목길 프로젝트'(이하 '미로미로 프로젝트')가 선정되는 성과를 거두었다. 마추픽추 프로젝트와 미로미로 골목길 프로젝트는 공공미술의 성격을 가지고 있다는 점에서는 유사한 듯하면서도 내용에 있어서는 차이를 가지고 있었다.

마추픽추 프로젝트는 단순히 조형미술작품을 설치하는 프로젝트였으며, 대부분의 작품이 마을 내부가 아닌 마을 입구나 주변 도로, 옹벽 등에 설치되었다. 마을미술 프로젝트에 대한 이해가 부족한 마을 주민들이 마을 내부 공간에 작품을 설치하는 것을 받아들이지 않았기 때문이었다.

하지만 미로미로 프로젝트는 '테마가 있는 빈집 프로젝트'를 통해 마을의 빈집을 '사진갤러리', '북카페', '평화의 집', '빛의 집'과 '어둠의 집' 등 소규모 전시공간과 예술체험공간으로 바꾸고, '골목길 프로젝트'를 통해 미로 같은 골목길 구석구석에 지역 대학생과 주민들이 직접 만든

Culture and artists from the Art Factory in Dadaepo sought to bring vitality into the village by installing 10 public artworks throughout the village, which artworks included 'People and Birds', 'Dreaming Fish', 'Good-Morning!', 'Balloons with Songs of Hope', 'Autumn Travel', and 'Rainbow Blooming Village'. Among them, four works were directly participated in by residents, including 'My Heart in Balloon', in which children in the community school drew pictures of their hopes, and 'Sweet Whispers of Dandelion', which contain the village residents' messages of hope.

Saha-gu and the experts of art and culture, who gained confidence by successfully carrying out the Machu Picchu Project to install public artworks, were selected by the Cooperation Project of Content & Tourism granted by MCST in 2010 by submitting the 'Miro (maze) Miro Alley Project' (the Miro Miro Project). The Machu Picchu project and the Miro Miro Project seemed similar, in that they had the characteristics of public art, but also had differences in content.

The Machu Picchu Project was simply a project to install public sculptural artworks, and most of the works were installed at the entrance of the village, on roads or retaining walls around the village, though not inside the village. This was because villager people who lacked understanding of the public art project did not accept the installation of works in the village's inner space.

However, through the 'Empty House Project with Theme', the Miro Miro Project changed empty houses in the village into small exhibition spaces and art experience spaces, such as the 'Photo Gallery', 'Book Café', 'Peace House', 'The House of Light', and 'The House of Darkness', and through the 'Alley Project', installed sculptures made by local university students and residents in every corner of the maze-like alleys. In addition, a village map showing such projects and alleys was made for visitors to explore various places of the village as if they were hunting for hidden treasure. In

조형물을 설치하였다. 또 이렇게 설치된 공간과 골목길을 표시한 마을지도를 만들어 관람객들이 보물찾기를 하듯이 마을 곳곳을 찾아다니도록 하였다. 또한 마을 꼭대기에 빈집을 활용하여 '하늘마루'라는 전망대와 입주작가 숙소, 게스트하우스, 체험공간 등으로 조성하였다.

그림 5. 마을의 빈집을 이용한 공공미술작품 '빛의 집' / 'The House of Light' using empty house

미로미로 프로젝트가 마을의 주거지 내부를 대상으로 하면서 주민들과의 협력을 확보하고 삶의 터전을 보호하기 위해 '문화마을 운영위원회'라는 이름의 협의체를 구성하였다. 마을에 살고 있는 주민대표 5명, 마추픽추 프로젝트에 참여했던 아트팩토리 인 다대포 운영위원 3명, 문화예술단체 전문가 그리고 사하구청 공무원으로 구성된 이 협의체는 "감천2동 천마산과 옥녀봉 사이의 산비탈면에 계단식으로 형성된 마을의 역사성과 문화예술적 가치 및 지역 특성을 살려 원도심의 보존과 재생이라는 기본개념을 바탕으로 주민과 예술가 및 지역사회단체의 참여와 협의를 통하여 생활친화적인 도심의 문화마을로 가꾸는 데 기여"(백영제·김다희·이명희, 2011)하려는 목적으로 결성되었다.

addition, an empty house at the top of the village was used to create an observatory called Haneulmaru, an accommodation for artists' residence, a guest house, and an experience space.

A governance named the 'Culture Village Steering Committee' was organized to promote cooperation with residents and protect the living environment from the projects intended to be installed inside the village's residential area. The purpose of this Committee, consisting of five representatives of village residents, three members of the Art Factory in Dadaepo, experts in culture and art organizations, and officials from Saha-gu District Office, was "to use the historical and cultural value and local character of the village on the terraced slope between two mountains for the preservation and regeneration of the old downtown area of Busan, and to make a downtown cultural village based on the participation and cooperation by local residents, artists and local organizations." (Baek, Young-je, Kim, Da-hee, Lee, Myung-hee, 2011)

Another noteworthy point is that the 1st Hanmadang Cultural Festival was held in June 2010, which was mainly organized by the Steering Committee. This festival was expanded to the 1st Gamcheon Culture Village Alley Festival in 2011, and has developed into a representative festival in Busan to this day, to the extent that in 2019, it was selected as the best festival representing Busan.

The Machu Picchu Project in 2009 and the Miro Miro Project in 2010 caused a tremendous butterfly effect for the village. As public artworks installed in the Village became gradually known through SNS, visitors with digital cameras began to visit the village. Visitors who were generally young brought lively energy into the village of mainly elderly people. As the number of tourists increased, the empty house became a café.

또 하나 주목할 점은 2010년 6월 운영위원회를 중심으로 제1회 문화한마당 축제를 개최한 것이다. 이 축제는 2011년에 제1회 감천문화마을 골목축제로 확대되었고, 2019년에는 부산을 대표하는 최우수 축제로 선정될 정도로 오늘날까지 부산을 대표하는 축제로 발전하였다.

2009년 마추픽추 프로젝트와 2010년 미로미로 프로젝트는 마을에 엄청난 나비효과를 불러일으켰다. 감천문화마을에 설치된 공공미술작품들이 SNS 등을 통해 조금씩 알려지면서 디지털 카메라를 든 방문객들이 마을을 찾기 시작하였다. 방문객들은 대체로 젊은 사람들이 많았으므로 노인들이 많은 마을에 활기를 불어넣었다. 관광객이 증가하자 비어있던 건물에 카페가 들어서기도 했다.

공공미술 프로젝트들이 여러 가지 미디어에 소개되면서 알려지자, 2012년 당시 인기 있던

그림 6. 공공미술작품과 초기 방문객(2013년 8월)
Public art and visitors in early stage

그림 7. 감천문화마을 최초 카페(2012년)
First cafe in Gamcheon Cultural Village

그림 8. 감천문화마을의 명물 '어린왕자'와 관광객들 / The famous 'Le Petit Prince' and tourists

As public art projects became known as they were introduced by various media, in 2012, a popular weekend entertainment TV program was filmed in Gamcheon Culture Village; when it was broadcast nationwide, the number of visitors to the village suddenly began to increase.

In May 2013, Le Monde, the renowned French newspaper, introduced the historical and religious background of the village's formation with detailed village photos under the title "Gamcheon: Art at the end of the alley", and explained the public art projects and the residents' reactions. In July, CNN of the United States introduced it in a travel-related article titled "Gamcheon: Is this the most artistic village in Asia?" As a result, along with famous Buddhist temples, beaches, and traditional markets visited by tourists coming to Busan, Gamcheon Culture Village became recognized as one of the unique tourist destinations of Busan.

Sanbok-road Renaissance Project,
the Urban Regeneration Project by BMC

With the start of the fifth mayor's term by civil election in 2010, BMC established a new department called the Creative City Headquarters, and promoted various urban regeneration projects. In addition to the national government's projects, such as the Urban Vitality Promotion Project, BMC's own regeneration projects, such as Gangdong Creative City Project, Happy Village Making Project, Hope Village Project, and Railroad-side Underdeveloped Village Regeneration, were implemented. Among them, the representative project is the Sanbok-road Renaissance Project. The Busan Research Institute, a municipal research institute of BMC, prepared the master plan for the

주말 예능프로그램이 감천문화마을에서 촬영하였고 전국 방송을 타면서 마을을 방문하는 사람들이 갑자기 증가하기 시작했다.

2013년 5월에는 프랑스의 저명한 신문인 르몽드가 스타일 관련 기사로 "감천 : 골목 끝의 예술"이라는 제목으로 상세한 마을 사진들과 함께 마을이 형성된 역사적 상황과 종교적 배경을 설명하고, 공공미술 프로젝트와 주민들의 반응까지 폭넓게 소개하였다. 7월에는 미국의 CNN 방송이 여행 관련 기사로 "감천 : 이곳이 아시아에서 가장 예술적인 마을일까?"라는 제목으로 소개하였다. 그 결과 부산에 오는 관광객들이 많이 방문하는 불교사찰, 해수욕장, 전통시장과 함께 감천문화마을은 부산만이 가지고 있는 독특한 관광지 중의 하나로 인식하게 되었다.

부산시의 도시재생, 산복도로 르네상스사업

2010년 부산시 민선 5기 시정이 시작되면서 부산시는 창조도시본부라는 부서를 새로 신설하고 다양한 도시재생사업을 추진하였다. 도시활력증진사업이나 뉴타운 해제지역 재생사업 같은 전국적인 사업 이외에도 강동권 창조도시사업, 행복마을사업, 희망마을사업, 철로변 낙후마을 재생 등 부산만의 재생사업을 시행하였다. 그 중에서 대표적인 사업은 '산복도로 르네

그림 9. 산복도로 르네상스사업의 비전과 유형 / The vision and types of Sanbok-road Reanaissance Project

Sanbok-road Renaissance Project by dividing Busan's Sanbok-road area into 8 areas of 3 zones, which are distributed across Jung-gu, Seo-gu, Dong-gu, Busanjin-gu, Saha-gu, and Sasang-gu, and to implement various projects for individual areas according to annual order. This was an extensive and ambitious urban regeneration project that invested 15 billion won annually, with a total of 150 billion won for the 10 years from 2011 to 2020.

The Sanbok-road Renaissance Project was a kind of model project for the national government supported urban regeneration project, since it set a vision of 'making creative communication spaces full of vitality of life', and attempted 'comprehensive self-help regeneration through space regeneration, living regeneration, and cultural regeneration'.

Gamcheon Culture Village, along with the adjacent Ami-dong, was included in the Ami-Gamcheon area in the Gudeok–Cheonmasan zone, and was implemented as a second-year project. In 2011, pilot projects were introduced; and in 2012, the main projects were implemented at full scale. Since resident interest and tourist visits increased due to the Machu Picchu Project and the Miro Miro Project, the Sanbok-road Renaissance Project was able to expand the village infrastructure, culture and arts space, and anchor facilities for residents.

The Sanbok-road Renaissance project increased the convenience of residents by expanding parking lots and roads, which had previously provided insufficient infrastructure for the village. In addition, a small museum containing the history of the village, a small art museum as an art and culture space, and the Cheondeoksu well site, which is a historical space around an old well, were created to provide attractions for tourists. The local bathhouse building, which has been closed for a long time due to population decrease, has been remodeled into the Gamnae Community Center with

그림 10. 폐업 목욕탕을 리모델링한 감내어울터 입구 / Gamnae Community Center renovated from closed bathhouse

그림 11. 목욕탕 욕조를 남긴 감내어울터의 전시실 Galley with bathtub in Gamnae Community Center

상스사업'이다. 부산시의 시정연구원인 부산연구원에서 마스터플랜을 수립하여 중구, 서구, 동구, 부산진구, 사하구, 사상구에 걸쳐 분포하는 부산의 산복도로 지역을 3개의 권역, 8개 구역으로 나누어 연차별로 도시재생사업을 추진하는 계획이었다. 2011년부터 2020년까지 10년 동안 매년 150억 총 1,500억원을 투입하는 광범위하고 야심찬 도시재생사업이었다.

산복도로 르네상스사업은 대부분 물리적 사업에 집중하던 당시로는 드물게 '생활의 활기가 넘치는 창조적 소통공간 함께 만들기'라는 비전을 설정하고 '공간재생, 생활재생, 문화재생을 통한 자력수복형 종합재생'을 시도하였다는 점에서 이후 국가가 지원한 도시재생사업의 선구적인 사례라고 할 수 있다.

감천문화마을은 인접하는 아미동과 함께 구덕천마산권역의 아미·감천구역에 포함되어 2차년도 사업으로 실시되었다. 2011년도에는 시범사업이 도입되었고, 2012년부터 본격적인 사업이 시행되었다. 마추픽추 프로젝트와 미로미로 프로젝트로 주민들의 관심과 관광객들의 방문이 높아지던 시기에 산복도로 르네상스사업이 실행되면서 마을의 기반시설과 문화예술공간, 주민들을 위한 거점공간을 확충할 수 있었다.

산복도로 르네상스사업의 성과로는 우선 마을의 부족한 기반시설인 주차장과 도로를 확충하여 주민의 편의를 높였다. 또한 마을의 역사를 담은 작은 박물관, 예술문화공간인 작은 미술관, 오래된 우물 주변을 역사공간으로 정비한 천덕수샘터 등을 조성하여 관광객들의 볼거리를 제공하였다. 주민 감소로 인해 오랫동안 폐업 중이던 동네의 목욕탕 건물은 마을주민과

café, small gallery, meeting spaces, guest house, and rooftop observatory for villagers and visitors. In addition, Gamnaegol restaurant and Gamcheon Culture Village coffee shop were created using empty houses to be managed by the Resident Council to earn economic profits from the increase in tourists.

Thanks to the Sanbok-road Renaissance Project, it has become a cultural village with various cultural facilities, convenience facilities, and commercial facilities to increase the volume of tourists due to public art projects, and has become a popular attraction for more and more tourists.

The Community Council, which has grown into a community business company, is making enough profit to provide jobs to residents by selling village maps or managing community stores, such as restaurants, cafés, and souvenir shops. In addition, various services are provided to return some of the profit back to residents. To alleviate the inconvenience of living for the elderly, laundromat and community bath are operated free of charge, and free shuttle buses only for residents of Gamcheon 2-dong have also been introduced to supplement public buses. Daily necessities, such as garbage bags, are provided to help low-income residents; and during seasonal holidays, gifts are also given.

For these achievements, Gamcheon Culture Village was honored with the Excellence Award of the Regional Traditional Culture Brand Awards organized by the MCST in 2012, followed by the 2016 Space Culture Award, and the 2017 Korea City Award. In addition, it won international awards, including the Asian Townscape Award in 2012, and the Metropolis Award in 2014.

With the popularity of Gamcheon Culture Village, the number of visitors increased exponentially. The number of visitors, which in 2011 was only 25,000, in 2012 exceeded 100,000, in 2017 exceeded 2 million, and in 2019 exceeded 3 million. Gamcheon Culture

방문객을 위한 카페, 작은 갤러리, 모임공간, 게스트하우스, 옥상전망대를 갖춘 커뮤니티 센터 성격의 감내어울터로 리모델링되었다. 또한 마을의 비어 있던 공간을 활용하여 감내골 맛집, 감천문화마을 커피숍을 조성하여 주민협의회가 운영함으로써 관광객 증가에 따른 경제적인 수익도 거둘 수 있게 되었다.

마을기업으로 성장한 주민협의회는 마을지도를 판매하거나 맛집, 카페, 기념품점 등 마을 상점을 운영하면서 주민들에게 일자리를 제공할 정도로 수익을 창출하고 있다. 뿐만 아니라 수익금의 일부를 주민에게 돌려주기 위해 각종 지원사업도 계속하고 있다. 어르신들의 생활 불편을 해소하기 위해 빨래방이나 마을목욕탕을 무료로 운영하고 있고, 마을버스 대신 감천2동 주민들만 이용하는 무료셔틀버스도 도입하였다. 저소득주민들을 돕기 위해 쓰레기봉투 등 생필품을 지원하고, 명절에는 선물을 돌리기도 한다.

이러한 성과로 감천문화마을은 2012년에는 문화관광부에서 주관한 지역전통문화브랜드 대상에서 우수상의 영예를 안은데 이어, 2016년 공간문화대상, 2017년 대한민국 도시대상 등을 수상하였다. 뿐만 아니라 2012년에는 아시아도시경관상, 2014년에는 메트로폴리스어워드 1위 등 국제적인 상도 수상하였다.

감천문화마을의 유명세와 함께 방문객은 기하급수적으로 증가하였다. 2011년 2만5천명에 불과하던 방문객 숫자는 2012년에 10만명을 넘었고, 2017년에는 200만명을 돌파하였으며, 2019년에는 300만명을 넘었다. 특히 방문객 중에는 외국인 관광객 비율이 60%에 이를 정도로 국제적인 명소로 자리매김하였다.

젠트리피케이션과 지구단위계획

만 명이 채 되지 않는 작은 마을에 주민수의 300배가 넘는 관광객이 찾아오면서 마을에는 부정적인 변화들도 나타났다. 평일과 주일, 낮과 밤 가리지 않고 몰려드는 관광객들의 소음과 사진촬영으로 주민들의 사생활은 점점 더 침해되었다. 장기적으로 관광객들이 둘러볼 수

Village has established itself as an international attraction, with the proportion of foreign tourists reaching 60 %.

Gentrification and District Unit Plan

Negative changes have also emerged in the village as tourist numbers of more than 300 times the number of residents visited a small village with population of less than 10,000. Resident privacy was increasingly violated by the noise and photography of tourists regardless of weekdays or weekend, day or night. The District Office has even considered limiting the specific area allowed for the access of tourists, or designating a Silent Zone for the area with many residents in the long-term (Busan Ilbo, February 2, 2020).

The public bus used by the residents was filled with tourists, so there were no seats for the residents to sit. As the number of shops dealing with tourists increased, real estate prices soared several times, and neighborhood shops that had been managed for a long time were forced to vacate one-by-one. Convenience stores, such as grocery stores and small shops near alleys, have turned into souvenir shops, restaurants, cafés, and craft shops, causing inconvenience to residents. The Ilgam Shop, the meeting place for residents at the entrance of Gamcheon Culture Village where residents left their house keys when they went out, had to be closed, due to the prospective opening of a franchise commercial shop.

In addition, commercial facilities for tourists affected the townscape of the village. To catch the eyes of tourists, there were many decorations of unknown nationality, and loud colors of signboards and facades; large-scale commercial buildings that

그림 12. 젠트리피케이션으로 폐업한 마을상점과 주민피해
Closed neighborhood shop by the gentrification and warning signs

있는 영역을 특정구간으로 제한하는 방안이나, 거주하는 주민이 많은 구간은 묵음존(Silent Zone)을 설치하는 방안까지 구청에서 고려할 정도였다(『부산일보』, 2020. 2. 2.).

주민들이 이용하던 마을버스는 관광객으로 가득차 주민들이 앉을 자리가 없을 정도가 되었다. 관광객을 상대하는 상점들이 증가하면서 부동산 가격은 몇 배로 치솟았고, 오랫동안 마을을 지키던 상점들은 하나둘 쫓겨 났다. 골목 가까이 있던 부식가게나 작은 슈퍼 등 생활편의 가게들이 기념품 상점, 식당, 카페, 공예품 판매점 등으로 변하면서 주민들의 생활불편이 가중되었다. 감천문화마을 입구에 있던 일감상회는 마을사람들이 외출할 때면 집열쇠를 맡겨놓던 마을사랑방의 역할을 했지만, 프랜차이즈 상업시설의 입주로 인해 문을 닫아야 했다.

뿐만 아니라 관광객 상대 상업시설들은 마을의 경관에도 영향을 주었다. 관광객들의 눈길을 끌기 위해 국적불명의 형태, 원색적인 간판과 입면이 난립했고, 여러 집을 합친 대규모 상업건물들이 들어서 기존의 소규모 주거지 경관과 어울리지 않았다.

이에 위기감을 느낀 사하구청에서 감천문화마을의 독특한 도시공간과 경관을 보전하고 마을의 정주환경을 보호하기 위해 건축물의 형태·높이·용도 등의 기준을 설정하고 계획적으로 관리할 수 있는 지구단위계획을 2017년 수립하였다.

combined several houses were built, which did not sit well with the existing small house townscape.

Saha-gu District Office, which felt a sense of crisis, established a district unit plan in 2017 to set standards for the shape, height, and use of buildings to preserve the unique urban space and townscape of Gamcheon Culture Village, and protect the settlement environment of the village.

Table 1. Contents of the district unit plan for Gamcheon Culture Village.

Location: Gamcheon Culture Village, Gamcheon 2-dong, Saha-gu, Busan (188,177 m²)

Period: 2015–2017 (Reference year, 2016; Target year, 2025)

Details of Plan:

- Establish zone designation of district unit plan, guidelines of plan, and operation manual reflecting related plans and implementation plans for Gamcheon Culture Village.
- Private sector: Strengthen the characteristics of each zone according to the zone division, and present three-dimensional guidelines and sustainable management plans through resident's participation.
- Public sector: Zone designation of district unit plan reflecting the locality of Gamcheon Culture Village, layout plan of new facilities for suggestion of maintenance of public spaces and viable public projects, and guidelines for reviewing unexecuted facilities of urban planning (The blog of Site Planning Ltd. / https://blog.naver.com/uahan/221238404634).

The district unit plan investigated the current situation of the townscape and the 188,177 m² space of Gamcheon Culture Village in detail, and divided the overall area into seven: key management area, general management area 1, general management area 2, Gamnae 1-ro area, Gamnae 2-ro area, Okcheon-ro area, and Okcheon-ro 75th Street area. After that, the height, form, and function of the building were regulated for each area according to the characteristics of the townscape and space.

표 1. 감천문화마을 지구단위계획 개요

위치 : 부산광역시 사하구 감천2동 6-1610 일원 (감천문화마을) 188, 177m²

기간 : 2015~2017 [기준연도(2016년), 목표연도(2025년)]

과업 내용 :

- 감천문화마을의 관련한 각종 계획, 실행사업계획 등을 반영한 지구단위계획 구역지정과 계획지침 및 운영계획을 수
  립한다.

- 민간부문 : 구역별 구분에 따라 구역별 특성을 강화하고 3차원적 지침과 주민 참여를 통한 지속가능한 관리방안을 제
  시한다.

- 공공부문 : 감천문화마을의 장소성을 반영한 지구단위계획구역의 지정과 공공공간의 정비 및 실행가능한 공공사업
  의 제안을 위한 신규시설 배치계획과 미집행도시계획시설의 재검토를 위한 가이드라인을 제시한다.[1]

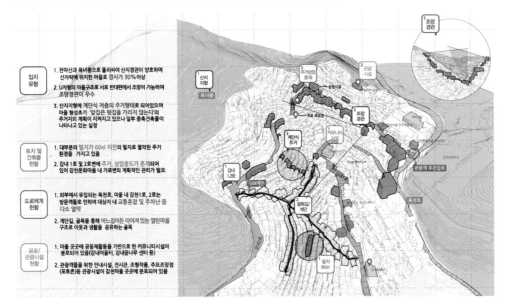

그림 13. 지구단위계획 수립을 위한 현황종합분석((주)싸이트플래닝) / Investigation of current situation for distric unit plan

지구단위계획은 감천문화마을 일원 188,177m²의 지역의 경관적·공간적 현황을 세부적으
로 조사하여, 특성에 따라 중점관리구역, 일반관리구역1, 일반관리구역2, 감내1로구역, 감내

---

1 싸이트플래닝 블로그 (https://blog.naver.com/uahan/221238404634)

To prevent large-scale construction activities through consolidation according to commercialization, the maximum development scale was limited to twice the average of all land parcels in each area. In the case of a key management area where the unique townscape of Gamcheon Culture Village is best preserved, the height of the building was limited to one story.

As infrastructure, roads between Goejeong and Gamcheon and two parking lots were newly designated to enhance the convenience of residents. The natural green area on the slope of Oknyeobong Peak behind the Saha-gu Social Welfare Center was designated as an urban agricultural park to prevent the encroachment of green space, as well as to prevent expedient development. Ten public open spaces were also designated in the Gamnae 2-ro area, situated along the main routes of visitors, to preserve the view of the village's townscape, and to protect in advance the right of view, which might be violated by the expected development of private buildings.

To improve the disorderly streetscape caused by the turmoil of commercial facilities, only one outdoor signboard in horizontal format is allowed for a shop, and the width of the signboard was limited to within 80 % of the width of the building, and up to 1m in height. The most unique characteristic of the district unit plan is that it included a regulation that prohibits franchise stores by large-scale capital in all areas.

Considering the district unit plan of Gamcheon Culture Village, it can be seen that it is very detailed, and the regulations and standards are very strict. The fact that the district unit plan with such strict regulations was able to be established shows that the sympathy and understanding of the residents, mainly of the Resident's Council played a major role. This is because the Resident's Council, which is making profits through community business companies as the number of tourists increases, feared that the disorderly townscape and franchise stores would reduce the attractiveness of the

2로구역, 옥천로구역, 옥천로75번길구역의 7개 구역으로 구분하였다. 이후 각 구역별로 경관과 공간의 특성에 맞게 건축물의 높이, 형태, 용도 등을 제한하였다. 상업화에 따라 합필을 통한 대규모 건축행위를 막기 위해 최대 개발규모를 각 구역 내 전체 필지 평균의 2배로 설정하였다. 감천문화마을의 독특한 경관이 가장 잘 보존되어 있는 중점관리구역의 경우 건축물 높이를 1층으로 제한하였다.

기반시설로는 괴정~감천 구간 도로와 주차장 2개소를 신설하여 주민들의 편의를 도왔다. 사하구종합사회복지관 뒤쪽 옥녀봉 자락의 자연녹지지역은 도시농업공원으로 지정하여 녹지의 훼손을 방지할 뿐만 아니라 난개발의 우려도 차단하였다. 방문객들이 주로 다니는 주요 동선을 따라 지정된 감내2로구역에는 마을의 경관조망 보전을 위해 공공공지 10개소를 지정해 건축물 진입으로 인한 조망권 침해를 사전 예방하였다.

상업시설의 난립으로 무질서한 가로경관을 정리하기 위해 옥외광고물은 점포당 1개소 이내로 가로형 간판으로 허용하고, 간판의 가로폭은 건물폭의 80% 이내, 세로폭은 최대 1m 이내로 제한하였다. 감천문화마을 지구단위계획에서 가장 독특한 점은 모든 구역에서 대규모 자본에 의한 프랜차이즈 점포의 입주를 불허하는 규정을 포함시켰다는 점이다.

이처럼 감천문화마을 지구단위계획을 살펴보면 매우 세부적일 뿐만 아니라, 규정이나 기준이 매우 엄격하다는 사실을 알 수 있다. 이처럼 엄격한 규제를 가진 지구단위계획을 수립할 수 있었던 것은 주민협의회를 중심으로 주민들의 공감과 이해가 큰 역할을 하였다. 관광객의 증가에 따라 마을기업을 통해 수익을 거두고 있는 주민협의회의 입장에서 무질서한 경관이나 프랜차이즈 상점의 입주는 마을의 매력을 감소시키고 결국 관광객의 감소로 이어질 것으로 우려했기 때문이다.

맺으며

2009년 마추픽추 프로젝트로 시작하여 2019년 관광객 300만명을 유치하기까지 약 10여

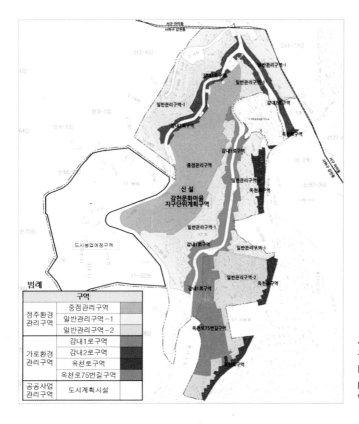

그림 14. 감천문화마을 지구단위계획 구역구분 (㈜싸이트플래닝)
Division of area for district unit plan for Gamcheon Culture Village (Site Planning Ltd.)

village, and eventually lead to a decrease of tourists.

Closing Remarks

Starting with the Machu Picchu Project in 2009, Gamcheon Culture Village has experienced revolutionary changes over a short period of about 10 years, before attracting 3 million tourists in 2019.

Gamcheon Culture Village became known as the most successful urban regeneration project in Korea, and has received many awards. On the other hand, as many experts and local media have pointed out, tourists have infringed on the privacy

년의 짧은 기간에 감천문화마을은 상전벽해의 변화를 경험하였다.

감천문화마을은 우리나라에서 가장 성공적인 도시재생사업으로 알려졌고, 많은 상도 받았다. 하지만 다른 한편으로 많은 전문가들과 지역의 언론들이 지적하는 바와 같이, 관광객들은 주민들의 사생활을 침해하였고, 난립한 상업시설들은 마을의 고유한 경관을 훼손하였으며, 부동산 가격이 폭등하여 주민들이 집을 팔고 마을을 떠나는 부작용들도 나타났다. 관광객은 증가하였지만, 인구는 계속 감소하기 때문에 마을공동체가 언제까지 지속될지도 불분명하다. 현재 감천문화마을에서 가장 필요한 일은 주민, 행정, 지역의 전문가 등이 함께 마을의 바람직한 미래상에 대해 진지하게 고민하는 것이다.

2020년부터 전세계를 휩쓴 코로나19로 인해 감천문화마을의 관광객도 급감하였고, 상점도 상당수 휴업이나 폐업상태이며, 떠들썩하던 골목축제도 연기하거나 조용한 행사로 변경되었다. 숨가쁜 변화 끝에 잠시의 멈춤이 마을의 미래에 대해 깊이 그리고 폭넓게 고민해 보는 시간이 될 수 있기를 기대한다.

참고문헌

1. 김형균 외, 도시재생 사업지역의 주민생활 및 상권변화 연구, 부산:부산연구원, 2015
2. 문화체육관광부, 마을미술 프로젝트, 서울: 2009 마을미술 프로젝트 추진위원회, 2010, 2009
3. 백영제·김다희·이명희, 감천문화마을 이야기, 도서출판 두손컴, 2011
4. 오재환 외, 산복도로 르네상스 사업 평가를 통한 부산형 도시재생 방향 설정, 부산: 부산연구원, 2021
5. 우신구, 아미동이야기 : 포개진 삶, 겹쳐진 공간, 서울: 국립민속박물관, 2020
6. 윤지영, 도시재생마을 브랜딩 전략 방안 : 감천문화마을을 중심으로, 부산:부산발전연구원, 2015

of residents, scattered commercial shops have damaged the unique townscape of the village, and real estate prices have soared, resulting in side effects of residents selling homes and leaving the village. Although the number of tourists has increased, it is unclear how long the village community will last, because the population continues to decline. Currently, the most necessary thing for Gamcheon Culture Village is for residents, administrative, and local experts together to discuss in depth the desirable future of the village.

Since 2020, the number of tourists to Gamcheon Culture Village has plummeted due to COVID-19, and many stores have been closed temporarily or permanently. The noisy alley festival has been postponed, or scaled down to small events. Many people hope that a brief pause after breathless change will be an opportune time to think deeply and broadly about the future of the village.

References

1. Kim, Hyung-gyun et al., Research on changes in residents' life and commercial business in urban regeneration project areas, Busan: Busan Research Institute, 2015

2. Ministry of Culture, Sports and Tourism, Community Art Project, Seoul: 2009 Community Art Project Promotion Committee, 2010, 2009

3. Baek, Young-je, Kim, Da-hee, Lee, Myung-hee, Gamcheon Culture Village Story, Book Publishing Dusoncom, 2011

4. Oh, Jae-hwan et al., Busan-style urban regeneration direction through the evaluation of the Sanbok-road Renaissance Project, Busan: Busan Research Institute, 2021

5. Woo, Shin-koo, Ami-dong Story: Overlapping Life, Layered Space, Seoul: National Folk Museum of Korea, 2020

6. Yoon Ji-young, Branding Strategy Plan for Urban Regeneration Village: Busan: Busan Development Institute, centering on Gamcheon Culture Village, 2015

Chapter 2

공공주도의 재생과 재개발

Public-led Urban Regeneration and Redevelopment

도시설계의 특성은 무엇보다 중앙 및 지방정부가 공공성을 목적으로 쇠퇴한 원구도심의 물리적 환경을 개선하고 경제활동의 활성화를 유도하는 주도적인 공공투자의 역할이다. 이 장에서는 2000년대 초부터 서울시가 추진해온 일련의 사례들인 청계천의 복원과 도심 수변공간의 조성, 굴뚝 산업단지인 구로공단의 첨단산업 G밸리로의 재개발, 쓰레기 매립지였던 상암동의 미디어산업 거점으로의 변화, 그리고 미군부대의 반환으로 진행중인 용산공원 조성을 살펴본다.

One of the characteristics of urban design is the central and local government's role as public investor to improve the physical environment of the declined old city area, and to induce revitalization of economic activities. This part introduced the four cases implemented by the Seoul Metropolitan Government since the early 2000s: the restoration of Cheonggyecheon stream in Seoul and the creation of a public waterfront space in the city center; the redevelopment of Guro Industrial Complex, a chimney industrial complex, into a high-tech industrial G-valley; the transformation of Sangam-dong, which used to be a waste landfill, into a media industry base; and the creation of Yongsan Park in progress upon the return of the US military base.

서울 청계천, 미래도시를 위한 복원과 재생

김형일 (삼육대 건축학과 교수)

들어가며

청계천복원사업은 서울 구도심의 낙후된 이미지와 환경개선을 위하여 새로운 도시관리 패러다임을 도입하여 시민들의 삶의 질을 향상시키고자 시도한 환경친화형 도시설계 프로젝트로 시작되었다.

자연건천에서 인공하천으로 또 복개천 상부 도심고속도로로 변화를 거듭했던 청계천에 복개된 하천과(1961년)과 고가구조물(1971년)을 복원하고 철거함으로써 맑은 물을 다시 흐르게 하여 생태계의 복원은 물론 서울이 자연과 함께 어우러지는 환경친화적 도시로 거듭나게 하는 프로젝트이다.

서울시의 공공사업으로서 2003년 7월부터 2년 3개월간 3867억원의 사업비를 투입하여 청계광장에서 성동구 신답철교 사이의 5.84㎞ 구간을 복원하여 250,2000m²의 녹지와, 283만9000본(本)의 식물을 이식, 12.04㎞의 산책로를 조성하여 대략 50년 만에 청계천이 도심하천으로 기능을 회복하고 역사문화를 회복하여 청계천 주변의 노후지역에 새로운 활력을 제공하도록 계획된 도심재생 프로젝트의 일환으로 시작되었다(그림 1).

Cheonggyecheon Stream, Seoul

Restoration and Regeneration for a Future City

Hyeongil Kim (Professor, Sahmyook University, Dept of Architecture)

Introduction

The Cheonggyecheon Restoration Project started as an environment-friendly urban design project that attempted to improve the quality of life of citizens by introducing a new urban management paradigm to improve the environment and develop the image of the old downtown of Seoul.

The restoration and dismantling of a covered structure (1961) and an elevated structure (1971) in Cheonggye Stream, which has repeatedly changed from a natural dry stream to an artificial stream and to an urban highway at the upper part of the covered stream, allowed a project that restores the ecosystem by allowing clear water to flow again, and reinforces Seoul as an eco-friendly city that lives in harmony with nature.

As a public project of Seoul, from July 2003, by investing KRW 386.7 billion for two years and three months, the 5.84 km section between Cheonggye Plaza and Sindap Railroad Bridge in Seongdong-gu was restored, 2,502,000 m² of green space was created, and 2,839,000 plants were transplanted. This began a 50-year urban

그림 1. 청계천 위성사진(출처: 청계천 백서, 서울시설관리공단) / Satellite image of Cheonggyecheon

주요 사업내용은 복원된 하천의 생태환경 조성, 이용자를 위한 산책로와 접근로, 양안도로의 설치, 진입부 상징광장, 교량설치 등이다. 청계천 복원사업을 통해 서울은 보행과 대중교통 중심의 도시교통정책과, 구도심 활성화와 지역균형발전정책, 하천중심의 도시환경복원정책을 확대하였다는 평가를 받았다.

청계천 복원사업 이후 실제로 인근 주변정비사업과 재개발 등이 이어지고 있으며 이로 인하여 구도심부에 지속적으로 상주인구가 늘어나고 새로운 도심주거지로서 변화하고 있다.

청계천의 역사적 배경

청계천은 서울의 도심인 종로구와 중구를 관통하여 흐르는 하천으로, 조선왕조가 시작하면서 도성건설과 함께 자연하천이었던 청계천을 정비하여 배수시설을 확보하고자 하였다. 조선의 도읍으로 시작한 한양은 도심의 동·서·남·북의 외곽에 낙산, 인왕산, 목멱산, 백악산의 4개 산으로 둘러싸여 있으며, 이들을 연결하는 도성을 쌓아 방어에 유리한 도시로 형성되었다(그림 2). 그러나 강우량이 많아지면 산에서 흘러내린 물로 인하여 도성 안이 홍수에 쉽게 노출되는 문제가 발생하여 도심을 관통하는 개천 조성사업으로 청계천이 시작되었다(그림 3).

regeneration project to restore the function of Cheonggyecheon as an urban river by creating a 12.04 km walking trail, restoring historical culture, and providing new vitality to the aging areas around Cheonggyecheon (Fig. 1). The main contents of the project are the creation of an ecological environment for the restored water stream, the installation of a walking path, access roads, and shore side roads for users, a symbol plaza at the entry point, and the installation of bridges. Through the Cheonggyecheon Restoration Project, Seoul expanded its urban transportation policy that was centered on pedestrian and public transportation, old downtown revitalization and regional balanced development policies, and river-centered urban environment restoration policy.

Historical Background of Cheonggyecheon

Cheonggyecheon river flows through Jongno-gu and Jung-gu, the downtown areas of Seoul. At the beginning of the Choseon Dynasty, along with the construction of the capital city, Cheonggyecheon - which was a natural river - was repaired to secure drainage facilities. Hanyang, which started as the capital of Choseon, is surrounded by four mountains: Naksan, Inwangsan, Mokmyeoksan, and Baekaksan to the east, west, south, and north of the city center, respectively (Fig. 2). However, when rainfall increases, the city is easily exposed to flooding from the water flowing down from the mountains, so the beginning of Cheonggyecheon was the construction of the stream that runs through the city center (Fig. 3).

During the Japanese colonial period, it was neglected and contaminated with excrement and garbage; due to mounting environmental and sanitation

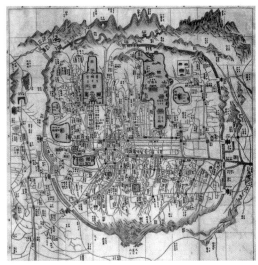

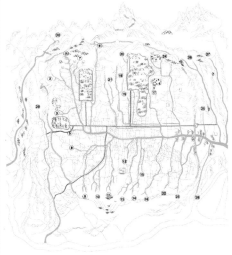

그림 2. 한양 도성전도(출처: 규장각 한국학 연구원)
Capital city map of Hanyang

그림 3. 청계천 수계도(출처: 청계천 백서, 서울시설관리공
단) / Cheonggyecheon water system map

일제강점기를 거치면서 방치되어 분뇨와 쓰레기 등으로 오염되고 환경과 위생문제로 1920년대부터 청계천 복개에 대한 논의가 시작되었으며 1934년 대경성계획 수립 후 청계천에 고가철도안 발표(1935.1. 매일신보), 청계천 운하계획 발표(1936.8. 매일신보), 태평로에서 광교까지 복개(1937) 등 잇단 계획이 발표되었다.

해방 후 한국전쟁을 거치면서 전국 주택의 1/3 정도가 파괴되었으며 주택을 잃은 사람들이 서울로 몰려와 청계천 인근에 무허가 불량주택이 대량으로 지어졌으며 50년대에서 60년대까지 대한민국 최대의 빈민가를 형성하였다. 이로 인하여 청계천 인근은 위생문제와 치안질서, 그리고 홍수피해와 화재사고 등 다양한 도시문제들이 야기되었다.

청계천 복개공사는 너비 50m로 1958년부터 본격적으로 시작하여 1978년까지 구간을 나눠서 진행되었다(그림 4). 그리고 서울시는 1967년 복개된 청계천 위로 5.4km의 청계고가도로(당시 3.1도로)를 착공하여 1971년에 완공하였다(그림 5). 공사 당시 청계천 주변의 판잣집들은 복개과정에서 철거되고, 그 자리에는 맨션이나 상점가가 건설되었다.

problems, discussions on the covering of Cheonggye Stream began in the 1920s. After the establishment of the Great Gyeongseong plan in 1934, a series of plans were announced, including the announcement of an elevated railroad plan on Cheonggyecheon (1935, Maeil Shinbo), the announcement of the Cheonggyecheon Canal Plan (1936), and the covering from Taepyeong-ro to Gwanggyo (1937).

After liberation, during the Korean War, about one-third of the houses in the country were destroyed. People who had lost their homes flocked to Seoul, and from the 1950s to the 1960s built many illegal houses near the Cheonggyecheon Stream, which formed the largest slum in Korea. Due to this, various urban problems, such as sanitation and public order, flood damage, and fire accidents, were caused in the vicinity of Cheonggyecheon.

The Cheonggyecheon Stream Covering Construction started in 1958 with a width of 50 m, and was divided into stages, with construction continuing until 1978 (Fig. 4). Seoul Metropolitan Government started construction of the 5.4 km Cheonggye elevated highway (later called the 3.1 Road) over the covered Cheonggyecheon Stream in 1967, and completed it in 1971 (Fig. 5). At the time of construction, the shacks around Cheonggyecheon were demolished during the covering construction, and mansions or shopping streets were built in their place.

After the Cheonggyecheon Covering construction, a market formed around Cheonggyecheon, and clothing factories were established. In the 1970s, the floating population increased, and Cheonggyecheon began to change into a center of commerce and industry in the city.

그림 4. 청계천 복개공사(출처: 서울역사박물관)
Cheonggyecheon Stream Covering Construction

그림 5. 청계 고가도로(출처: 서울역사박물관)
Cheonggye elevated highway

청계천 복개공사 이후 청계천을 중심으로 시장이 형성되었으며, 의류공장 등이 자리를 잡으며 1970년대에는 유동인구가 많아지며 도심내의 상공업의 중심으로 변화하기 시작하였다.

청계천 복개와 청계고가도로

1980년대와 90년대를 거치며 청계천 주변지역은 각 권역별로 주거, 업무, 산업, 도·소매 활동 등 다양하게 발전되었다. 태평로~삼일로 구간의 대로변 중심의 도심권역은 업무중심 및 상업서비스지역으로 동대문권역은 대형 시장 위주의 도매 판매시설과 대형 도소매 상업시설이 군집하여 있으며 외곽권역은 주로 주거지역으로 개발되었다. 특히 청계천 주변지역은 전국을 대상으로 하는 거대한 도매상권으로서 전문화, 특화된 기능이 서로 얽혀 움직이는 거대한 골목산업생태계를 형성하였다(그림 6, 7).

2000년대에 와서는 건설된 지 30~40년이 지난 청계고가 및 복개도로와 지하구조물의 안전성 문제가 제기되었으며 주변 건물들의 노후화로 지역환경이 낙후되고 상주인구가 줄어들어 지역발전이 저해되었다.

그림 6. 청계고가도로 주변(출처: 연합뉴스)
Around Cheonggye elevated highway

그림 7. 청계천 복원 전(출처: 사진작가 박영기)
Before restoration of Cheonggyecheon

Covered Cheonggyecheon Stream and Cheonggye Elevated Highway

Through the 1980s and 1990s, the area around Cheonggyecheon has developed in various ways, including residential, business, industrial, and wholesale/retail activities for each district. The downtown area centered on the roadside between Taepyeong-ro and Samil-ro has developed into a business-oriented and commercial service area, while the Dongdaemun area has wholesale and retail facilities centered on large markets and large wholesale and retail commercial facilities, with the outer area being mainly developed as a residential area. In particular, the area around Cheonggyecheon now forms a huge alley industrial ecosystem, where various specialized functions are intertwined as a huge wholesale shopping district for the whole country (Figs. 6 & 7).

In the 2000s, safety issues were raised about the Cheonggye elevated highway, covered roads, and underground structures that had been built for 30 to 40 years.

청계천 복원 이전 상황

복원 이전 청계천은 복개공사로인 한 역사유적 훼손과 교통혼잡, 통과교통 과다, 대기오염, 소음 등으로 도심 쇠락이 일어나고 있었으며 이로 인하여 동북권 경쟁력이 최하위권으로 분석되어 세심한 정비가 요구되고 있었다.

청계천 복개구조물 아래에 광교 등 역사유적의 일부가 방치되어 있음이 확인되었는데, 장충단 공원으로 이전되어 있는 수표교는 석축에 대한 기록이 있어 석축이 남아 있으리라 예측되었으며 그 모습의 일부가 확인되어 역사유적 발굴과 복원이 논의되고 있었다.

복원계획 이전 청계로와 청계고가도로에서는 하루 16.8만대 이상의 교통량(약 62.5%는 단순 통과목적)으로 교통부하량이 과다하게 집중되어 있었다.

또한 청계천 도로변의 대기환경수준을 측정한 결과, 미세먼지(PM10) 오염항목을 제외한 일반기준 오염물질은 서울 평균치를 상회하였으며 질소산화물의 경우 서울시 대기환경기준을 초과했고, 발암성 물질인 휘발성유기화합물질(VOC) 가운데 특히 벤젠의 농도가 높게 측정되어 대기오염이 매우 심각한 수준이었다.

청계천 지역의 교통문제, 대기오염, 주택 및 건물의 노후화 등으로 강북도심의 경쟁력이 현저히 저하하여, 이전 10년간 상주인구 5만, 고용인구 8만이 감소하고, 사업체 본사의 수는 강남의 63% 수준에 불과하다는 연구결과가 나왔으며, 도심재개발은 민간자본을 끌어들이는데 어려움을 겪고 있었으며, 특히 도심의 금융 및 비즈니스 기능(12.5%)이 강남 부도심(27.0%)에 비해 크게 미흡한 것으로 나타났다.

그리고 비도심형 소규모 전통제조업이 과잉 집중되고 지식기반산업의 성장이 기대에 미치지 못한 것으로 평가되어, 낙후되고 취약한 도심산업구조는 동북아 중심지로서 수도 서울의 경쟁력을 저하시키는 중요 요인으로 여겨지고 있었다.

Before the Restoration of Cheonggyecheon

Prior to restoration, the Cheonggyecheon Stream was undergoing urban decline due to congestion, excessive traffic, air pollution, noise, etc., along with damage to historical remains due to the covering construction. As a result, the competitiveness of the Northeast region was analyzed as being of the lowest rank, and careful maintenance was required.

It was confirmed that some of the historical remains, such as Gwanggyo, had been neglected under the covering structure of Cheonggyecheon. Supyo Bridge was relocated to Jangchungdan Park, with the intention that its stone structure would remain as a record of stone construction. Excavation and restoration of historical assets were being discussed as the existence of a part of it was confirmed.

Before the restoration project, the traffic load was excessively concentrated on Cheonggye-ro and Cheonggye elevated highway, with more than 168,000 vehicles a day (about 62.5 % of which was for simple transit purposes). In addition, as a result of measuring the atmospheric environment level along the Cheonggyecheon roadside, the general standard pollutants, except for fine dust (PM10) pollution, exceeded the Seoul average; the nitrogen oxides and volatile organic compounds (VOC), which are carcinogenic, exceeded the Seoul air environment standards. In particular, the concentration of benzene was measured to be high among VOC, and the air pollution was very poor.

The competitiveness of the old downtown Gangbuk (River North Area) has significantly decreased due to traffic problems in the Cheonggyecheon area, air pollution, and the aging of houses and buildings, resulting in a decrease over the past 10 years in the permanent population by 50,000, and in the employed population by

청계천 복원사업 추진배경

2000년대에 들어서며 환경에 대한 관심이 높아지고 무분별한 성장과 개발보다 환경과 문화를 중요하게 생각하는 사회 분위기가 조성되었으며 지속가능한 개발이라는 새로운 개념이 보편화되었다. 서울시에서도 환경문제를 중심으로 새로운 도시개발의 전략 : '자연과 인간 중심인 환경도시로 거듭나는 사업'의 필요성이 제기되었다.

청계천지역은 하수관로 정비가 이루어지지 않은 채 복개공사가 이루어져 하수 역류와 악취 발생 등 생활환경 개선이 시급하였으며, 따라서 서울시는 청계천을 자연하천으로 복원하고, 수변을 생태공원으로 조성하여 도시개발 패러다임을 환경친화적으로 전환하고 깨끗한 환경과 도심지 내 친수공간을 제공하고자 청계천 복원사업을 추진하게 되었다. 아울러 도심 경쟁력을 잃고 슬럼화된 채 방치된 청계천 주변을 청계천 복원사업을 통하여 도심 재활성화를 추구하여 지역간의 경제·문화적 불균형을 극복하고자 하였다.

청계고가 철거시에 나타난 '교통마비 반대여론'

복원사업을 위하여 꼭 필요했던 청계고가도로의 철거 결정은 서울시가 서울도심의 교통체계를 대중교통 중심으로 전환하고자 하는 계획을 내세웠지만 시민, 학계, 중앙정부, 시민단체로부터 심각한 반대에 부딪혔으며, 시민들이 불편을 겪게 될 것이라 주장하며 반대하였다.

서울시는 청계천복원에 따른 교통분야업무를 효율적으로 추진하고 교통국, 청계천복원추진본부, 청계천복원지원연구단을 구성하고 청계천복원 교통분야대책은 청계천복원추진본부 추진단장 주재의 교통분야 주요 관계관회의를 통해 업무를 분담하고, 사업의 중요성과 교통에 미치는 영향을 고려해 도심교통체계 개편사업을 시행하여 교통영향을 최소화하도록 사업을 추진하였다.

청계천복원추진본부와 시민위원회, 연구지원단 등 청계천복원을 위한 거버넌스 체계를

80,000; a study found that the number of corporate headquarters is only 63 % that of Gangnam. Urban redevelopment had difficulties in attracting private capital, and in particular, the financial and business functions of the old downtown area (12.5 %) were significantly insufficient, compared to the sub-center of Gangnam (27.0 %).

The over-concentration of small-scale traditional manufacturing industries that were not suitable for the downtown area, especially the growth of knowledge-based industries, was evaluated as not meeting expectations; and the backward and weak urban industrial structure was considered as an important factor that was undermining the competitiveness of Seoul as the capital of Northeast Asia.

Background of the Cheonggyecheon Restoration Project

In the 2000s, interest in the environment increased, a social atmosphere was created to consider the environment and culture more important than reckless growth and development, and the new concept of sustainable development became common. In Seoul, a new urban development strategy centered on environmental issues raised 'the necessity of a project to be reborn as an environmental city centered on nature and humans'.

In the Cheonggyecheon area, the sewage pipeline was covered without maintenance, and there was an urgent need to improve the living environment from detractions such as sewage reverse flow and odor generation; therefore, the city of Seoul initiated the Cheonggye Stream restoration project to restore Cheonggye Stream to a natural stream, and to transform the urban development paradigm into an environment-friendly paradigm by creating an eco-park on the waterfront, and providing a clean

적극 활용하여 사전대책 마련, 지속적인 의견수렴을 통해 이해 당사자들의 협조를 얻을 수 있었다.

청계천 복원사업의 기본구상

청계천 복원사업은 도시생태계의 중요한 역할을 하는 물순환체계를 회복시켜 자연의 자생 능력에 대한 생태계를 복원하고, '자연이 있는 도시하천' 조성에 기본의의를 두고 '서울 도심의 생동감과 친근함'을 제공하는 공간으로 탄생하는 '공간 창조' 사업으로 구상되었다(그림 8).

복원구간의 다양성을 부여하기 위하여 역사(전통), 문화(현대), 자연(미래)이라는 3개의 큰 시간축으로 구상하여 복원 시점부로부터 2km까지는 역사와 전통을 중시하고, 2km 지점부 터 4km 지점까지는 문화와 현대를 중심 테마로, 4km 지점부터는 자연과 미래의 개념을 도

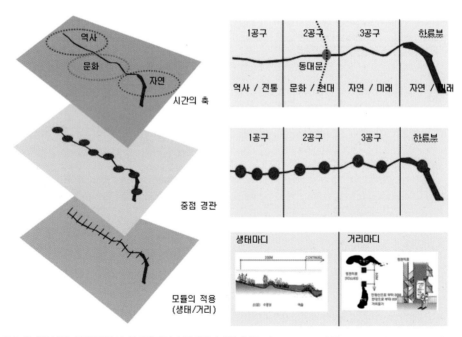

그림 8. 청계천 복원 기본구상(출처: 청계천 백서, 서울시설관리공단) / Basic concept of Cheonggyecheon restoration

environment and water-friendly space in the downtown area.

In addition, through the Cheonggyecheon restoration project, the area around Cheonggyecheon, which had lost its competitiveness in the city center and was neglected, was revitalized to overcome economic and cultural imbalances between regions.

When Cheonggye Elevated Highway was Demolished, 'Opposition to Traffic Paralysis' Appeared

The decision to demolish the Cheonggye elevated highway, which was essential for the restoration project, was agreed upon by citizens, academia, the central government, and civic groups. The Seoul Metropolitan Government proposed a plan to transform the city's transportation system into a public transportation-oriented one, but citizens objected, arguing that it would cause inconvenience to the citizens.

The Seoul Metropolitan Government efficiently promotes transportation tasks in accordance with the Cheonggye Stream Restoration, and organizes the Transportation Bureau, Cheonggye Stream Restoration Promotion Headquarters, and Cheonggye Stream Restoration Support Research Group. In consideration of the importance of the project and its impact on transportation, the urban transportation system reform project was implemented to minimize the transportation impact.

By actively utilizing the governance system for the restoration of Cheonggye Stream, such as the Cheonggye Stream Restoration Promotion Headquarters, Citizens' Committee, and Research Support Group, they were able to obtain cooperation from stakeholders through proactive measures and continuous opinion gathering.

입한 설계가 진행되도록 하였다. 3개의 시간축 구간은 총 8개의 중점경관을 포함하도록 구성하고, 여기에 다시 생태/거리의 모듈을 적용하였다.

또한 청계천복원 기본구상은 청계천복원사업이 단순히 하천의 복원뿐만 아니라 주변지역의 활성화를 목표로 구상되었다. 복원사업의 마스터플랜은 청계천과 청계천 일대의 도시 관리계획을 재정립하고 복원된 청계천은 매력적인 도심지 내 수변공간으로 재탄생시켜 주변지역 정비를 위한 마중물이 되어, 도심의 경제 활성화를 이루고자 하였다(그림 9).

청계천 복원사업 마스터플랜을 통하여 지역별 특성을 고려한 도심부 개발의 세심한 관리, 사대문안 역사문화성 회복, 도심 공동화현상을 해결하고 보행자 중심 도심 교통체계를 새롭게 정립하고자 하였다.

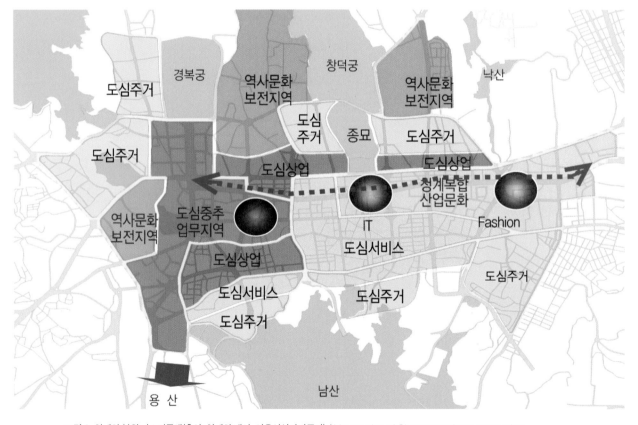

그림 9. 청계천 복원 마스터플랜(출처: 청계천 백서, 서울시설관리공단) / Master plan of Cheonggyecheon restoration

Basic Concept of the Cheonggyecheon Restoration Project

The Cheonggyecheon restoration project was conceived as a 'space creation' project by restoring the ecosystem to allow the self-sustainability of nature. By restoring the water cycle system, which plays an important role in the urban ecosystem, it was intended to create a space that provides 'the vitality and friendliness of downtown Seoul' with the basic significance of creating an 'urban river with nature' (Fig. 8).

To give diversity to the restoration section, the project was conceived along three major time axes: history (tradition), culture (modern), and nature (future). From the point of restoration to the 2 km mark, history and tradition were emphasized; from the 2 km to 4 km mark, culture and modernity were the central themes; and from the 4 km mark, the concept of nature and the future was introduced, so that the design proceeded according to this sequence. The three temporal axis sections were composed to include a total of eight key landscapes, and the ecology/street modules were applied here again.

In addition, the Cheonggyecheon Restoration basic concept was conceived with the goal of revitalizing the surrounding area, as well as simply restoring the stream. The master plan of the restoration project re-established the urban management plan for Cheonggyecheon and the surrounding area, and the restored Cheonggyecheon was reborn as an attractive waterfront space in the downtown area to serve as a pilot project for the maintenance of the surrounding area, thereby revitalizing the urban economy (Fig. 9).

Through the master plan of the Cheonggyecheon Restoration Project, it was intended to carefully manage the development of the downtown area in consideration of regional characteristics, to restore the historical and cultural properties within the four main

청계천 복원사업의 계획요소 및 디자인 이슈

청계천 하천복원 기본계획은 청계천 주변의 도시계획 기본구상을 중심으로 복개시설 및 청계고가 철거, 기존시설의 이설 및 공사 중 교통처리계획 수립(보행계획 포함) 등 복원 이전 선행되어야 할 사항들을 정리한 후 청계천복원을 위한 시설기본계획(하천, 하수도, 도로, 환경, 조경, 교량 등)을 수립하고 주변 녹지축[동·서 수경녹지축 복원 및 연결 (덕수궁~청계천~중랑천~한강)과 남·북 녹지축(종묘~세운상가~남산) 연결의 거점을 확보하여 약 8만평 규모의 선형녹지체계를 수립하는 계획]과 연계하는 방안으로 진행되었다(그림 10).

청계천 복원 및 생태계 복원

복개구조물과 청계고가 철거로 청계천에 맑은 물이 흐르도록 하고 녹지를 복원시켜 각종 생물의 서식공간을 제공하여 전체 생태계를 복원하도록 계획되었다. 상류지역은 친수공간을 중심으로 하류지역은 자연성이 살아있는 공간으로 복원의 방향성이 설정되었다(그림 11). 이에 따라 어류 및 조류들이 청계천으로 회귀할 수 있도록 하였으며, 하천을 따라 바람길이 만들어지면서 도시 내 기온이 떨어지도록 하여 도심 열섬현상의 감소를 유도하였다.

그림 10. 청계천 인근 그린 네트워크 개념도(출처: 청계천 백서, 서울시설관리공단)
Green network concept map near Cheonggyecheon

gates, to solve the urban hollowing phenomenon, and to establish a new pedestrian-oriented urban transportation system.

Planning Elements and Design Issues for the Cheonggyecheon Restoration Project

The Cheonggyecheon stream Restoration basic plan is based on the master plan around the Cheonggyecheon stream, after arranging the priorities prior to restoration, such as the removal of the covered structure and Cheonggye elevated highway, the relocation of existing facilities, and the establishment of a traffic treatment plan during construction (including the pedestrian plan). A basic plan for the restoration of Cheonggyecheon (streams, sewers, roads, environment, landscaping, bridges, etc.) was established, and the plan was carried out in connection with the surrounding green axis (Fig. 10).

a. 청계천 상류지역 / Upstream areas   b. 청계천 하류지역 / Downstream areas

그림 11. 청계천 상류지역과 하류지역(출처: 청계천 백서, 서울시설관리공단)
Upstream and downstream areas of Cheonggyecheon

복원사업의 가장 큰 요소인 도심하천으로의 복원을 위하여 국지성 집중호우 대비용 200년 빈도의 통수단면(제방어유고)을 확보하고 도시하천임을 감안하여 여유고는 50~80년 빈도를 적용하여 설계되었다. 그리고 기존 복개구조물 철거 시 일부분을 활용, 양안의 도로 및 통수단면으로 사용하도록 계획되었다(그림 12).

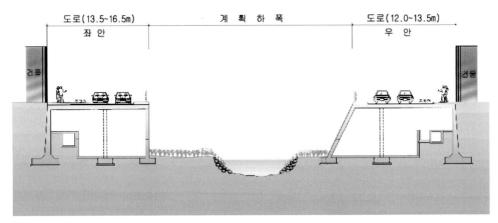

그림 12. 청계천 하천 단면 개념도(출처: 청계천 백서, 서울시설관리공단)
Conceptual Cross Section of Cheonggyecheon River

청계천은 한강원수를 채취하여 1b등급 이상의 수질로 정수 처리한 용수를 하루 4만톤(최대 12만톤) 공급하여 하천수심을 평균 40cm로 유지시키며서 맑은 물을 공급하여 상시 물이 흐르는 생태하천으로 복원하였다.

복원된 청계천은 하류지역에서 중랑천을 거쳐 한강과 연결되며 청계천과 중랑천이 만나는 지점에는 물고기의 산란지 등 생태습지를 조성하여 생태보존지역으로 지정하였다.

Cheonggyecheon Stream Restoration and Ecosystem Restoration

It was planned to restore the entire ecosystem by dismantling the cover structure and Cheonggye elevated highway to allow clear water to flow through Cheonggye Stream, and to restore green areas to provide habitat for diverse wildlife. The direction of restoration was set as the upstream area centered on a waterfront space, and the downstream area as a space where nature was alive (Fig. 11). This allowed fish and birds to return to the Cheonggyecheon Stream, and also reduced the urban heat island phenomenon by creating a winding path along the river, causing the temperature in the city to drop.

For the restoration of urban rivers, which is the biggest element of the restoration project, a flow surface (bank clearance) with a frequency of 200 years was secured to prepare for localized torrential rain; and considering that it is an urban river, the free clearance was designed with a frequency of 50 to 80 years. It was devised to utilize a part of the existing covering structure when demolishing it, and it was planned to be used as a road and passage section for both banks of the river (Fig. 12).

In Cheonggyecheon, 40,000 tons (up to 120,000 tons) of water that is collected from the Hangang River and treated with water quality of grade 1b or higher is supplied to maintain the depth of the river at an average of 40 cm.

The restored Cheonggyecheon is connected to the Han River through Jungnangcheon in the downstream area, and an ecological wetland, such as a spawning ground for fish, was created at the point where Cheonggyecheon and Jungnangcheon meet, and designated as an ecological conservation area.

Waterside Park Plan

To restore environmental functions and provide a habitat for living things, the

수변공원계획

환경적 기능을 회복하고 생물서식기반을 제공하기 위해 생태적 호안을 설계하고 자연형 하천 조성을 위해 하천시설물과 하도기반을 조성하여 경관의 절제 및 조화, 고풍스러운 하천경관의 연출, 녹색경관을 도입하도록 하였다. 특히 하천 주변에 경화토 산책로와 녹지대, 징검다리 등을 조성하고 고수부지, 분수, 수변데크를 설치하여 다양한 계층의 이용자들이 쉽게 접근할 수 있는 도심 속 수변공원이 되도록 하였다(그림 13).

상류에서 중류지점까지는 휴게시설과 보행동선을 분리(고수부지, 저수부지)하여 도시형 호안형태로 계획하여 이용자들의 친수성을 증진시키고 주변경관과의 조화를 이루도록 하고

a. 청계천 징검다리 / Stepping stone bridge   b. 청계천 분수 / Cheonggyecheon fountain

그림 13. 청계천 징검다리와 분수(a 출처: https://blog.daum.net/sansol/4095)(b 출처: 서울시설공단, 뉴스와이어)
Cheonggyecheon stepping stone bridge and fountain

그림 14. 청계천 식재개념도(출처: 청계천 백서, 서울시설관리공단) / Conceptual image of Cheonggyecheon planting

ecological shoreline was designed, and river facilities and necessary bases were created for the creation of natural rivers, so that the landscape was restrained and harmonized, a genteel river landscape produced, and a green landscape introduced.

In particular, it was designed to become a waterfront park in the city that users of various classes could easily access by creating a light soil promenade, green zones, and stepping-stones around the river, and installing terrace lands, fountains, and waterside decks (Fig. 13).

From the upstream to the midstream point, the rest facilities and the pedestrian circulation are separated (terrace lands, reservoir site), and planned in the form of an urban shoreline to enhance the water site so that it is friendly to users, and harmonizes with the surrounding landscape. It also planned to use the walls and vegetation of the terrace land to create a green landscape (Fig. 14).

The landscape was relieved by reproducing the image of traditional stone stacking (bank stone stacking, castle stone stacking, gwanggyo stone wall) on the vertical wall, greening the wall and planting trees on the terrace land, and reeds, pampas grass, succulents, and greenery on the slopes of the soil section. In the case of the terrace land, where mats were planted, it was planned in connection with the (1.5 – 3.0) m wide promenade as a space for users to learn about and experience nature (Fig. 15).

Cheonggye Plaza and Open Space Plan

The plaza was created by enhancing the meaning and symbolism of the restoration of Cheonggyecheon in the area where Taepyeong-ro and Cheonggye-cheon-ro meet. Cheonggyecheon starting point Plaza, a symbolic place at the start of the first waterway, was planned to be reborn as an open civic plaza with changes in line with the needs of citizens for leisure culture and the flow of the times [7].

벽체 및 고수부지의 식생을 통하여 녹색경관이 연출되도록 계획하였다.

하류구간은 자연형 하천의 조성을 위한 하도기반을 조성하고 생물서식공간 조성을 위한 생태적 호안으로 설계하였으며 하천식물을 식재하고 왼쪽 기슭은 산책로를 설치하지 않고 사람의 접근이 없는 자연회복구간으로 조성하였다(그림 14).

수직벽면에 전통적 돌쌓기(제방돌쌓기, 성돌쌓기, 광교석축) 이미지를 재현하여 경관을 완화시켰으며, 벽면녹화 및 고수부지 상단에 수목을 식재하고, 토사구간의 사면에는 갈대, 물억새, 수크링, 녹화매트 등을 식재하고 고수부지의 경우 이용자가 자연학습과 체험할 수 있는 공간으로 폭 1.5~3.0m의 산책로와 연계하여 계획하였다(그림 15).

a. 청계천 석벽 / Stone wall

b. 벽면녹화 / Wall greening

그림 15. 청계천 석벽과 벽면녹화(출처: https://blog.daum.net/sansol/4095) / Cheonggyecheon stone wall and wall greening

청계광장 및 오픈스페이스계획

청계천이 시작되는 태평로와 청계천로가 만나는 접점지역을 청계천복원의 의미 및 상징성 제고하여 조성하였다. 첫 물길을 여는 상징적인 장소인 청계천 시점부 광장은 시민들의 여가문화에 대한 요구와 시대 흐름의 변화를 담아 열린 시민광장으로 재탄생되도록 계획하였다.

청계광장은 광장부와 하천부로 구성되어 있으며 지상부의 친수 및 휴식공간은 쾌적한 오픈스페이스로 만들어 광화문광장과 인근 주변 관광자원 등과 연계시켜, 가변적이고 각종 문화행사를 비롯한 다양한 이용행태를 수용하는 열린 마당으로 구성하였다(그림 16).

Cheonggye Plaza is composed of a plaza section and a river section, and the water-friendly and resting space on the ground is made into a pleasant open space and linked with Gwanghwamun Square and nearby tourist resources. It is composed of an open yard that is flexible, and can accommodate various cultural events and usage patterns (Fig. 16). A candle fountain with three-color lighting and a two-stage 4 m waterfall were created to present a dynamic scene. On both sides of the waterfall, 'Palseokdams' made of 8 stones imported from different regions of the country were created to give visitors a unique waterfront experience.

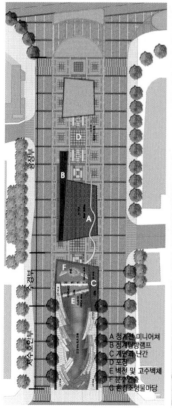

a. 청계광장 배치도 / Plaza Site Plan   b. 청계광장 분수와 팔석담 / Fountain and Palsukdam

그림 16. 청계광장 개념도(출처: 한국조경학회지, 신현돈) / Conceptual image of Cheonggye Plaza

삼색 조명이 어우러진 캔들 분수와 4m 아래로 떨어지는 2단 폭포를 구성하여 역동적인 모습을 연출하였으며, 폭포 양 옆에는 전국에서 가져온 돌로 8도석으로 제작된 '팔석담'을 조성하여 방문객에게 독특한 수변을 경험하도록 하였다. 특히 야간에는 빛과 물이 어우러지는 환상적인 모습을 연출하여 다양한 야경을 제공하도록 하였다.

문화재 발굴과 교량복원

서울의 구도심지를 관통하는 장소성을 고려하여 청계천복원사업은 서울의 역사성 회복과 문화공간의 창출이라는 과제와 함께 수행되었다. 이에 따라 문화재 지표조사, 문화유적 시굴조사, 문화재 발굴조사 및 공학적 정밀실측조사를 거쳐 전문가 자문을 통한 역사문화의 복원방안의 기본방향을 설정하였다.

조사를 통해 발견된 유적들은 일제강점기와 1960년대 청계천 복개공사 과정에서 콘크리트속에 매몰되었던 유적들로, 청계천 복원구간에 있던 광통교, 수표교 및 오간수문 등 다리와 수문 그리고 호안 석축 등이며, 이는 조선시대의 건축기술과 도시계획을 짐작할 수 있는 중요한 자료로 일부를 복원하도록 결정되었다.

복원과정에서 서울시는 주변환경을 고려한 역사복원을 추진하였으나 문화연대 등 시민단체들은 원형, 원위치 복원을 주장하여 다양한 의견 수렴과정을 거치고 문화재청의 결정에 따라 복원이 진행되었다.

문헌자료를 통해보면 청계천복원구간의 옛 다리는 상류쪽으로부터 모전교, 광통교, 장통교, 수표교, 하랑교, 효경교, 태평교, 오간수다리, 영도교 등 9개소인 것으로 파악되었는데 옛 다리에 대한 복원은 원형 고증이 가능한 것에 따라 장기적이고 단계별로 시행하는 것으로 계획되었다.

Cultural Heritage Excavation and Old Bridge Restoration

Considering the sense of place that penetrates the old downtown of Seoul, the Cheonggyecheon restoration project was carried out with the task of restoring Seoul's historic nature and creating a cultural space. Accordingly, the basic direction of the restoration plan for historical and culture was established through expert advice after conducting surface surveys of cultural properties, prospecting surveys of cultural relics, excavation surveys of cultural properties, and detailed engineering surveys.

The relics discovered through the investigation are the bridges and sluice gates, including the Gwangtong Bridge, Supyo Bridge, and Ogan Sumun Gate, which were in the restoration section of the Cheonggye Stream, and the stonework of the shoreline. It was decided to restore some of the ruins buried in concrete during the Japanese colonial period and during the construction of Cheonggyecheon covering structure in the 1960s, as important data that can infer the architectural technology and urban planning of the Choseon dynasty.

During the restoration process, the city of Seoul pursued historical restoration in consideration of the surrounding environment, but civic groups, such as Cultural Action, insisted on restoring the original form and original location, and after collecting various opinions, restoration was carried out according to the decision of the Cultural Heritage Administration.

According to the literature materials, it was identified that there were 9 old bridges in the Cheonggyecheon Restoration Section from the upstream side, namely Mojeon Bridge, Gwangtong Bridge, Jangtong Bridge, Supyo Bridge, Harang Bridge, Hyogyeong Bridge, Taepyeong Bridge, Ogansu Bridge, and Yeongdo Bridge. It was planned in the long term to restore or rebuild them in stages, according to the possibility of

광통교

광통교는 폭 15m, 길이 12m 규모로 도성에서 가장 규모가 큰 교량이었으며 경복궁-육조거리-운종가-(광통교)-숭례문 등 도성을 남북으로 연결하는 중심통로에 위치하여 어가행렬이나 사신행렬 때 이용되었다고 한다(그림 17).

이는 잔존유규가 잘 남아있고 사진자료 등 사료들이 충분하여 원형복원을 하는 것으로 결정되었지만 교통, 하천 단면, 안전 등의 문제로 원위치 복원에는 어려움이 있었다. 원 위치에서 170m 상류지점에 보도교로 형태복원이 되었다(그림 18).

a. 광통교 과거 모습 / Gwangtong Bridge in the past   b. 복원 후 광통교 / Gwangtong Bridge after restoration

그림 17. 광통교, 과거와 현재(출처: 문화재청, 위키백과) / Gwangtong Bridge, past and present

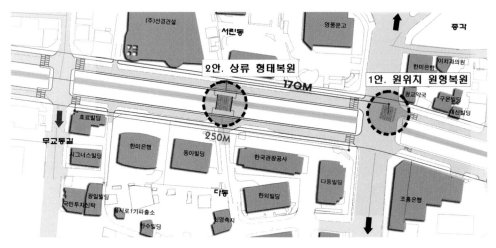

그림 18. 광통교 복원 개념도(출처: 청계천 백서, 서울시설관리공단) / Restoration concept image of Gwangtong Bridge

ascertainment of their original form.

Gwangtong Bridge

Gwangtong Bridge was the largest bridge in the city with a width of 15 m and a length of 12 m, and was located in the central passage connecting the city on its north—south axis, such as Gyeongbok Palace – Yukjo street – Unjong street – (Gwangtonggyo) – Sungnyemun gate, etc., and was used for the procession of the King and envoys (Fig. 17).

It was decided to restore the original shape because the remaining relics were well preserved and materials such as photographs were sufficient, but there were difficulties in restoring it in its original position, due to traffic, river cross-section, and safety issues. As a result, the footbridge was rebuilt 170 m upstream from the original location (Fig. 18).

Supyo Bridge

Supyo Bridge was said to be 8.3 m wide and 27.5 m long and have 45 piers, and was called Supyo Bridge after a Supyo (water depth finder) was erected next to the bridge to measure the water level of the stream. In 1959, when the covering of Cheonggyecheon started, Supyo Bridge and Supyo were moved to Jangchungdan Park, and Supyo was then moved to the King Sejong Memorial Hall in Cheongnyangni-dong, Dongdaemun-gu, Seoul (Fig. 19).

As with Gwangtong Bridge, there was no problem with the original restoration; however, when the original location and original shape were restored, the planned river width was narrow, so the possibility of maintaining the original shape, such as changing the planned road, emerged as a problem. Therefore, it was planned as a pedestrian bridge, and its shape was restored to its original location in consideration of

수표교

수표교는 폭 8.3m, 길이 27.5m 규모의 교량으로 45개의 교각을 가지고 있었으며 다리 옆에 개천의 수위를 측정하기 위해 수표를 세운 이후 수표교라고 불렀다고 한다. 수표교와 수표는 청계천이 복개공사가 시작되던 1959년 장충단공원으로 옮겨졌으며, 이후 수표는 다시 동대문구 청량리동에 있는 세종대왕기념관으로 이전되었다(그림 19).

a. 수표교 과거 모습 / Supyo Bridge in the past   b. 복원 후 수표교 / Supyo Bridge after restoration

그림 19. 수표교, 과거와 현재(출처: 서울시설공단, 한국관광공사) / Supyo Bridge, past and present

광통교와 마찬가지로 원형복원에는 문제가 없었으나 원위치 원형복원 시 계획된 하천의 폭이 좁아 계획도로 변경 등의 원형보존 유지 가능성이 문제가 대두되었다. 보도교로 계획하고 교통 및 주변 상황과 교각, 다리 높이 등을 고려하여 원위치에 형태복원되었다.

그러나 현재 청계천에 복원된 수표교는 석교가 아닌 어설픈 나무다리로 원형의 모양새만 갖췄다는 비판을 받고 있다(그림 20).

traffic and surrounding conditions, piers, and bridge heights.

Nevertheless, the currently restored Supyo Bridge in Cheonggyecheon is criticized for having only a gentle circular arc side elevation as a clumsy wooden bridge, rather than a stone bridge (Fig. 20).

Ogansumun Gate

Although there are no remnants of the Ogansumun Gate, it was classified as a relic whose shape can be restored, because its original form can be known through data such as photographs and drawings.

It was called Ogansumun because the sluice gate consisted of 5 sections, that is, 5 sluice gates, as it was a water gate from the Choseon Dynasty at 6-ga, Cheonggyecheon 6-ga street under the wall from Dongdaemun to Eulji-ro. Each gate was a 1.5 m sluice gate installed to allow the stream flowing under the southern wall of Dongdaemun to flow out of the city wall.

During the excavation of the Cheonggyecheon ruins, an end support, a base of the hongye (rainbow-shaped structure), and a stone turtle were excavated, and restoration was decided on. A newly created Ogansugyo Bridge was built at the original location for reasons such as city traffic, and the 5 sluice gates and a rainbow-shaped Hongye arch were reproduced using the traditional shape (Fig. 21).

There are a total of 22 bridges in Cheonggyecheon, starting at Mojeon Bridge, and ending at Gosanja Bridge. Among them, the bridges dedicated to the pedestrian are Gwangtong Bridge, Jangtong Bridge, Supyo Bridge, Sewoon Bridge, Saebyok Bridge, Narae Bridge, Malgeunnae Bridge, and Dumul Bridge, while the rest of the bridges are planned as driveway and sidewalk bridges (Fig. 22).

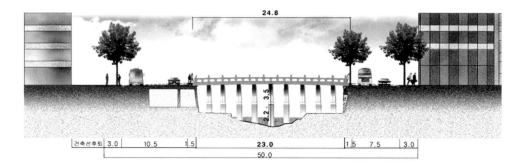

그림 20. 수표교 복원 개념도(출처: 청계천 백서, 서울시설관리공단) / Restoration concept image of Supyo Bridge

오간수문

오간수문은 잔존 유적은 없으나 사진이나 그림 등의 자료를 통해 원형을 알수 있어 형태복원이 가능한 유적으로 분류되었다. 동대문에서 을지로로 가는 성벽 아래 청계천6가에 있던 조선시대의 수문(水門)으로 수문이 5칸, 즉 5개의 수문으로 이루어졌다고 하여 오간수문으로 불렸으며 동대문 남쪽 성벽 아래로 흐르는 냇물이 도성 밖으로 잘 빠져나갈 수 있도록 설치한 1.5m 크기의 수문이다. 청계천 유적 발굴 시 끝받침과 홍예(虹霓 : 무지개 모양의 구조물) 기초부, 돌거북 등이 발굴되어 복원이 결정되었으나, 도심 교통 등의 문재로 원위치에는 새롭게 만들어진 오간수교를 세웠으며 전통적인 모양을 살려 5개 수문과 무지개 모양의 홍예 아치를 재현하였다(그림 21).

a. 오간수문 과거 모습 / Ogansumun in the past　　　　b. 복원 후 오간수문 / Ogansumun after restoration

그림 21. 오간수문, 과거와 현재(출처: 서울시설공단, 문화재청) / Ogansumun Gate, past and present

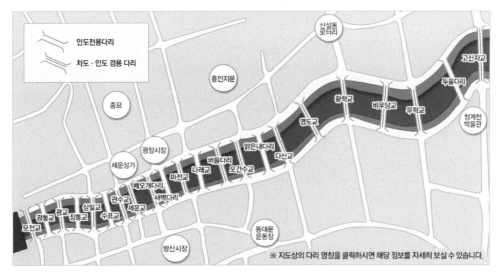

그림 22. 청계천의 교량들(출처: 서울시설공단) / Bridges of Cheonggyecheon

Evaluation and Limitations of the Cheonggyecheon Restoration Project

The Cheonggye Stream restoration project is evaluated by considering the performance of the Cheonggye Stream restoration as an urban regeneration project, and the limitations of the Cheonggye Stream restoration as an ecological/historical restoration project.

Restoration of Cheonggyecheon as an Old Downtown Regeneration Project

From the perspective of the downtown regeneration project, evaluation shows that the safety problem was solved by the demolition of the old, elevated highway passing through the city center, and this resulted in drastic improvement of the city landscape.

청계천에는 시작지점인 모전교에서 끝지점인 고산자교까지 총 22개의 다리가 있으며, 이 중 인도전용 다리는 광통교, 장통교, 수표교, 세운교, 새벽다리, 나래교, 맑은내다리, 두물다리는 보도용 다리이고, 나머지 교량들은 차도·인도겸용 다리로 계획되었다(그림 22).

청계천 복원사업의 평가와 한계

청계천 복원사업은 도심재생사업으로 청계천 복원의 성과와 생태·역사복원사업으로 청계천 복원의 한계를 나누어서 평가되어지고 있다.

도심재생사업으로 청계천 복원

도심재생사업의 관점으로는 도심을 관통하는 오래된 고가도로의 철거로 안전문제를 해결했으며 이로 인한 도심경관의 획기적 개선을 이루었다는 평가이다. 또한 하천의 복원으로 도심 내의 친수공간을 제공하고 도심자연을 일정부분 회복하였다.

청계천지역의 대기오염도와 기상자료를 복원사업 전후를 비교하여 분석한 연구결과를 살펴보면 전반적으로 복원 후 가스상 오염물질이 4~7ppb 수준으로 개선된 것으로 나타나 도심 대기환경 개선에 기여한 것으로 평가되었다. 청계천은 복원 후에는 풍속이 다소 빠르게 나타나는 것으로 분석되었으며 분석결과에 따르면 평균풍속은 최소 2.2%, 최대 7.1%가 증가한 것으로 나타났다. 복원 전후의 온도분포 조사결과에 따르면 청계천 주변지역의 온도는 낮아지고 풍속은 빨라지는 등 이 지역의 미기후가 주변지역에 비해 상대적으로 쾌적하게 변화한 것으로 분석되었다.

또한 청계천 주변지역에서는 복원사업 이후 도심재생사업이 시행되었으며 미시행재개발지구의 활성화를 유도하고 도심활성화를 이끌어온 것으로 평가되었다. 재개발 추진과 토지이용변화에 따른 전반적인 건축행위의 증가로 이어져 도심 환경개선과 상주인구의 증가에 기여한 것으로 평가되었다. 2006년 발표된 2020 서울도시기본계획에 따르면 도심권 활성화 방

In addition, the restoration of the river provided a water-friendly space in the city center, and to a certain extent restored the nature of the city.

Looking at the results of a study comparing the air pollution level and meteorological data of the Cheonggyecheon area before and after the restoration project, it was found that the overall level of gaseous pollutants after restoration was improved to (4-7) ppb, which was evaluated as contributing to the improvement of the urban air environment. The wind speed of Cheonggyecheon was analyzed to appear rather faster after restoration, the average wind speed increased by at least 2.2 % and at the maximum by 7.1 % according to the analysis results [9]. In accordance with the temperature distribution survey results before and after restoration, analysis showed that the microclimate of the area around Cheonggyecheon has changed to be become relatively more comfortable when compared to the surrounding areas, such as lowered temperature and faster wind speed, which are important in the heat of summer [10].

In addition, in the area around Cheonggyecheon Stream, the downtown regeneration project was implemented after the restoration project, and evaluation found that this has led to the vitalization of the unimplemented redevelopment district, and led to urban revitalization. The evaluation showed that the project has contributed to the improvement of the urban environment and the increase of the resident population, as the redevelopment promotion and land use change have led to an increase in overall construction activity [11]. According to the 2020 Seoul Master Plan announced in 2006, a plan to revitalize the downtown area through restoration of Cheonggyecheon is suggested as a way to revitalize the downtown area, and as the tasks of the living area plan policy goals and implementation strategies, the planning maintenance and management of the area around the Cheonggye Stream, and guidelines for the management of buildings along the Cheonggye Stream, have been prepared and

안으로 청계천 복원을 통한 도심기능 활성화 방안이 제시되어 있으며 생활권계획 정책목표 및 추진전략의 과제로 청계천 주변지역의 계획적 정비와 관리, 청계천변 건축물 관리유도지침을 마련하여 전략적으로 관리하고 있다.

생태·역사복원사업으로 청계천 복원

생태적 측면에서의 복원효과는 식생의 종조성이 자연하천과 유사해지고 다양성이 늘어났으며, 외래종 비율이 낮아지고 있는 점과 일정수준의 수질 개선효과도 관찰되었다. 반면 생태복원이라는 측면에서는 과다한 인공적인 시설과 장치들로 인하여 정체성의 문제와 함께 한강에서 양수한 물을 흘려보내는 방식의 복원으로 생태적 기능을 완전히 고려한 자연하천이 아니라는 부정적 평가도 함께 받고 있다.

역사복원의 경우는 고가도로와 복개구조물 제거로 인하여 하천의 옛모습을 어느 정도 찾았다는 긍정적인 평가도 있지만 도심 내의 현실적 조건으로 발굴된 유적인 석축을 복원하지 못하였고 광통교, 수표교 등 옛 교량을 원래의 모습과 다르게 복원하거나 복원하지 않았다는 부정적 평가를 함께 받았다.

맺으며

청계천 복원이 당초부터 생태환경적 개념이 아닌 도심정비를 위한 개발사업의 일환으로서 진행되었다는 비판도 있지만 복원 이후 방문객 현황을 살펴보면 다양한 산책코스의 제공하였으며 점심시간에는 주변 직장인들의 산책장소로, 또 일몰 후에는 청계천의 야경과 주변환경을 즐기는 가족·연인·친구 단위의 방문객들로 서울 도심 속 오아시스 역할을 하며 휴식, 여가 및 관광지로 변신하여 새로운 도시환경을 창출하였다는 평가를 받고있다.

managed strategically.

Restoration of Cheonggyecheon as an Ecological and Historical Restoration Project

As for the ecological restoration effect, it was observed that the species composition of vegetation became similar to that of natural rivers, the diversity increased, the ratio of exotic species decreased, and a certain level of water quality improvement effect was also observed [12]. On the other hand, in terms of ecological restoration, taking into account the excessive number of artificial facilities and devices, the problem of identity, and the restoration of the method of flowing water pumped from the Han River, the project is being negatively evaluated as not presenting the features of a natural river, and full consideration has not been given to its ecological functions.

In the case of historical restoration, positive reviews observe that due to the removal of the elevated highway and the covering structure, the old shape of the river has to some extent been restored, but the stone structures of the ruins excavated in the city center could not be restored under realistic conditions in the city center, and negative reviews have also received that observe that old bridges, such as Gwangtong bridge and Supyo bridge, have either been rebuilt differently from their original appearance, or not restored at all.

Closing Remarks

Although there has been criticism that the restoration of Cheonggyecheon was carried out from the beginning as a part of a development project for urban renewal,

참고문헌

1. 청계천 복원사업, 서울연구데이터서비스, https://data.si.re.kr/node/190

2. 청계천복개공사, 1958 ~ 1978, 서울기록원, https://archives.seoul.go.kr/authority/TOPIC-00273

3. 굴곡진 역사와 함께한 서울의 얼굴: 청계천, 국가기록원, https://theme.archives.go.kr//next/koreaOfRecord/cheonggyecheon.do

4. 청계천 복원사업 백서, 서울 특별시, 2006

5. 청계천 복원사업 기본배경, 서울 정책아카이브, 2015, https://www.seoulsolution.kr/ko/content/3250

6. 강수학 외 2인, 청계천 복원 전·후의 식물상 변화 연구, 한국환경재생녹화기술학회지, 2007

7. 신현돈, 서울 청계광장 설계 연구, 한국조경학회지 40(3), 2012

8. 허영일, 청계천 복원에 따른 도시대기환경 개선효과에 관한 연구, 서울시립대학교, 2006

9. 복원 전·후 바람길 변화, 청계천 소개, 서울시설관리공단, https://www.sisul.or.kr/open_content/cheonggye/intro/effect.jsp

10. 김운수, 청계천.서울숲 조성애 따른 미기후 및 생태변화, 서울연구포커스, 서울시정개발연구원, 2006

11. 임의지, 청계천 복원사업의 도심재생효과, 서울경제, 2005

12. 이창석, 되살아난 생태하천 '청계천' 복원 효과, 이미디어, 2020

rather than an ecological environment concept, evaluation has found that it has created a new urban environment by providing a variety of walking courses, serving as a walking area for nearby office workers during lunchtime and as a visiting place for families, lovers, and friends to enjoy the night view of Cheonggyecheon and the surrounding environment after sunset, and it serves as an oasis in downtown Seoul, transforming downtown into a rest, leisure, and tourist destination.

References

1. Cheonggyecheon Restoration Project, The Seoul Research Data Service, https://data.si.re.kr/node/190

2. Cheonggyecheon Stream Coverage Construction, 1958 ~ 1978, Seoul Metropolitan Archives.

3. The face of Seoul with its winding history: Cheonggyecheon Stream, National Archives of Korea https://theme.archives.go.kr//next/koreaOfRecord/cheonggyecheon.do

4. Cheonggyecheon Restoration Project White Paper, Seoul Metropolitan Government, 2006

5. Basic Background of Cheonggyecheon Restoration Project, Seoul Solution, 2015

6. Kang Suhak et al., A study on changes in flora before and after restoration of Cheonggyecheon, Journal of the Korean Society for Environmental Regeneration and Greening Technology, 2007

7. Hyundon Shin, A study on the design of Cheonggye Plaza in Seoul, Journal of the Korean Landscape Architecture Society 40(3), 2012

8. Heo Young-il, A Study on the Improvement Effect of Urban Air Environment by Restoring Cheonggyecheon, University of Seoul, 2006

9. Changes in wind path before and after restoration, Introduction to Cheonggyecheon, Seoul Facilities Corporation, https://www.sisul.or.kr/open_content/cheonggye/intro/effect.jsp

10. Kim Woon-su, Changes in microclimate and ecology following the creation of Cheonggyecheon and Seoul Forest, Seoul Research Focus, The Seoul Institute, 2006

11. Im, Euiji, Urban regeneration effect of Cheonggyecheon restoration project, Seoul Economic Daily, 2005

12. Changseok Lee, The restoration effect of the revived ecological river 'Cheonggyecheon', Emedia, 2020

서울 용산공원, 대형공원의 조성과정과 시민참여

최혜영 (성균관대학교 건설환경공학부 교수)

들어가며

용산공원은 서울시 용산구에 위치한 미8군 기지가 평택으로 이전한 뒤, 반환될 부지 위에 지어질 대형 도시공원이다.

용산 미군기지의 공원화는 1988년 3월, 노태우 대통령의 지시로 시작되었다. 그러나 '공원'의 조성보다 장시간 서울의 중심을 자치해온 '주한 미군의 이전'이 논의의 핵심이었다.

이렇다 보니 공원화 과정은 용산 미군기지의 이전 계획에 큰 영향을 받을 수밖에 없었다. 기지 이전은 한-미 두 국가간 정치적, 군사적, 외교적 상황 변화에 의해 복잡하게 진행되었다. 공원 조성과정도 마찬가지였다. 기지 이전 논의가 중단되면 공원화 논의도 수그러들고, 기지 이전 협상에 진척이 있으면 공원 조성과정도 탄력을 받아 진행되었다.

지난 30여 년간 공원화과정은 멈춤과 재개를 거듭하며 진행되어오고 있다. 그러나 그 과정의 불확실함을 야기한 것은 비단 기지 이전만의 문제는 아니었다.

도시계획시설 중 하나인 공원은 남녀노소, 지위, 빈부의 차이를 막론하고 어느 누구나 보편적으로 사용할 수 있는 공간이다. 공원은 도시민의 세금으로 조성되며, 그렇기에 시민들의 의견을 들어 민주적으로 조성해야 마땅하다. 이러한 시민 소통 과정은 대형 공원을 조성할 때 더욱 필요하다. 대형 공원은 규모가 주는 복잡성이 크고, 조성이 완료되기까지 시간이 오래

Seoul Youngsan Park, Public Engagement in the Large Urban Park Development

Hyeyoung Choi (Professor, Sungkyunkwan University, School of Civil, Architectural Engineering, and Landscape Architecture)

Introduction

Yongsan Park is a large urban park to be built in Yongsan-gu, Seoul, on the site of Yongsan Garrison which is currently being relocated to the city of Pyeongtaek and to be returned following the relocation.

The Yongsan Park project was initiated as of March 1988 when President Roh Tae-woo ordered the relocation of the US military base in Yongsan. At first, it was not intended to create a 'park.' The 'relocation of U.S. Forces in Korea,' which has been at the center of Seoul for a long period, was the focus of the discussion.

As a result, the process of turning the military base into a park had to be greatly affected by the relocation schedule of Yongsan Base. The relocation of the base was complicated by changes in the political, military, and diplomatic terrain between the two countries. The park development process has also been affected. Thus, the discussion as to relocating the base was paused, and so was the park. If progress was made for the base relocation, the park creation process also gained momentum.

For the past 30 years, the park development process has been stopped and resumed.

걸리며, 다양한 이해관계자가 존재하기 때문이다.

따라서 지속적으로 시민들에게 사업의 당위성을 알리고 공감대를 형성하여, 이들의 의견을 경청하고 목소리를 모아 나가는 것이 중요하다. 시민들의 생각이 쌓여 하나의 단단한 여론을 형성할 때 공원 조성을 위협하는 불확실한 상황이 와도 유연하게 대응해 나갈 수 있다.

용산공원 조성과정에서 시민 참여의 중요성은 지속적으로 언급되었다. 이에 따라 진행 과정마다 시민들의 의견을 수렴하고 소통할 수 있는 장치를 두었다.

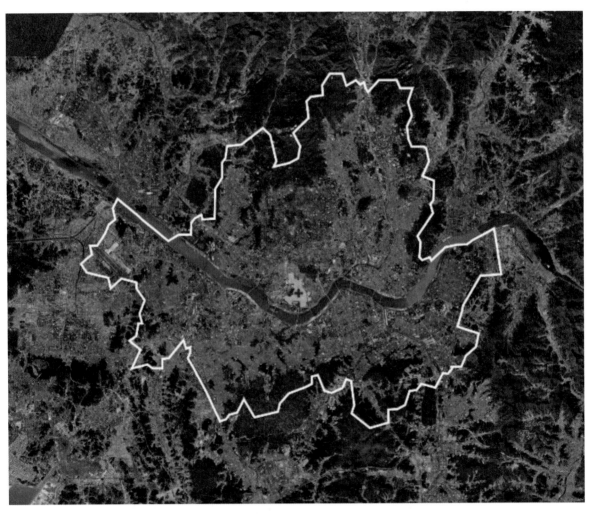

그림 1. 서울시 경계와 용산공원의 위치 / Location of the Yongsan Park

However, it was not just the base relocation that caused the uncertainty of the process.

A civic park, one of the urban planning facilities, is a space which can be used universally by anyone, regardless of gender, age, or financial status. Parks are built with city residents' taxes, so they should be created democratically in view of the citizens' opinions. The process of citizen participation is even more necessary when creating a large park. This is because large parks are complex in scale, take a long time to complete, and have diverse stakeholders. Therefore, it is important to continuously inform citizens of the project, form a consensus, listen to their opinions, and gather their voices. When citizens' ideas accumulate to form a solid public opinion, they can respond flexibly to uncertain situations which threaten the park's creation.

However, these attempts were not enough in the process of developing Yongsan Park. The importance of public participation was mentioned throughout the creation process. Accordingly, there was a device to collect citizens' opinions in each process of the Yongsan Park development project. However, these attempts were not enough in the process of developing Yongsan Park. The task force team (TF) under the Ministry of Land, Infrastructure, and Transport (MOLIT) has tried to engage with the citizens since 2017 actively.

In this writing, the efforts made to gather citizens' opinions and to form a discourse on the complex and uncertain large-scale park development process will be reviewed through the lens of the Yongsan Park case. It is also explored the necessity of sharing information with the public and how to engage with them during the park development process.

그러나 실질적인 시민 소통이 부족했다는 비판 또한 대두되었다. 이러한 상황을 극복하고자, 용산공원 조성을 담당하고 있는 국토교통부 용산공원조성추진기획단(이하 기획단)은 2017년 이후 보다 적극적인 방식으로 시민들의 의견에 귀 기울이기 시작했다.

이 장에서는 용산공원을 사례로, 복잡하고 불확실한 대형 공원 조성과정에서 시민들의 의견을 모으고 담론을 형성해 나가기 위해 어떠한 노력을 기울였는지 살펴보고, 정보의 공유와 시민 참여의 필요성을 모색해보고자 한다.

용산공원 조성사업의 태동

1988년 8월 정부는 서울 용산에 위치한 백만평 규모의 미군부대를 한강 이남으로 이전하고 이를 공원으로 조성한다고 발표했다. 이어 1988년 12월 중순, 노태우 대통령은 용산기지의 공원화계획 수립을 지시했다. 서울시는 1989년 프로젝트에 착수해 1991년 '용산 미8군기지 활용계획'을 발표한다. 당시 서울시가 제안한 공원화 기본 구상안은 뉴욕 센트럴파크와 같

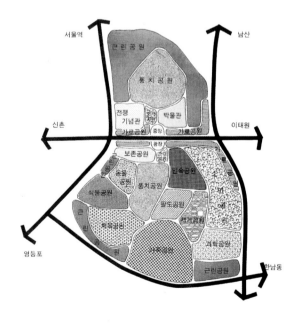

그림 2. 서울시 용산 군이적지 활용방안과 기본계획 구상안(1991)
출처: 『용산 군이적지 활용방안과 기본계획 구상안』 (서울특별시, 1991), 『용산공원정비구역 종합기본계획』(국토해양부, 2011), p. 33에서 재인용.
Initial Plan of the Yongsan Park (1991)

The Beginning of the Yongsan Park Development Project

On August 12, 1988, the government announced that the U.S. military bases, including the U.S. Forces Korea Command in Yongsan, Seoul, would be moved to another city, and that the city center of Seoul with an area of 330 hectares, where the U.S. military base was located, would be turned into a park. Seoul Metropolitan Government began to set up the re-utilization plan of the base and announced the 'Plan for the Relocation of the 8th US Army in Yongsan' in May 1989.

The basic park master plan announced by Seoul Metropolitan Government was a picturesque-style civic space akin to New York's Central Park, and a symbolic space for the restoration of national self-esteem. However, there was also an opinion that it should be developed as an affordable housing area to support low-income families instead of creating a park.

After that, as the relocation of the base became unclear due to various international issues occurring during the 1990s, such as the North Korean nuclear crisis in 1994 and the bailout crisis in 1997, the establishment of the park plan was also suspended.

Site relocation discussions resumed as of 2000. The two countries began to discuss the relocation of the base in earnest according to the LPP (Land Partnership Plan). For the US military, consolidating military bases in one place was useful for modern military operations, and for the Korea government, it was necessary to develop cities which had been occupied by the US military for an extended period.

Both countries initiated the 'Yongsan Base Relocation Steering Committee' to discuss the relocation, and at the Korea-US summit in May 2003, they reached an agreement addressing the 'Yongsan Base Relocation Agreement (UA/IA),' explaining that the entire Yongsan Base, including the UNC and CFC, would relocate to

은 자연풍경식의 시민휴식공간, 민족자존의 회복을 위한 상징성을 가진 공간이었다. 그러나 당시 서울시의 주택난을 해소하기 위해 저소득층을 위한 임대주택지로 개발해야 한다는 의견도 있었다.

이후 공원계획 수립은 잠정 중단되었다. 90년대에 발생한 다양한 국제적 이슈 - 94년 북핵사태, 97년 구제금융사태 등 - 로 인해 기지 이전이 불투명해지자 공원 조성 또한 더 이상 추진될 수 없었다.

기지 이전 논의는 2000에 다시 시작되었다. 한미 양국은 연합토지관리계획(LPP)에 따라 전국에 산재한 미군기지를 한 곳으로 통폐합하고자 했다. 미군은 현대적 군사작전을 위해서 군기지를 한곳으로 집중하는 것이 유용했고, 한국 정부는 오랫동안 미군기지에 내주었던 땅을 되찾아 적절한 도시개발을 하는 것이 필요했다.

이러한 배경 하에 한미 양국은 '용산기지이전추진위원회'를 구성하여 기지 이전을 협의해 나갔다. 2003년 5월 한미정상회담에서 2008년 말까지 유엔사 및 연합사를 포함한 용산기지 전체를 평택으로 이전하는 데 합의하고 '용산기지이전협정(UA/IA)'을 체결했다. 2004년 12월에는 회담 결과에 대해 국회 비준을 얻음으로써 비로서 기지 이전이 본격화되는 발판을 마련하였다.

기지 이전이 가시화되면서 반환기지 활용방안에 대한 연구도 다수 이루어졌다. 2005년 대한국토·도시계획학회와 한국조경학회의 '용산기지 공원화연구'의 결과를 토대로, 2006년 노무현 대통령은 '용산 국가공원 선포식'을 개최하여 '국가 주도'로 공원을 조성한다고 발표했다.

한편, 1990년 한미간 기지 이전 협의에 따라 미8군 골프장이 먼저 반환되었다. 당시 서울시의 계획에 따르면 이곳에는 '가족공원'이 들어설 예정이었다. 그러나 공원 조성 도중 김영삼 대통령의 지시로 국립중앙박물관이 이곳에 지어졌다. 공원은 '비어 있는 땅', '시설을 위한 유보지'라는 인식으로 인해 온전한 공원이 조성되지 못하고 박물관에게 땅을 내어준 것이다. 계획한 대로 공원 조성이 어렵다는 것을 보여준 사례라 할 수 있다.

Pyeongtaek by the end of 2008. Then, in December 2004, the agreement was ratified by the Republic of Korea National Assembly.

As the relocation of the base became visible, many studies were conducted addressing how to utilize the return base. Based on the results of the study to transform Yongsan Garrison into the civic park in 2005, President Roh Moo-hyun held the 'Yongsan National Urban Park Proclamation Ceremony' in 2006 to announce that the park would be built by the 'state-led' approach.

Meanwhile, according to the agreement to relocate the base between Korea and the US in 1990, the 8th Army golf course was returned first. It became a 'family park' developed by Seoul Metropolitan Government per the park master plan in 1991. In the meantime, there was a need for a new National Museum of Korea. In consideration of various sites, the National Museum of Korea was established in this place, which was recognized as 'empty and free to use' land. This is an example of how difficult it is to create a civic park exactly as planned.

History of Yongsan Garrison

Yongsan-gu is the geographical center of Seoul. In the past, Yongsan received attention as it was adjacent to the city center, the center of politics, and the Han River, the hub of logistics.

Historically, various government offices were established such as Itaewon, the first lodging facility on the way from Hanyang to Yeongnam, Waseo, baking the roof tile, Jeonsangseo, raising livestock using for the royal rites, and Seobinggo, storing ice were located this area.

용산기지의 역사

현재 서울시 지도를 살펴보면 용산구는 지리적으로 서울의 중심이다. 이러한 용산은 과거에도 정치의 중심인 사대문 안 도심과 물류의 중심인 한강과 인접하여 주목을 받았다.

한양에서 영남으로 가던 첫 번째 숙박시설인 이태원, 기와를 굽던 와서, 궁중 제사에 쓸 가축을 기르던 전생서, 얼음을 저장하던 서빙고 등 다양한 관청이 이곳 주변으로 위치하고 있었다.

지리적 이점으로 인해 외세의 전략적 요충지가 되기도 했다. 고려시대에는 몽고의 병참기지였으며 청일전쟁(1894년) 때는 청나라군과 일본군이 주둔했다. 그러다 러일전쟁(1904년)을 계기로 일제의 군기지로 변모하기 시작하였다. 일제는 한일의정서 제4조 규정을 근거로 용산 일대 115만평을 강제로 수용한 뒤 위수지역으로 선포했으며, 1906년에서 1914년까지 이곳에 군사시설을 집중적으로 조성하였다. 한편, 일제가 헐값에 땅을 강제수용하면서 둔지미 마을 등 이곳에서 살던 원주민들은 인근 보광동 일대로 쫓겨날 수밖에 없었다.

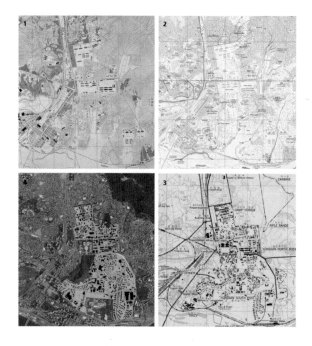

그림 3. 용산 미군 기지의 변화상
출처
1: 1927년 용산시가도(서울역사박물관 소장) /
2: KOREA CITY PLANS 1:12,500(미 텍사스대학교 도서관 소장)/ 3: 1950년대 중반 지도
(서울역사박물관 소장)/ 4: 2019년 현재 지도
(연구자 작성)
Changes in the Yongsan Military Base

It was also used as a strategic point by foreign militaries. During the Goryeo Dynasty, it was a logistical base for Mongolia. During the Sino-Japanese War (1894), the Qing dynasty and Japanese forces were stationed in Yongsan area. Then, since the Russo-Japanese War (1904), it began to be transformed into a military base for the Japanese Empire. Japan forcibly expropriated the area of appropriate 380 hectares in Yongsan based on Article 4 of the Korea-Japan Protocol and declared it a garrison. From 1906 to 1914, the Japanese Empire concentrated on building military facilities at the site. As the Japanese imperialists plundered the land at a low price, the natives living here, including in Dunjimi Village, had no choice but to face eviction to nearby Bogwang-dong.

Following liberation in 1945, the US army was stationed in Yongsan base. They left the base soon, but came back again due to the Korean War. Since then, the US army has been stationed Yongsan area for more than 60 years. Many of the buildings and roads constructed during the Japanese colonial period were used by the US military. However, the current base layout has been completed during the US army era.

Yongsan Park Site

The US military base consists of two parts. The main site, the Main Post and the South Post, and the surrounding sites, the Camp Kim, the UN Compound (UNC), and the Transportation Motor Pool (TMP). To the northern end of the Main Post is the Camp Coiner. The total size is about 2.8 million square meters. Considering the area of Yeouido, the financial district in Seoul is about 2.9 million square meters, it is equivalent to the size of a main functional zone in the city. There are about 1,000

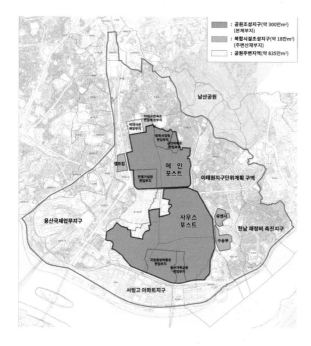

그림 4. 용산공원 정비구역 현황 (2021년)
출처: 『용산공원 정비구역 종합기본계획 (2차변경)』(국토교통부, 2021)
Zoning of the Yongsan Park and its Vicinity (Updated in 2021)

1945년 해방 이후 미군은 비어 있는 일군기지를 사용하다가 일시 퇴거했으나 한국전쟁으로 인해 다시 용산에 주둔하기 시작했다. 미군은 일제가 조성한 상당수의 건물, 도로, 시설물 등을 그대로 수용했다. 그러나 현재까지 사용하고 있는 대다수의 건물과 기지 내부의 형태는 미군에 의해 조성되었다고 할 수 있다.

용산공원 부지

용산 미군기지는 본체 부지인 메인 포스트(Main Post)와 사우스 포스트(South Post), 주변으로 산재한 캠프 킴(Camp Kim), 유엔사(UN Compound), 수송부지(TMP)로 이루어져 있다. 메인 포스트의 북쪽에는 캠프 코이너(Camp Coiner)가 위치해 있다. 전체 크기는 약 280만m²에 달한다. 여의도 면적이 약 290만m²임을 감안하면 하나의 도시 크기와 맞먹는

buildings within the site, and at one time more than 30,000 US soldiers and their families lived there.

In 2011, the central government of Korea designated and announced the Yongsan Park site and its vicinity. First, it was decided to subsidize the cost of relocating the US military base to Pyeongtaek by implementing the mixed-use development of the Camp Kim, the UNC, and the TMP. The Main Post and the South Post will be developed into a park of about 243 hectares, of which the sites for the US embassy in Korea (approximately 49,000 m^2), Dragon Hill Hotel area (84,000 m^2), and the heliport (57,000 m^2) area were excluded.

The area of the park has changed once in 2021. First, the site of the former Defense Acquisition Program Administration (about 95,000 m^2) and the military apartment complex (45,000 m^2) located between the northern end of the Main Post and Haebangchon were incorporated into the park. In addition, the War Memorial Museum (120,000 m^2), the National Museum of Korea and Yongsan Family Park (335,000 m^2), which functioned as urban cultural and green facilities at the boundary of the park site, were also designated as park sites. As a result, the area of the park has increased to about 3 million square meters. This is roughly equivalent to the area of Central Park in New York (3.41 million square meters).

The War Memorial Museum and the National Museum of Korea were incorporated into the park area, but they cannot be regarded as a single and integrated park because the department of operation and management in charge is separate from that of the park. It can be said that the expansion of the park has been accomplished in a symbolic sense, which aims to establish an integrated plan by tying these facilities and parks with the characteristics of 'public functions' in the city.

규모이다. 대상지 내에는 1,000여 동의 건물이 있으며, 한때 3만명 이상의 미군 및 군속가족들이 거주했다.

2011년, 기지 이전과 공원 조성을 위해 정부는 용산공원 정비구역을 지정·고시했다. 먼저, 3개의 산재부지는 복합개발을 추진하여 미군기지의 평택으로의 이전비용을 보조하는 것으로 결정했다. 본체부지인 메인 포스트와 사우스 포스트 일대는 공원으로 조성하되, 이 중 미군잔류부지(드래곤힐 호텔 일대 8.4만m²), 헬기장(5.7만m²), 캠프 코이너 일대로 이전해 오는 주한미국대사관(약 4.9만m²) 부지를 제외한 약 243만m²를 공원으로 조성하고자 하였다.

공원의 면적은 2021년 한차례 변경되었다. 먼저, 메인 포스트 북단과 해방촌 사이에 위치한 구 방위사업청(약 9.5만m²)과 군인아파트(4.5만m²) 부지가 공원으로 편입되었다. 또한 공원 대상지 경계에서 도시의 문화, 녹지시설로 기능하던 전쟁기념관(12만m²), 국립중앙박물관, 용산가족공원(33.5만m²) 또한 공원부지로 지정되었다. 이로 인해 공원 면적은 약 300만m²로 늘어나게 되었다. 용산공원 북쪽에 들어설 예정이었던 주한미국대사관 직원숙소까지 공원으로 편입될 경우 면적은 300만m² 이상으로 넓어진다. 이는 미국 센트럴파크 면적(341만m²)과 맞먹는 크기이다.

그러나 전쟁기념관과 국립중앙박물관은 용산공원과 관리·운영의 주체가 달라 하나의 공원으로 보기 어렵다. 도시에서 '공적 기능'을 하는 이들 시설과 공원을 한데 묶어 통합적으로 계획을 수립하고자 하는, 상징적 의미에서 공원의 확대가 이루어졌다고 할 수 있을 것이다.

용산공원 계획 및 설계

일반적으로 공원은 '도시공원 및 녹지 등에 관한 법률'에 의해 지방자치단체에서 조성하고 관리·운영한다. 2016년, 도시공원법이 개정되면서 제15조 1항에 '국가도시공원'이란 공원의 종류가 생겨났지만 아직까지 국가가 지정하거나 조성을 한 공원 사례는 없다. 그러나 용산공

Yongsan Park Planning and Design

In general, parks in Korea are created, managed, and operated by local governments in accordance with the 'Act on Urban Parks and Green Areas.' In 2016, with the revision of the Urban Parks Act, a type of park called 'national urban park' was created in Article 15 (1), but there has been no case of park designated or created by the state yet. However, in Yongsan Park, the central government takes the lead in creating and managing the park according to the 'Special Act on Yongsan Park Creation' enacted in 2007. At the time of declaring national park development in 2006, the central government insisted that the government take the initiative to create parks to prevent the development of historically important base in a haphazard manner. Accordingly, in 2007, the 'Special Act on the Creation of Yongsan Park' was enacted.

In 2011, the first statutory plan specified in the Special Act, the 'Comprehensive Master Plan with Basic Design Guidelines for Yongsan Park Development Area,' was established and announced. The Comrehensive Master Plan is a plan covering the inside of the park and surrounding areas. Based on this, the two-stage international design competition was held at the end of 2011, and the West8+Iroje team's 'Healing: The Future Park' plan was selected. The winning team started the 'Yongsan Park Basic Design and Park Master Plan' stage at the end of 2012, but due to various issues such as delay of the relocation of the US military base to Pyeongtaek, the Korean government's request to postpone the transfer of wartime OPCON, domestic political issues, and budget cuts, the project could not be completed until the end of 2018. However, the second statutory plan was placed in the cabinet, not being announced.

The main concept of the winning scheme at the design competition and of the basic design and park master plan phase is a 'healing.' Based on the concept of healing that

원은 2007년 제정된 '용산공원조성특별법'에 의해 중앙정부, 즉 국가가 공원 조성과 관리·운영을 주도한다. 2006년 국가공원화를 선포할 당시 중앙정부는 민족의 아픔이 담긴 거대한 기지가 '난개발' 되는 것을 막기 위해 국가가 나서서 공원을 조성해야 한다고 주장하였으며, 이에 따라 2007년 '용산공원조성 특별법'이 제정되었던 것이다.

2011년, 특별법에 명시된 첫 번째 법정계획인 '용산공원 정비구역 종합기본계획'이 수립·고시되었다. 종합기본계획은 용산공원 조성지구 전체에 대한 계획으로 공원 내부와 공원 주변 지역을 망라한다. 이를 바탕으로 2011년 말 2단계 국제설계공모전이 실시되었으며 West8+ 이로재 팀의 'Healing: The Future Park'안이 당선되었다.

당선팀은 2012년 말부터 '용산공원 기본설계 및 공원조성계획' 수립단계에 착수했다. 그러나 미군기지의 이전 연기, 한국 정부의 전시작전권 이양 연기 요청 및 이로 인한 한미연합사 일대의 기지 내 잔류 결정, 국내의 정치적 이슈 발생, 예산 삭감 등 여러 문제로 인해 2018년 말에 와서야 과업이 종료되었다. 그럼에도 두 번째 법정계획인 '공원조성계획'은 고시되지 못한 채 캐비닛 속에 방치되게 되었다.

공모전 당선작과 2018년 말 완료된 기본설계 및 공원조성계획안의 기본개념은 치유의 공원이다. 지형의 치유, 자연의 치유, 역사의 치유, 연결의 치유 등 대상지를 아우르는 네 가지 치유의 개념을 바탕으로 군사적 용도로 사용되어 오던 기지를 작동하는 대형 도시공원으로 탈바꿈시키고자 했다(한국도시설계학회, 2020).

시민 참여

30년 넘게 진행되고 있는 용산공원 조성사업에서 다양한 방식의 시민 참여 기회가 있었지만 실질적인 시민 참여는 드물었다. 이로 인해 소통이 부족하다는 지적이 계속되었다. 서울시나 중앙정부에 의해 '선 계획 수립', 공청회를 통한 '후 결과 통보' 방식은 용산공원 조성과정에서 지속적으로 반복되었다. 이러한 상황은 2011년 '용산공원 정비구역 종합기본계획'을 수립

encompasses four sub themes: the healing of the terrain, the healing of nature, the healing of history, and the healing of connection, it was intended to transform the site that had been used as a military base into a operative large urban park.

Public Engagement

There were various opportunities for citizen participation in the Yongsan Park development project, having taken place for over 30 years, but substantial citizen participation was rare. As a result, criticism continued to be made about the lack of communication.

The approach of 'plan establishment in top-down manner' by Seoul Metropolitan Government or central government and 'notifying the plan to the general public' by the public hearings was repeated continuously in the process of Yongsan Park development thereafter. This attitude was kept while establishing the comprehensive master plan for the Yongsan Park development area in 2011, which eventually led to the change of the plan in 2014.

In fact, there were quite a number of attempts to collect citizens' opinions or induce citizen participation in the park development process. Surveys, seminars, debates, and symposiums were opportunities to investigate citizens' opinions. Exhibitions and contests were intuitive participation platforms. However, in general, there were many professional-oriented events, which made it difficult for citizens to access them.

Park slogan event (2008), posting the journal about urban parks (2008), Yongsan Park storytelling contest (2010), advertising contest (2011), UCC contest (2012), citizen reporter recruitment (2013), Yongsan National Park children's drawing contests (2013-

하는 과정에서도 마찬가지였으며, 결국 미흡한 시민 참여를 빌미로 정치권과 지역사회의 요구에 의해 2014년 종합기본계획이 한차례 변경되었다.

사실 공원 조성과정에서 시민들의 의견을 수렴하거나 시민 참여를 유도하려는 시도는 꽤 많았다. 시민들의 의견을 파악하기 위한 설문조사, 세미나, 토론회, 심포지엄 등의 의견 교류 기회, 참여를 유도하는 전시나 공모전 등이 열렸다. 그러나 대체로 전문가 중심의 행사가 많아 시민들의 접근성이 떨어졌다. 용산공원 슬로건 만들기 이벤트(2008), 국내외 공원탐방기 올리기 이벤트(2008), 용산공원 스토리텔링 공모전(2010), 광고 공모전(2011), UCC 공모전(2012), 국민기자단 모집(2013), 용산국가공원 어린이 그림그리기 대회(2013~2015) 등은 시민들이 직접 참여할 수 있는 방식이었지만 일회성으로 이루어져 공원 조성의 핵심 이슈를 다루기에는 무리가 있었다.

무엇보다 시민들의 의견이 계획이나 설계안으로 연결되기가 어려웠으며 시민들은 자신들의 생각이 공원 조성과정에 어떻게 반영이 되고 있는지 확인할 수 없었다.

종합하면, 용산공원 조성과정에서 논의의 대상이 되는 모든 범주 - 기지 이전, 정치·외교·군사에 관한 사항, 공원화, 부분개발, 공원의 성격, 공원의 계획 및 설계, 공원의 관리 및 운영 프로그램 등 - 에서 시민들이 의견을 낼 수 있는 플랫폼이 제한적이었으며, 의견이 받아들여져 공원 조성과정에 반영될 수 있는 시스템이 미비했다.

다 함께 논의할 수 있는 공론의 장 없이 공원의 성격에 대해 대립되는 주장들이 개별적으로 등장했으며, 이들의 충돌로 공원 조성과정이 불안정해졌다. 탑-다운방식으로 성급하게 추진하다 보니 시민 참여과정과 공원 설계과정이 통합적으로 작동되지 못하고 병렬적으로 진행되었으며, 때로는 복잡하게 얽혀 혼란을 야기하기도 했다.

이런 상황에서 2016년 6월 개최된 공청회는 지금까지의 공원 조성과정의 패러다임을 바꾼 계기가 되었다. 공원설계팀은 1,000여 동에 달하는 기지 내 건물 중 일부를 공원의 실내 프로그램 공간으로 보존 및 재활용하고자 하였다. 공원 조성과정을 이끄는 국토교통부 산하 기획단은 기존 건물의 재사용 기준을 마련하기 위해 심의기구인 '용산공원조성추진위원회' 내 소

2015) were held, but most of the events were single occassion so it was difficult to deal with the pertinent key issues of park development. Above all, it was hard for citizens' opinions to be connected to a plan or design proposal, and citizens could not know how their thoughts were being reflected in the park development process.

In other words, the platform by which the public could express their opinions was limited on all issues to be discussed in the process of developing Yongsan Park - base relocation, political/diplomatic/military matters, creating a large park, phasing, characteristics of the park, planning and design, operations and management, etc.. Even if they could express themselves, there was insufficient system through which these opinions could be accepted and reflected in the park development process. Eventually, conflicting opinions regarding the park development continued to arise, and caused instability in the park development process. Due to the hasty top-down approach, the citizen participation process and the park design process were not integrated and proceeded with in parallel, sometimes leading to confusion and complication.

The public hearing held in June 2016 changed the paradigm of the development process so far. The park design team planned to preserve and reuse some of the 1,000 buildings in Yongsan Base for the indoor program spaces in the park. The TF for the Yongsan Park development in the MOLIT, who leads the park development process organized the sub committee under the 'Yongsan Park Steering Committee' to set up the guidelines for the building reuse. They discussed the issue several times, and finally generated the potential and feasible building program guidelines in the park by preserving and reusing the existing building in the base. At the same time, a survey of demand was conducted for some of the buildings to be preserved and reused. In accordance with the guidelines, the TF and the sub committee first looked at the

위원회를 구성하여 수차례 논의 끝에 공원 내 수용 가능한 건물 프로그램의 기준을 마련하였다. 동시에 보존 및 재사용 대상 건물들 중 일부에 대해 입지 수요조사를 실시했다. 마련한 기준에 따라 도입 가능한 프로그램(콘텐츠) - 가령 박물관, 미술관 등 - 의 입지를 원하는 정부부처의 의견을 우선 살폈으며 콘텐츠 수용기준, 수용 가능 시기, 입지를 원하는 각 정부부처의 예산계획, 공원설계와의 정합성 등을 살펴 최종 8개의 콘텐츠가 잠정 수용되었다.

그러나 이러한 내용을 공청회에서 발표하자마자 언론과 서울시로부터 큰 비난이 쏟아졌다. 공원 전체가 자연녹지가 아닌 건물로 뒤덮이고, 특히 용산공원과 맥락이 닿지 않는 정부부처의 시설물이 들어온다고 이해한 언론에서 연일 비판의 기사를 내보냈다. 결국 기획단은 2016년 말, 8개의 콘텐츠를 백지화하겠다는 보도자료를 냈다. 또한 100년을 내다보고 공원을 조성하겠다고 발표하며 긴 시간에 걸쳐 시민들의 의견을 수용해 가며 공원을 만들 것을 선언했다.

2017년, 기획단은 설계과정을 잠시 중단하고 '용산공원 라운드테이블 1.0' 프로그램을 시작했다. 이는 6개월간 용산공원과 관련된 다양한 주제 - 생태, 역사, 문화, 예술, 도시 등 - 에 대해 전문가와 시민들이 함께 의견을 나누는 자리로 기획되었다. 여전히 시민 참여가 다소 수동적이고 제한적인 측면이 있었지만, 총 8회에 걸친 행사 동안 이전과는 다른 방식으로 공원 조성에 대해 민과 관, 전문가가 함께 담론을 만들어 나갔다.

라운드테이블의 핵심은 청년 프로그래머 프로그램이었다. 향후 용산공원 조성과정에 관심을 가지고 이끌어 나갈 수 있는 젊은 세대를 키우는 것이 이 프로그램의 목적이었다.

청년 프로그래머들은 6개월 동안 적극적으로 행사에 참여하여 공원 조성과정을 둘러싼 다양한 이해관계들을 살펴보았으며, 이를 바탕으로 공원에 대한 스스로의 의견을 여러 매체로 표현하는 창작작업을 진행했다. 이들은 마지막 행사인 '공원 서평'에서 각자의 결과물을 시민들과 공유했다.

opinions of central government agencies that want to locate programs (contents) in the park - for instance, museums and art galleries - and reviewed the availableness of those items. The final eight items of content were tentatively accepted after examining the budget capability of proposed agencies and consistency with the park development schedule and design.

However, as soon as these intended inclusions were announced at the public hearing, a great deal of criticism poured out from the media and the Seoul Metropolitan Government. Criticism from the media was harsh, which understood that the entire park was covered with buildings rather than nature, and in particular the intended inclusion of government agency structures out of context with Yongsan Park. Eventually, as of the end of 2016, the TF issued a press release stating that eight buildings would be cancelled. In addition, the TF declared the long-term park development as a principle while including citizens' opinions.

In 2017, the TF paused the design process and began the Yongsan Park Roundtable 1.0. This was planned as a place for experts and citizens to share opinions on various topics related to Yongsan Park for a period of 6 months - ecology, history, culture, art, city, etc. Although civic participation was still somewhat passive and limited, the public, government, and experts collaborated to create a discourse revolving around the park in a different way than before during the total of eight events.

The essence of the Roundtable 1.0 was the youth programmer program. The purpose was to nurture a younger generation who could take an interest in and lead the development of Yongsan Park in the future. The young programmers actively participated in the event for six months to examine the various interests and opinions surrounding the park development process, and based on this, they proceeded with creative work to express their thoughts about the park through various modes of

그림 5. 라운드테이블 1.0 시민 참여 / Citizen Participation in Roundtable 1.0

　2018년에는 국토교통부와 서울시, 용산문화원의 협력으로 '용산기지 버스투어'가 실시되었다. 용산기지는 서울 한복판에 위치한 땅이지만 높은 담이 에워싸고 있다. 지금까지 일반시민들은 기지 안으로 들어가 볼 수 없었기에 이곳은 그들에게 미지의 공간이며 도시의 섬과 같은 존재였다. 공원의 미래상을 논하기 앞서 대상지에 대한 이해가 시급했다.

　미군과의 협의 문제로 기지 전체를 걸으며 둘러보는 것은 가능하지 않았다. 그러나 용산문화원 해설사의 설명을 들으며 하는 버스투어만으로도 시민들은 기지에 대해 충분히 이해할 수 있게 되었다. 2020년 전 세계를 강타한 코로나로 인해 버스투어는 잠정 중단되었지만,

media. Their ideas were presented and exhibited at the last event and shared with citizens.

In 2018, in cooperation with the Yongsan Cultural Center and Seoul Metropolitan Government, the MOLIT launched the Yongsan Base Bus Tour. Because of the high concrete walls along the base, it has been an unknown space. There was no chance to get into the base for the public. Before discussing the vision of the park, people need to understand the site's condition.

Although it was not possible to take a walking tour for potential security issues, it was a good opportunity for individuals to understand the base by listening to the explanations of the commentators at the Yongsan Cultural Center. Citizens left fruitful thoughts on the park development. The bus tour was paused due to the COVID-19 in 2020, but more than 2,000 people visited the site.

In 2019, as the return of some areas of the U.S. military base became visible, a 'research on temporary use of the U.S. military base in Yongsan, etc.' was conducted to obtain ideas for using the returned base temporarily. In this study, as part of the citizen participation program, an idea workshop – 'Yongsan Lab: Alpha' – was held with college students and experts for period of 3 days. As a result of active communication between younger generations and experts under the theme of the temporary opening of Yongsan Park, many creative and ingenious ideas were derived.

Throughout 2020 and 2021, the TF held more active citizen participation events. Before this event, as a boom-up, Public contests such as 'naming contests' and 'photo contests' of Yongsan Park to induce people's interest and participation in the park development process were held. In addition, a direct communication channel was planned for citizens to express their opinions on the park planning and design process. The 5th officer's dormitory complex located at the southeast end of the base was

2,000명이 넘는 시민들이 기지를 다녀갔으며 그 과정에서 시민들은 공원의 조성 방향에 대해 많은 의견을 남겨주었다.

2019년, 미군기지의 일부지역 우선 반환이 가시화되면서 기획단은 반환기지를 임시로 사용할 수 있는 아이디어를 얻고자 시민 참여 프로그램을 시행했다. 아이디에이션 워크숍 - '용산랩 : 알파' - 을 통해 3일 동안 청년과 전문가가 용산공원 임시개방을 주제로 상호 소통한 결과 창의적이고 기발한 아이디어들이 다수 도출되었다.

2020년과 2021년 두 해에 걸쳐 기획단은 보다 적극적인 시민 참여 행사를 개최했다.

용산공원의 네이밍 공모전, 사진 공모전 등 국민 참여 공모전을 개최해 공원 조성과정에 대한 시민들의 관심과 참여를 유도했다. 동시에 시민들이 공원계획 및 설계과정에 의견을 전달할 수 있는 직접적인 소통창구를 기획했다. 기지 동남단에 위치한 장교숙소 5단지를 개방하여 물리적인 참여 공공간을 마련했다.

310명의 '국민참여단' 또한 모집했다. 용산공원에 관심 있는 전국의 남녀노소로부터 지원을 받았으며, 자기소개서, 지원사유서 등을 면밀히 검토하고 대면과 비대면 인터뷰를 진행하여 참여단원을 선발했다.

국민참여단은 본격적인 활동이 시작되기 전, 생태·역사·도시 등 용산공원과 관련된 여러 주제를 다루는 전문가들의 특강을 6차례에 걸쳐 온라인으로 수강했으며, 2021년 3월부터 6월까지 실질적인 참여 활동에 임했다.

국민참여단의 운영은 10개 분과로 나눠 진행되었다. 각 분과에는 분과장인 General Manager(GM) 1인, 코디네이터 4인이 배치되어 시민들의 소통 및 참여활동을 도왔다. GM은 공원, 역사, 도시 등의 분야 전문가로서 분과활동을 기획·운영하였으며 도출된 시민들의 의견을 조율·정리·종합하는 역할을 맡았다. 코디네이터에게는 시민들의 참여활동을 기록하는 사관의 역할이 주어졌다. 소셜미디어를 다루는 20명의 청년 크리에이터 그룹도 운영되었으며, 이를 통해 참여단의 활동을 알리고 공원의 전반적인 이슈들을 창의적인 콘텐츠로 재생산해 일반 국민들과 소통할 수 있도록 하였다. 30명의 연구그룹도 선정했다. 이들은 국민참여단의 일반 그룹과 밀접히 교류하며 용산공원을 소재로 향후 필요한 연구주제를 발굴하는

opened so as to provide a space for public engagement.

Then 'National Public Participation Group (NPPG)' was organized. To select participants, it was called for from people of all ages and genders across the country who were interested in Yongsan Park. 300 participants were selected through face-to-face and online interviews after carefully reviewing the statement of purpose. Before the major activity began, NPPG took online courses offered by experts dealing with various topics related to Yongsan Park, such as ecology, history, and urban issues, and participated in practical participation activities from March to June 2021.

NPPG was divided in 10 sub groups. Each group had a single 'General Manager (GM),' and 4 'Coordinators' to effectively handle citizens' communication and engagement. As an expert in the fields of parks, history, and cities, GM planned and operated the sub group activities, and played a role in mitigation, organizing and synthesizing the opinions of citizens. The coordinator took a charge of documenting citizens' opinions and thoughts. Together along with them, 20 'Creators' familiar with the social media were operated to inform the general public about the activities of citizens and to communicate issues of the park with creative contents. 30 'Research Groups' were also selected. They worked alongside the general public in the NPPG and proposed the necessary research themes related to Yongsan Park.

The result of the activity was derived under the name of 'Yongsan Park NPPG's proposal' in July 2021, with seven primary objectives: 1. A park that is always convenient and safe to use, 2. A park that preserves historical and cultural values, 3. A park that balances conservation and utilization, 4. A park that embraces diverse values and new possibilities, 5. A park that operates cultural and artistic programs flexibly, 6. A park that coexists with the surrounding area, 7. A park where the public participation process becomes history.

그림 6. 용산기지 버스 투어 / Bus Tour to the Yongsan Army Base

역할을 맡았다.

활동의 결과는 2021년 7월, '용산공원 국민참여단의 제안'이라는 보고서로 발표되었으며, 이 중 7대 제안 - 1. 언제나 편리하고 안전하게 이용할 수 있는 공원, 2. 역사·문화적 가치를 지키는 공원, 3. 보존과 활용의 균형을 이루는 공원, 4. 다양한 가치와 새로운 가능성을 포용하

These goals and directions were reflected in the 'Comprehensive Master Plan with Basic Design Guidelines for Yongsan Park Development Area,' which was altered once again due to the expansion of the site area in 2021. The 'Yongsan Park Basic Design and Park Master Plan' is also being revised from the second half of 2021 in order to incorporate the opinions of the public participation group and create a park both desired and imagined by citizens.

Closing Remarks

Yongsan Park is a grand space located in the middle of Seoul, and it is a very complex project not only in scale, but also in character in that various stakeholder such as the central government, local governments, and USFK are intertwined. It is a project which will go on for a long time as it has been carried out for over 30 years thus far, and there is a high probability that the social, political, military, and diplomatic terrain surrounding the park will evolve. Each time this happens, the park master plan may change or the construction process may falter.

As a representative example, the 'Amendment to the Special Act on the Creation of Yongsan Park' was proposed as a way to allow 20% of Yongsan Park's area to be purposed the affordable housing when the housing problem became a social issue in 2021. This isn't the first time the move has appeared in 2021. The Youth Architects Association insisted on constructing public housing on half of the base instead of a park in 1989. From the 1990s to the mid-2000s, when the national urban park was decided, there was a series of movements movement to introduce facilities other than parks, such as the construction of a new city hall in Seoul, a museum complex, a

는 공원, 5. 문화·예술 프로그램이 유연하게 운영되는 공원, 6. 주변지역과 상생하는 공원, 7. 국민 참여과정이 역사가 되는 공원 - 은 2021년 부지 면적 확대로 또 한 차례 변경된 '용산공원 정비구역 종합기본계획'의 비전, 목표, 세부전략 등에 반영되었다. '용산공원 기본설계 및 공원조성계획' 또한 국민참여단의 의견을 받아들여 시민들이 원하고 상상하는 공원으로 한걸음 나아갈 수 있도록 2021년 하반기부터 수정 중에 있다.

맺으며

용산공원은 서울시 한복판에 위치한 대형의 공간이며 중앙정부, 서울시, 용산구, 주한미군 등 다양한 주체가 얽혀 있다는 점에서, 규모면에서뿐만 아니라 성격면에서도 매우 복잡한 프로젝트이다. 지금까지 30년 넘게 장기적으로 진행되어 왔지만 앞으로도 긴 시간 동안 지속될 프로젝트이며 공원 조성을 둘러싼 사회적, 정치적, 군사적, 외교적 지형이 바뀌어 나갈 확률이 높다. 그럴 때 마다 공원 계획 또한 바뀌거나 조성과정이 휘청거릴 수 있다. 대표적으로 2021년 주택 문제가 사회적 이슈가 되면서 용산공원 면적의 20%를 공공임대주택을 지을 수 있도록 하는 '용산공원 조성 특별법 개정안'이 발의되기도 했다.

이러한 움직임이 2021년에 처음 등장한 것은 아니다. 1989년 청년건축인협의회의 서민주택 건설 주장이 있었고 90년대부터 국가주도로 공원을 조성하기로 결정한 2000년대 중반까지 서울시 신청사 조성, 박물관 콤플렉스, 영빈관, 외국인타운, 6·25전쟁 납북피해기념관 등 공원 이외의 것을 도입하려는 움직임이 지속적으로 있었다. 비교적 최근인 2018년에도 국민 임대주택 건설에 대한 청와대 국민청원이 개진되었다. 공원의 본질적 성격 - 비어 있는 땅이라는 인식 - 은 어떠한 변화가 필요하거나 닥쳐왔을 때, 손쉬운 해결책으로 부담 없이 사용할 수 있는 공간으로 이해되어 왔다.

그러나 이럴 때마다 공원 조성을 계획대로 이끌어 나가는 힘은 결국 시민들로부터 나온다. 시민들과 지속적으로 정보를 공유하고 함께 헤쳐나가기 위한 소통의 장을 활성화한다면 어떠

용산공원
국민참여단의 제안
용산공원을 위한 국민의 바람

2021. 7.

그림 7. 국민참여단 활동 / Activities of the National Public Participation Group (NPPG)

VIP guesthouse, a foreigner town, and the construction of the Korean War Abduction Memorial Hall. In 2018, a petition was filed to the Blue House to construct public rental

한 충격이 오더라도 이겨내고 시민들이 원하는 공원을 조성할 수 있을 것이다. 용산공원은 비록 지금까지 탑-다운방식 중심으로 달려왔지만 변화해 나가는 시대의 흐름에 발맞추어 시민들을 점진적으로 계획의 영역 안으로 끌어들이고 있다. 여전히 국민 참여방식과 수준에 대해서는 반론이 있을 수 있지만 그 첫걸음을 내디뎠다는 데 의의가 있을 것이다.

국민참여단의 7대 제언 중 마지막 구절인 '국민 참여과정이 역사가 되는 공원'처럼 장기간 진행될 프로젝트에서 다양한 사람들이 참여해 의견을 개진하고, 그 의견이 켜켜이 쌓여 공원이라는 결과물로 다가오길, 그리고 그렇게 조성되어 나갈 공원이 우리가 미래세대에게 주는 큰 선물이 되길 기대해 본다.

참고문헌

1. 국토교통부. 용산공원 국민참여단의 제안. 세종: 국토교통부, 2021a

2. 국토교통부. 용산공원 정비구역 종합기본계획(2차). 세종: 국토교통부, 2021a

3. 국토해양부. 용산공원정비구역 종합기본계획. 과천: 국토해양부, 2011

4. 김한배 외. 보이는 용산 보이지 않는 용산. 파주: 마티, 2012

5. 최혜영. 용산공원 조성 과정의 리질리언스 연구. 서울대학교 박사학위 논문, 2020

6. 한국도시설계학회. 도시설계의 이해 - 실무편. 고양: 도서출판 대가, 2020

housing in the park site. The essential character of the park, i.e. the perception of it being empty land, has been understood as a space that can be used without burden as an easy solution when any change is needed or impending.

However, whenever this happens, the power to guide the development process on the planned path ultimately emerges from the citizens. If it is sought to share information with citizens and communicate to get through together actively, it will be able to overcome any shock and create a park the citizens want. Although Yongsan Park has been running mostly with a top-down approach so far, it is gradually bringing citizens into the realm of the planning and designing to keep pace. There may still be objections to the method and the extent of public engagement, but it will be meaningful in that it has taken the first step.

I hope that various individuals and groups will participate in a long-term project akin to 'a park where the public participation process becomes history,' which is the last phrase found in the 7 proposals of the NPPG, and that their opinions pile up and come to a result called a park. I wish that the park will be a great gift we can deliver to future generations.

References

1. MOLIT. Proposals of Yongsan Park National Public Participation Group. Sejong: MOLIT, 2021
2. MOLIT. General Basic Plan for the Creation and Zoning of the Yongsan Park (2nd Revision). Sejong: MOLIT, 2021
3. MLTM. General Basic Plan for the Creation and Zoning of the Yongsan Park. Gwacheon: MLTM, 2011
4. Kim, H. B. et al. Visible Yongsan, Invisible Yongsan. Paju: Mati, 2012
5. Choi, H. The Development Process of Yongsan Park and Its Resilience. Ph.D. Thesis, Seoul National University, Seoul, 2020
6. Understanding of Urban Design - Practice. Goyang: Daega, 2020

서울 상암 디지털미디어시티(DMC)

쓰레기매립지에서 창조산업의 메카로의 변신

오 다니엘 (고려대학교 건축학과 교수)

상암 디지털 미디어 시티(DMC)는 서울 서북부 경계에 위치한 미디어, IT, 콘텐츠산업을 중심으로 하는 복합업무지구로 1990년대에 급속도로 진행된 도시화로 인한 환경오염과 주택문제를 함께 해결하면서, 도심 밀도를 낮추고 서울 전역으로 균형적인 발전을 위한 서울시 주도의 첫 부도심지 공공개발사례이다. 쓰레기매립지를 대형공원으로 조성하며 환경오염을 완화시키고, 지역에 필요했던 녹지와 여가, 레저시설을 제공하는 데 큰 역할을 하였고, 중심업무단지의 보행우선 가로 조성과 광장을 중심으로 서울의 스마트도시의 시초를 알리는 '스마트 스트리트'부터 최근 서울의 첫 자율주행자동차 지역 지정까지 스마트도시 도입의 테스트베드 역할을 하고 있다.

들어가며

아름답고 평화로운 섬인 난지도는 조선시대의 저명한 예술가들의 많은 역사적인 시와 그림에 등장했다. 그러나 1960년대 이후 수도권 확장에 속도가 붙고 전국의 많은 이주민들이 도시로 밀려들면서 판자촌이 형성되고 쓰레기 무단투기가 시작되면서 이런 그림 같은 모습은 서서히 사라졌다. 서울의 급속한 도시화로 생활쓰레기, 1,000만명이 생산한 건축용 폐자재

Sangam Digital Media City, Seoul

From Landfill to Mecca for Creative Industry

Daniel Oh (Professor, Korea University, Dept of Architecture)

Sangam Digital Media City (DMC) is a central business district planned specifically for the media, IT, and content industries located on the northwestern border of Seoul. It aims to reduce the density of the city center and achieve balanced development throughout Seoul while simultaneously solving environmental pollution and housing problems caused by rapid urbanization in the 1990s. This is the first case of public development in sub-centers led by Seoul. The waste landfill site was turned into a large park to alleviate environmental pollution, and it played a significant role in providing green space, leisure, and leisure facilities necessary for the region. From 'Smart Street,' announcing the beginning, to the recent designation of Seoul's first autonomous vehicle region, it is serving as a testbed for the introduction of smart cities.

Introduction

The site, a beautiful and peaceful island, Nanjido, appeared in many historic poems

그림 1. 1740년경 겸재 정선의 금성평사에 표현된 난지도, 왼쪽 상단부 (출처: 한국콘텐츠진흥원)
1740 Painting Portraying Nanjido, Top left

그림 2. 1990년도 난지도 쓰레기 섬의 모습 (출처: 서울시)
Nanjido Landfill Island in 1990

등 폐기물이 압도적으로 많이 발생했다. 1978년부터 15년이 넘는 1993년까지 한때 아름다운 섬은 높이 100m, 길이 2km, 약 1억2,000만톤에 이르는 거대한 쓰레기 산으로 변했다.

서울시는 1996년 매립지 환경오염 방지 및 안정화사업에 착수했고, 이듬해 서울시가 상암 신도시 개발 종합계획인 '서울신도시개발'을 발표했다. 이 계획은 서울의 주택난에 대응한 대부분 주거지 개발로 이뤄져 있었다. 하지만 1997년 발생한 한국 금융위기로 인해 모든 산업 활동은 몇 년 동안 정지상태에 직면하면서 이 계획 역시 어려움을 겪게 되었다. 이 역사적인 사건으로 인해 사회적인 분위기가 국제화만이 국가 경제를 재건하는 열쇠로 여겨졌다. 그리고 2000년도에 당선된 고건 시장은 디지털 미디어산업을 선도하는 도시인 '정보도시', 자연과 인간이 공존하는 도시 '생태도시', 그리고 서울과 대한민국의 관문인 '게이트 시티'를 세계에 알리기 위해 '상암 새 천년 신도시 계획'을 제시했다.

2001년에는 국가 대표 방송사 중 하나인 문화방송(MBC)과 함께 케이블TV 대기업들의 미디어·엔터테인먼트 특화 클러

그림 3. 상암 새천년 신도시계획 (출처: 서울시)
Sangam New Millennium New Town Plan

and paintings by notable artists from Chosun Dynasty. However, since the 1960s, as the expansion of the Seoul Metropolitan Area gained speed and many migrants from all over the country flooded into the city, this picturesque island slowly disappeared as shanty settlements started to form and illegal waste dumping began. Seoul's rapid urbanization produced an overwhelming amount of waste such as household waste and building waste materials produced by 10 million people. From 1978 to 1993, over 15 years, once beautiful island transformed into a huge mountain of waste measuring 100 meters high, 2 kilometers long and approximately 120 million tons.

In 1996, the Seoul Metropolitan Government launched a stabilization project to prevent environmental pollution in landfill and to withhold development, and the following year, Seoul Metropolitan Government announced a comprehensive plan to develop Sangam called 'New Seoul Town Development'. The urban development proposal included mostly housing development in response to the housing shortage in Seoul; however, due to the Korean financial crisis of 1997, the plan could not be realized.

The crisis affected all levels of socio-economic status and all industrial activities had to face a standstill for a couple of years. This historic event led to consensus seeking to become an active role in the global markets. As a society, globalization was seen as the key to rebuilding the nation's economy. And the newly elected mayor, Goh Kun, presented 'Sangam Millennium New Town Plan' in 2000, aiming to develop an 'Information City' – a city leading digital media industry, 'Ecological City' – a city with nature and human coexisting, and 'Gate City' – a city as a gate of Seoul and South Korea to the world.

In 2001, the 'Digital Media City Plan' was proposed with more details about specific programs and anchor tenants including relocation of one of the main national

스터 조성 등 구체적인 프로그램과 앵커 입주사들에 대한 보다 구체적인 내용이 담긴 '디지털미디어시티(DMC) 플랜'이 제안되었다. 이로써 상암 DMC는 서울의 도시개발과 경제발전을 통합한 최초의 시도이며, 국내 최초의 유비쿼터스 도시로 조성됐다는 점에서 의미가 있다. 또한 상암 디지털미디어시티(DMC)는 서울이 국제적인 도시로 도약하는 준비하는데 있어서 서울을 위한 중요한 도시개발 프로젝트를 대표한다. 이 사례는 기존에 주거지 개발에 초점이 맞춰진 택지개발과는 완전히 새로운 도시개발 모델로 미래산업과 새로운 일자리를 유도하는 사례로서 서울시는 어려운 정치와 경제적 맥락을 배경으로 사업의 공론화를 이끌어낸 사례이다.

개발개념

2000년도 서울 서북권 상암지역이 서울의 부도심화 가능한 유일한 미개발지역으로 난지도 매립지를 선정하여 훼손된 환경을 복원하기로 한다. 서울의 주택 부족은 90년대부터 빠르게 증가했고, 서울시는 이 지역에 주택지와 함께 서울의 균형적인 발전을 위해 부도심으로 조성하는 계획에 힘을 얻게 되었다. 또한 서울시는 이 일대가 도심과 쉽게 연결되는 관문도시가 되고, 정보기술(IT)기반도시인 생태도시로 거듭날 것으로 기대했다. 이 개념에 따라 이 지역은 크게 생태주거지, 정보와 미디어 중심의 상업중심지 그리고 이전에 유치한 월드컵 주경기장으로 세 가지의 주요 프로그램으로 나뉘었다.

경의선은 남북통일에 대비해 DMC가 북한을 거쳐 중국과 러시아로 이어지는 '국경횡단철도'의 출발점 역할을 할 수 있도록 한다. 그리고 상암역에서 경의선 신공항고속철도로 연결되는 신공항 이전이 잘 진행되고 있었다. 두 열차가 DMC에서 합병되면서 서울과 아시아를 연결하는 '게이트웨이' 도시를 만드는 것은 합리적이고 논리적인 도시계획 전략이었다.

broadcasting companies, Munhwa Broadcasting Company (MBC) and other cable television major companies successfully creating a cluster specialized in media and entertainment. DMC is the first attempt to integrate Seoul's urban development and economic development and is meaningful in that it was created as the nation's first Ubiquitous City.

Development Concept

Sangam, the north-west region of Seoul, was the only lasting undeveloped place in Seoul that could be turned into a sub-center of Seoul, restoring the damaged environment by Nanjido landfill. Besides, the shortage of housing in Seoul gradually increased in the 90s, so the Seoul city government planned this area for a new residential town. Therefore, the Seoul city government expected this area to be a gateway city that is easily connected to the center of the city, an information techno-based city, and an ecological city. Based on this principle, the area is divided into three main programs: World-Cup Park, Ecological residential development and Digital Media City.

In preparation for the reunification of South and North Korea, the Gyeongui Line allows DMC to function as a starting point for the 'cross-border railway' that runs through North Korea to China and Russia. And the new international airport relocation was well underway which was connected by the new Airport Express Train shares the Gyeongui Line from Sangam Station. With both trains merging at DMC, creating a 'gateway' city connecting Seoul with the rest of Asia was a rational and logical urban planning strategy.

그러나 주요 중심업무지구를 조성하려면 공공부문과 국가정부의 복잡하고 장기적인 약속이 필요하다. 경제위기를 감안하고 국가의 미래경제를 대비하기 위해 신도시 부도심으로서의 심도있는 타당성 검토가 이뤄졌다. 오늘날 미디어산업은 우리나라의 중요한 산업으로 자리잡았고, 해외로 수출하고 그 규모가 급격하게 증가하고 있는 고부가가치 산업이지만 이 개발계획 당시만 하더라도 미디어산업은 내수시장에 의존하고 있던 산업에 불과했다. 또한 미디어산업은 양질의 인적자원과 관련 기업 간 조율이 매우 중요한 요소였다. 이런 점에서 DMC 개발은 서울의 부도심지로서 적절한 투자였다고 할 수 있다.

개발개념은 매립지에 부지가 어떻게 위치할 것인지를 고려하여 생태적으로 지속 가능한 개발을 완전히 수용했고 새로운 도시 개발과 대부분 낙후된 환경 사이에서 신중하게 협상하고 균형을 맞추었다. 인간과 자연이 친환경적으로 공존하면서 첨단정보기술과 조화를 이루자는 취지로 쓰레기매립지를 타 지역으로 이전하는 대신 첨단환경공학으로 매립지를 안정화를 통해 대형 생태공원과 공공 골프장으로 조성하였다. 또한 수도권에서 가장 필요로 하는 공공주택을 공급하는 택지개발도 포함되었다.

상암 DMC의 개발은 궁극적으로 서울 도심 밀집을 완화하는 서울의 진정한 부도심지 역할을 하고 있다. 또한 산재한 선도기관들과 민간부문을 통합하는 산업별 CBD로, 관련 기업과 전문가들로 구성된 생태계가 해당 산업의 성장을 보다 효율적이고 생산적인 도시환경을 재편성할 수 있도록 도움을 주었다.

마스터플랜 기본개념

DMC는 기본적인 도시활동과 첨단기술이 결합된 미래형 도시생산공동체이자 산업생태계 구축에 초점을 두었다. 보행자 중심의 다목적화를 위해 창의적인 근로자들이 활기찬 도시환경에서 생활하고 일하고 즐길 수 있는 첨단산업 클러스터의 새로운 모델을 지향하였다. 이를 실현하기 위해 다음과 같은 개념이 수립되었다.

Creating a major central business district, however, requires a complex and long-term commitment from the public sector and the national government. Considering the economic crisis and preparing for the future economy of the nation, the program for the new urban sub-center took some serious investigation and many major feasibility studies took place. Today, Korea's media industry is a high value-added industry; however, when the master plan was drafted, media industry mostly relied on domestic consumption. Investing on such industry required high-quality human resources and coordination among related companies. Furthermore, linking media and entertainment sector to be an attractive investment for City of Seoul took some time for the public to realize.

And the development concept fully embraced ecologically sustainable development considering how the site will be located on landfill and carefully negotiate and strike a balance between new urban development and mostly underdeveloped surroundings. The plan called for humans and nature to coexist environment-friendly while creating harmony with innovative information technology. Instead of relocating the landfill, cutting-edge environmental engineering consolidated waste and turned it into an ecological park and a public golf course. And the plans also included residential development which a major percentage of the housing provided much-needed public housing in the Seoul Metropolitan Area.

Ultimately, the development of Sangam DMC serves as a true sub-center of west Seoul which alleviates densification of Seoul's city center. And an industry-specific CBD consolidating scattered leading institutions and private sectors which also benefit from having an ecosystem of related businesses and professionals to regroup for more efficient and productive urban environment for the sector's growth.

첫째, 방문객의 특성과 선호도에 맞는 '지식'으로 정보를 처리해 즉시 공급하는 '맞춤형 지식 제공'이 가능하도록 하고자 하였다.

둘째, '투과형 스트리트 엣지' 개념을 도입해 기존 공공·민간지역의 이질화된 구조에서 벗어나 건물 내외부의 투과성을 높여 거리와 건물 간 상호작용을 높인다. 건물 1층을 민간과 공공이 공존하는 공간으로 만들고, 건물 내부에서 일어나는 생산과 정보, 지식의 혁신이 외부, 특히 길거리와 공유될 수 있는 프로그램을 마련하고자 하였다.

셋째, 단일 공간 내에서 기능 간 혼합 사용을 유도해야 하는데, 이를 위해서는 DMC 관리자가 토지 및 건물 소유자에게 바람직한 혼합 사용을 위한 인센티브를 제공하는 동시에 다양한 사용을 수용할 수 있는 지상층에 공간 확보가 가능하도록 하였다.

넷째, ICT 등 첨단기술을 통해 보행자를 실시간 관리하면서 다양한 도시활동을 지원하고 보행자의 요구에 대응할 수 있는 프로그램이 가능한 도시경관이 구축되었다. 일반 거리에서는 경험할 수 없는 특별한 경험을 제공해 공적인 공간을 차별화하여 실시간 관리가 가능한 가로시설을 통해 DMC의 정숙성을 높일 예정이다. 예를 들어, 네트워크를 통해 새로운 ICT 도시 인프라에 접근할 수 있도록 일반 고정 간판을 피하고, 프로그래밍 가능한 파사드 미디어를 사용하고, 모든 간판과 가로등에 IP 주소를 부여하여 미래 확장성에 대비하였다.

토지 이용 계획

개발 초기 목적인 미디어 첨단산업 중심지를 조성하기 위해 지구 대부분이 첨단기술과 미디어 콘텐츠산업에 유리하도록 구획되어 있다. 서울특별시는 더 많은 IT 및 미디어 관련 기업들이 이 지역에 진출할 수 있도록 DMC 지역 전체에 앵커 용도를 배치했다. 그리고 주요지

Master Plan

DMC is a futuristic urban production community and industrial ecosystem that combines basic urban activities with high technology. It aims for a new model of the high-tech industrial cluster where creative workers can live, work, and enjoy in a vibrant urban environment for pedestrian-centered, multi-purpose. To realize this, the following concepts have been established:

First, the need for 'Provision for Customized Knowledge', which processes information into 'knowledge' tailored to the characteristics and preferences of visitors and immediately supplies to them.

Second, the concept of 'Permeable Street Edge' is introduced to enhance the permeability between the inside and outside of the building by blurring the boundary between the existing public and private areas, thereby increasing the interaction between the streets and buildings. It is to make the first floor of the building as s space where the private & public features coexist and to produce a program that allows the production, innovation of information, and knowledge taking place inside of the building to be shared with the outside, especially the streets.

Third, it is necessary to induce Mix of uses between functions within a single space, which requires DMC manager to provide incentives to land and building owners for a desirable mix of uses, while at the same time securing a single floor facing space that can accommodate a variety of uses.

Fourth, Programmable Urban Landscape that can support various urban activities and respond to the demands of pedestrians while managing them in real time through cutting-edge technologies such as ICT will be established. It provides a special experience that cannot be experienced on ordinary streets, making public

역에 대한 배정과 입찰은 주요 플롯에 대한 구매자의 적합성과 적합성을 검토하기 위해 특별히 구성된 검토위원회의 엄격한 검토 과정을 거치도록 하였다. 또한 IT·미디어 산업과 연관성이 높을수록 기업의 우선순위에서 높아져 위치선정과 선택권이 주어지도록 하였다.

그림 4. 상암 새천년 신도시의 토지이용구상 (출처: 서울시)
Sangam New Millennium New Town Concept Land Use Diagram

각 거점의 특성을 설정하기 위한 토지용도 할당 외에도, 중앙 핵심지역과 공공서비스 건물은 넓은 개방공간을 가지고 있으며, 다양한 특성을 가진 총 4개의 공원이 주요 축을 따라 배치되어 주변 프로그램과 용도에 대응하는 다양한 유형의 활동을 제공한다.

도시설계 전략 1 : DMC의 축

DMC에는 '미디어 & 컬처 축'과 '디지털 & IT 축'의 두 가지 주요 축이 제안되었다. 이 두 축은 L자형 부지를 연결하면서 중요한 장소에 중앙광장을 만드는 계획이다. 2개의 축 모두 내부 보행자 친화적 환경 도입이 돋보인다. 명확한 수직 축을 세움으로써 전 DMC 지역의 가독성은 물론 원래 계획했던 클러스터들도 강화되었다. 그러나 실행과정에서 미디어와 문화 프로그램은 원래의 개념을 희석시키는 두 축 모두에서 우세했다. 이에 비해 고층빌딩에 할당된 소포와 함께 종료될 예정인 디지털 & IT 축은 지난 20년 동안 여전히 비어있기 때문에 개발자들의 많은 수요를 끌어모으지 못한 것으로 보인다. 미디어 앤 컬처 액시스는 길이와 공공 영역 프로그래밍 측면에서 잘 작동한다.

DMC는 초기 개발 이미지가 개발 파급효과에 영향을 미치는 도시개발사업으로 사업의 신뢰성을 확보하고 계획개념 조기 실현을 가시화하기 위해 부지 중앙에 개척시설을 집중 배치

spaces different. Placeness of DMC will be enhanced through street facilities that can be managed in real time. For example, avoid ordinary fixed signboards, use programmable media facade media, and give IP addresses to all signboards and streetlights to make new ICT urban infrastructure accessible over the network.

Land Use Plan

In order to fulfill the objective and goals of the development, most of the district is zoned to favor hi-tech and media content industry. To have more IT and media related companies in the area, SMG distributed anchor uses throughout the DMC area. And the assignment and bidding for key plots had to go through a stringent review process by a review committee set up specifically for reviewing appropriateness and fitness of purchaser for the key plots. The more related to the IT and media industry, the sooner the company will have priority.

Besides allocating land uses for establishing a character of each node, the central core area and the public service buildings have generous open spaces; a total of four parks of varying characteristics are located along the main axis providing different types of activities responding to the surrounding program and use.

Urban Design Strategy 1 : Axes of DMC

Two main axes are proposed for DMC, Media & Culture Axis, and Digital & IT Axis. These two axes are meant to link the L-shaped site available at the time of planning

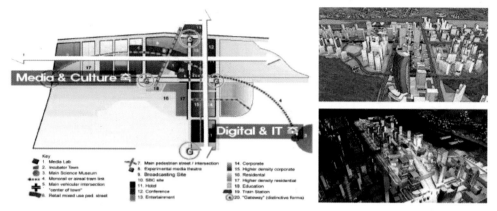

그림 5. 디지털미디어시티의 2개의 축 (출처: 서울시) / DMC's Two Axis

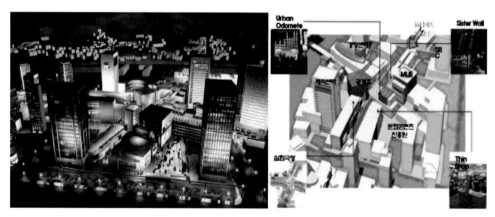

그림 6. 문화진흥원, 서울산업진흥원과 같은 공공기관과 핵심 시설이 두 축이 만나는 블럭에 위치하고 있음(출처: 서울시)
Public agencies like KOCCA and SBA are located at the block where two axis converge

하는 것을 매우 중요하게 생각했고, 거점을 형성하기 위해 2개의 축이 만나는 곳에 공공기관과 핵심역할을 할 수 있는 공간의 배치로 거점 형성에 노력했다.

도시설계 전략 2 : 보행자 전략

디지털 미디어 스트리트(DMS)는 디지털 스트리트 환경과 미디어 콘텐츠를 결합한 DMC의 중심가이다. 거리 활동, 물리적 공간, 사이버 공간이 상호 작용하며, 이러한 상호작용은 첨

to meet at the junction creating a clear central plaza. The introduction of the two axes was highlighted by introducing an inner pedestrian-friendly environment. Setting up clear perpendicular axes strengthened the legibility of the district and as well the different clusters the plan originally intended. However, over the course of execution, the media and culture program predominated in both axes diluting the original concept. By comparison, the Digital & IT Axis which is supposed to terminate with a parcel allocated for a skyscraper does not seem to have attracted much demand from developers as it is still sitting empty over the last two decades. Media & Culture Axis works well in terms of its length and the public realm programming.

At the junction of the two axes, the master plan team paid extra attention to locate public institutions and public facilities in order to secure the reliability of the project and to visualize the early realization of the concept of the plan so that it could be the basis for the subsequent development of the surrounding area.

Urban Design Strategy 2 : Pedestrian Strategy

Digital Media Street (DMS), is the main street of DMC which combines a digital street environment with media content. Street activities, physical space, and cyberspace interact with each other, and this interaction reflects DMC's characteristics as the Advanced Digital Technology and Media Entertainment Cluster. With four nodes that reflect the features of the surrounding area, media street triggers pedestrians' more active participation and street activities. Interfaces expand and supplement Street's function, and the façade of skyscrapers is considered as the place for the Media board, area for showcase, and space to express DMS visually. DMS can be evaluated as DMC's

단 디지털 기술 및 미디어 엔터테인먼트 클러스터라는 DMC의 특성을 반영하며, 쇼케이스를 위한 리빙랩, 에어리어 기능을 한다. 주변 지역의 특징을 반영한 4개의 거점으로 이루어진 미디어 스트리트는 보행자의 보다 적극적인 참여와 거리 활동을 촉발한다. 인터페이스는 스트리트 기능을 확장·보완하고, 고층빌딩의 파사드는 미디어보드, 쇼케이스 공간, DMS를 시각적으로 표현할 수 있는 공간으로 꼽힌다. DMS는 DMC의 특성을 충실히 반영하고 있어 DMC의 창의적인 거리 및 도시 디자인 전략으로 평가받을 수 있다.

DMC의 동서남북 중심부에 보행자 중심의 활동거리를 조성하여 DMC 도시활동의 중심지이자 보행환경 네트워크의 핵심이 되도록 계획하였다. 가로변에 3~5층의 낮은 층이 형성하고, 고층부가 후퇴하여 인적 저울의 거리 환경을 조성하였다. DMC 내 남북방향의 운행로를

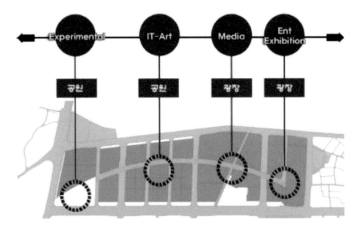

그림 7. DMC 중요 거점장소의 Concept (출처: 서울시)
Major Public Places of DMC

그림 8. DMC 중심 보행자 전용거리 모습 (출처: 저자)
DMC's Main Pedestrianized Street

creative street and urban design strategy as it faithfully reflects DMC's characteristics. Below images are the planning of DMS, and the photo of the Media Board on the MBC Building.

It was planned to create a pedestrian-centered activity street in the east-west and north-south centers of DMC to become the center of DMC urban activities and the core of the pedestrian environment network. The low floors of the third to fifth floors on the side of the street were formed, and the high-rise areas were retracted to create the street environment of the Human Scale.

A service road in the north-south direction within the DMC was constructed to induce vehicles to movement on the road, and the pedestrian traffic line was separated from the vehicle as much as possible by allowing pedestrian-oriented streets on the east-west road.

Urban Design Strategy 3 : Eco-friendly Planning

One of DMC's goals is to provide development in harmony with the environment, and the technology to consider humans first. As a result, large-scale environmental ecological parks such as Noeul Park, Hanuel Park and Peace Park, Nanjicheon Park, and Nanji Hangang Park were created while stabilizing Nanjido garbage landfill in DMC. In particular, the Noeul Park was constructed with a public golf course with considering nature as well as a resting place for citizens. The ecological golf course applied eco-friendly design and management technologies, such as transforming the infrastructure, minimizing the use of pesticides, and providing space for biological habitat.

조성해 도로에서 차량이 이동하도록 유도하고, 동서 차로에 보행자 중심의 도로를 허용해 보행자 통행선을 차량과 최대한 분리했다.

도시설계 전략 3 : 친환경계획

DMC의 목표 중 하나는 환경과 조화를 이루는 개발과 인간을 먼저 고려하는 기술을 제공하는 것이다. 이에 따라 DMC 내 난지도 쓰레기매립장을 안정화하면서 노을공원, 하늘공원, 평화공원, 난지한강공원 등 대규모 환경생태공원을 조성했고, 특히 노을공원은 시민 휴식공간으로 자연을 고려한 공공골프장을 조성했다. 생태골프장에는 기반시설 전환과 살충제 사용 최소화, 생물서식지 공간 제공 등 친환경 설계와 관리기술을 적용했다.

DMC는 넓은 녹색 네트워크를 가진 '생태 축'으로 형성되었다. 녹색 네트워크는 2개의 생태학적 축에 의해 구축되었다. 한 축은 북쪽의 봉산과 상암 매봉산의 관계이고, 다른 축은 서쪽의 대덕산과 상암 매봉산의 관계이다. 축을 기반으로 중앙 녹지로 방사형 녹지계획이 수립

그림 9. DMC 입구 광장(출처: 저자) / DMC Exhibition Plaza

DMC was formed with the 'Ecological Axes With a Wide Green Network.' That green network was established by two ecological axes. One axis is the relationship between the north's Bongsan and Sangam Maebongsan, and the other axis is between the west's Daedeoksan and Sangam Maebongsan. Based on the axes, a radial green area plan was established with a central green area. And it formed base green areas for each complex in response to the existing ones and linked them with Nanjicheon ecological park, Hangang, and Bongsan, which are green areas outside the complex.

A sustainable residential complex, combining nature, people and technology, tried to create an eco-friendly, human-friendly and landscape-friendly village. The green network, which harmonizes with the natural environment and is organically connected with the surroundings, and the network from Nanjicheon to the Hangang was established with the eco-friendly plan. A human-friendly plan created a participatory living space for residents and established a network that considers the convenience for all users and the safety of pedestrians. The landscape-friendly plan created a skyline that is harmonized with the landscape and made the street furniture design that can integrate the images of the DMC residential district.

Place Branding and Smart City

In order to brand DMC as an entertainment and med hub, Seoul Media has a public promotion agency dedicated to facilitate and grow the ecosystem necessary to build a strong hub. Periodically, media festivals are held to promote and celebrate DMC and the entertainment industry. These events include not not only music and entertainment programs, but also job fairs to attract high-quality human resources to DMC, and these

되었다. 그리고 기존 단지에 대응해 단지별 기반녹지를 조성하고 단지 밖 녹지인 난지천 생태 공원, 한강, 봉산과 연계했다.

자연과 사람, 기술이 어우러진 지속 가능한 주거단지가 친환경, 인간친화, 경관친화 마을을 만들기 위해 노력했다. 자연환경과 조화를 이루며 주변과 유기적으로 연결되는 녹색 네트워크와 난지천에서 한강까지의 네트워크가 친환경계획으로 구축됐다. 인간친화적인 계획으로 주민참여형 생활공간을 조성하고 모든 이용자의 편의와 보행자의 안전을 고려한 네트워크를 구축했다.

장소 브랜딩과 스마트시티

상암디지털미디어시티를 방송과 엔터테인먼트산업의 중심지역으로 만드는 데 필요한 생태계를 구축하고 지원하기 위해 서울시는 전담 공공기관을 이 장소로 이전하였다. 정기적으로 수많은 행사를 기획하여 대중으로 하여금 이 장소를 방송산업의 중심지로 브랜딩하였고, 행사 중에는 해당 산업군에 종사하는 전문가들을 대상으로 하는 취업, 창업페어도 지역을 알리는 데 많은 도움이 되었다. 또한 DMC를 연구개발단지로 발전시키기 위해 유럽에서 연구 클러스터로 유명한 프랑스 소피아 안티폴리스와 파트너십을 맺었다. 두 클러스터 간 정보 및 기술교환이 이루어지고 서로 다른 두 기업 간의 파트너십을 지원하는 노력을 하였으나 결과적으로는 연구기관들의 참여와 원동력이 되지 않았다.

DMC는 기획단계부터 계획된 서울의 최초 스마트도시 단지이다. 이를 통해 지속적으로 혁신적인 도시환경을 조성하기 위해 서울시는 꾸준한 노력을 하고 있다. 그 중 하나가 서울시의 스마트도시 기술을 도입단계에 부분적으로 지역에 적용해보는 테스트베드로 적극 활용하고 있다. 국토교통부는 2019년 6월 KT와 함께 '상암 자동운전 5G 페스티벌'을 열고 일반인을 대상으로 '자동운전 버스'를 선보였다. DMC는 세계 최초의 '5G 융합형 도시 자율주행 테스트베드'로 자율주행이 가능한 시스템 구현을 위한 모든 인프라를 갖추었다. DMC는 이러한

contribute competitiveness of the DMC complex. In addition, it also has several lectures about technology and future science for citizens. Public artworks are displayed along the street of DMC which are mostly about inspiration from technology and science. Some media artworks are interactive, attracts people, especially children.

In order to establish DMC as a research and development cluster, the mater plan team also facilitated a partnership with Sophia Antipolis, France, a high-quality research and development cluster in Europe. However, the interchanges between two institutions did not attract enough research institutions to create a fruitful partnership.

DMC was planned and constructed as the first smart cities in Seoul. In order to maintain this status as a smart city, SMG continues to invest on ICT infrastructure. One of the examples is to make DMC as a test bed for new technology. Sangam-dong and DMC area are assigned to be Demonstration District for Automated Driving. In June 2019, Ministry of Land, Infrastructure, and Transport held 'Sangam Automatic Driving 5G Festival' with KT and showcased 'Automatic Driving Bus' for the public. DMC, as the World's test self-driving vehicles. first '5G Converged Urban Self-Driving Testbed', boasts one of the most advanced infrastructure Through these projects DMC continues to be innovative center of Korea, and tries to be the front-runner of smart city.

Funding and Investments

Seoul city government has been supporting this area to attract private media entertainment and IT companies to DMC. So, it allocated city budget to develop and maintain this area. To start this development, the Seoul city government invested 1.7 trillion Won for purchasing land and complex construction which is being compensated

프로젝트를 통해 DMC가 스마트도시의 혁신을 이끄는 지역으로 조성하고자 노력하고 있다..

자금 및 투자

서울시는 민간 미디어 엔터테인먼트와 IT 기업을 DMC로 유치하기 위해 이 지역을 지원해 왔기 때문에 이 지역을 개발하고 유지하기 위해 시 예산을 할당했다. 서울시는 이를 위해 토지 공급과 투자로 보전되고 있는 토지 매입과 복합건설에 1조7,000억원을 투입했다. 게다가 매년 약 4억원을 기업의 개선과 지원을 위해 쓰고 있다.

이와 함께 문화관광부의 문화콘텐츠센터(Culture Content Center)와 공익법인에 의한 소프트월드(Soft World)도 이곳에 새로 설립되었으며, 대학의 연구센터도 설립되었다. 공영방송 MBC가 사옥을 이곳으로 옮겼고, SBS와 KBS도 자체 미디어센터를 건립했다. TBS는 IT 복합체도 구축했다. YTN, JTBC를 비롯한 많은 언론사들이 미디어센터를 이전하거나 새로 지었다.

외국인 투자를 장려하기 위해 DMC의 토지공급정책은 외국인 투자에 따라 가격이 조정될 수 있다는 예외를 두고 있다. DMC의 기업수와 외국인 인구는 2008년부터 2013년까지 거의 두 배가 되었다. 외국인 투자를 유치하기 위해 단기 및 장기 방문객 모두를 위한 다양한 숙박시설을 추가했다. 스탠포드 호텔과 DMC 빌(외국인을 위한 아파트)이 지어졌다. 이밖에도 외국인 학교[드와이트 외국인학교, 서울 일본인학교]가 설립되었다.

DMC 계획에는 우선시설 및 추천시설을 유치하기 위해 주요 인센티브로 시세의 50~70% 정도인 저렴한 토지가를 제공하였다. 이러한 유형의 시설들은 토지에 대한 분할납부를 허용하고 시가 재정적으로 지원할 수 있도록 하였다.

for by land supply and investment. Moreover, every year it spent about 0.4 billion won to improve and support enterprises.

With this, Culture Contents Center(문화콘텐츠센터) of Ministry of Culture and Tourism, and Soft World(소프트월드) by a public interest corporation were also newly founded here, and research center of universities as well. Public broadcast system MBC moved its head office building here, and SBS and KBS also built their own media center. TBS established IT complex as well. YTN, JTBC and many other media companies moved or newly built their media center.

DMC's land supply policy allowed land price adjustments in order to stay competitve and encourage foreign investments. The number of companies and the population of foreigners in DMC almost doubled from 2008 to 2013. Apart from direct incentives like land price, the plan included a range of accommodation for both short-term and longterm visitors: Stanford Hotel and DMC Ville (apartment for expats) were built. In addition, foreigners' schools (Dwight foreigners' school and Seoul Japanese school) were located near the housing area for expats.

The plan specified how priority facilities and recommended facilities can be incentivized with affordable land price which is about 50%-70% of market price. Addition to the attractive land price, these types of use were allowed to make partial payments for the land and even allowed the city to provide finanacial support acknowledging its benefit to the public.

맺으며

20년이 지난 지금, 상암 DMC는 여전히 개발 중에 있지만 한국 연예와 미디어의 메카로서의 위치는 확고하다. 그리고 한국 콘텐츠에 대한 세계적인 수요가 계속해서 빠르게 증가하고 있는 가운데, 이 클러스터가 시기적절하고 중요한 투자이자 정치적 결정임을 증명하고 있다.

성공적인 중심업무지구(CBD) 개발이 지역경제와 산업에 미치는 영향을 입증했으며, 특히 경제 성장과 일자리 창출을 위한 새로운 도시개발에 대한 수요가 높은 도시들에 대해 오늘날에도 유효한 많은 선진적인 도시계획기법과 설계원칙을 확인할 수 있다.

그림 10. DMC의 중심 보행자 전용 도로 전경 (출처: 저자) / DMC's Main Pedestrian Street

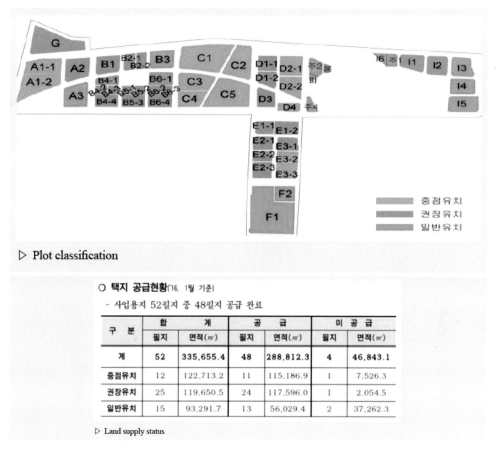

▷ Plot classification

○ 택지 공급현황(16. 1월 기준)
- 사업용지 52필지 중 48필지 공급 완료

구 분	합 계		공 급		미 공 급	
	필지	면적(㎡)	필지	면적(㎡)	필지	면적(㎡)
계	52	335,655.4	48	288,812.3	4	46,843.1
중점유치	12	122,713.2	11	115,186.9	1	7,526.3
권장유치	25	119,650.5	24	117,596.0	1	2,054.5
일반유치	15	93,291.7	13	56,029.4	2	37,262.3

▷ Land supply status

그림 11. 필지 유형과 투자유치유형 (자료: 서울시)
Attracting Priority and Recommended Facilities as Industry-building Businesses.

Closing Remarks

After 20 years, Sangam DMC is still under development; however, it secured its position as a mecca for Korean entertainment and media. And as the global demand for Korean content continues to rapidly grow, the cluster proves to be the timely and vital investment and political decision.

The case's general direction and concept for the CBD demonstrated the impact of

도시설계의 경우 보행자 전략이 하이라이트임이 입증됐다. 보행자 주축은 CBD가 자동차보다 보행자를 우선시함으로써 이익을 얻을 수 있다는 것을 증명했다. 산책로는 보행자 밀도에 비해 규모가 커보여 항상 사람이 많은 것은 아니다. 다만 시간이 지날수록 입주기업이 늘고 밀집되면서 영향력이 개선될 것으로 보인다. 보행자 도로를 따라 정면의 정렬과 활동은 충분한 활동성과 참여를 제공하는 것처럼 보이지만, 파사드의 디자인은 보다 인간적인 규모로 매력적으로 개선될 수 있다. 공공영역 디자인은 제공되는 좌석과 녹지가 매우 제한적으로 보여 환영과 따뜻함을 자아낼 수 있어 참여도를 높일 수 있다.

계획의 전반적인 방향과 컨셉은 CBD의 성공적인 조성을 보여주었지만, 지하철역과 기차역의 위치가 CBD와 편리하게 연결되지 않아 접근성에 한계가 있어 확장성과 수도권과의 한정적인 연결성은 미래의 제한요소로 작용하고 있다. 하지만 SMG의 헌신과 지속적인 구의 지원으로 시간이 흐르면서 이러한 우려를 극복할 수 있을 것이다. 그리고 탄탄한 협치구조와 조직화된 이해관계자들이 그 구의 성장과 지속 가능한 미래에 도움을 줄 것이다.

참고문헌

1. 디지털미디어시티 10년사 및 발전방향, 서울산업진흥원, 2013
2. 서울 상암디지털 미디어시티 (DMC)의 조성, 서울정책아카이브, 2015
3. Seoul's Digital Media City, Seoul Policy Archive, https://seoulsolution.kr/sites/default/files/ policy/1%EA%B6%8C_ 09_Urban%20Planning_Seoul%E2%80%99s%20Digital%20Media%20City.pdf, 2015
4. 서울특별시, 상암 새천년 신도시 기본계획, 2000
5. 김도년, 김정훈, 상암 Digital Media City 지구단위계획, 도시설계, 제5권 제1호, 2004

urban development on the regional economy and industry, and it reflects all the urban planning and design principles that are still relevant today especially for cities with high demand for new urban development for economic growth and job creation.

As for urban design, its pedestrian strategy proves to be the highlight of the case. The pedestrian main axis proved that a CBD could benefit from prioritizing pedestrians over cars. The pedestrian promenade is not always full of people as its scale seems oversized for the density of pedestrians. However, its impact will improve as more companies move in and densify over time. The alignment and.

Although the general direction and concept of the plan demonstrated a successful creation of a CBD, there is plenty of debate for the locations of subway stations and national rail station as they do not conveniently connect to the CBD. And its expandability and limited connections to the rest of the Seoul Metropolitan Area do pose a serious threat to its future. However, SMG's commitment and continued support for the district will overcome such concerns over time. And its strong governance structure and organized stakeholders will also help the district's growth and sustainable future.

References

1. Seoul's Digital Media City, Seoul Policy Archive, https://seoulsolution.kr/sites/default/files/ policy /1%EA%B6%8C_09_Urban%20Planning_Seoul%E2%80%99s%20Digital%20Media%20City.pdf, 2015

2. Song, Jun-Min & Joo, Yumin, From Rubble to the Korean Wave Hub: The Making of the New Digital Media City in Seoul, 10.1007/978-3-030-46291-8_10, 2020

3. https://use.metropolis.org/case-studies/sangam-digital-media-city-dmc-city-of-tomorrow

4. https://en.wikipedia.org/wiki/Digital_Media_City

5. https://www.risk.tsukuba.ac.jp/~ussrl/public_html/InterExchangeSeminar/materials/DMC.pdf

서울 G 밸리

21세기를 준비하는 도심산업단지

오 다니엘 [고려대학교 건축학과 교수]

한국전쟁 이후 폐허가 된 국가 경제를 살리고 산업화를 선도하는 국가산업단지로서 제조산업단지에서 첨단지식산업단지로 거듭나는 데 필요한 도시조직의 재구성에 대해 논의하는 적합한 사례이다. 도심지의 고부가가치에 필요한 산업구조와 생태계 재구성을 위한 공공의 역할과 정책의 변화를 위한 도구로써 도시설계의 역할을 확인할 수 있다. 제조업을 위한 공장에서 첨단산업을 위한 지식산업거점으로 변신에 필요한 용도계획부터 민간개발을 통해 쾌적한

그림 1. G 밸리를 상징하는 조형물과 광장의 모습 (출처: 저자) / G Valley Monument at the central plaza

G Valley, Seoul

Reinventing Urban Industry Complexes in the 21st Century

Daniel Oh (Professor, Korea University, Dept of Architecture)

As a national industrial complex that revived the devastated nation after the Korean War and led the country's industrialization, this case discusses the reorganization of the urban industrial complex transformed as a high-tech knowledge industrial complex from a manufacturing industrial complex. The role of urban design as a tool for policy change and the role of public institutions in the reconstruction of the industrial structure and ecosystem necessary to create a value-added industry core in the downtown area are the core findings of this case study. The urban design approach used to promote a comfortable and safe urban space through reinforcing urban design measures for the new private developments had shown its shortcomings. And the transformation from a factory for manufacturing to a base of knowledge industry for high-tech industry required more than just physical transformation. On the other hand, the public sector's effort to commemorate the past through archiving the city's past and especially the achievements of those who accomplished the reconstruction of the city and the country added layers of intangible quality to the unique place genuine to the public.

도시공간을 조성하는 데 사용된 도시설계적 접근의 결과를 확인할 수 있다. 또 한편 도시 재구성을 통해 사라지는 도시의 과거를 기억하고 특히 도시와 국가의 재건을 일궈낸 사람들의 업적을 기억하려는 공공의 노력이 인상적이다.

들어가며

전쟁으로 황폐해진 나라를 재건하려면 일자리뿐만 아니라 대중의 마음과 마음을 꿈으로 옮길 신중하면서도 대담한 전략적 계획이 필요하다. 인류의 역사를 통해 많은 성공사례가 있었지만, 그러한 사건을 경험한 사람들의 세대를 넘어 다음 세대에게도 같은 꿈과 희망을 제공하는 도시개발사례는 많지 않다. G 밸리가 놀라운 것은 지역의 모습은 완전히 새롭게 변하였으나, 여전히 젊은 세대를 위한 꿈과 기회를 위한 장소이자 장소의 맥락을 유지하며 차기 산업의 선도역할을 하는 지역이라는 점이다.

이러한 변화는 도시계획 및 도시설계를 통한 새로운 도시개발 및 규제 그리고 도시설계 지침의 수립을 통해 가능하다. G 밸리는 시대와 변화를 앞서는 정책기반에 맞춘 적절한 도시설계의 표본으로 한국의 첨단 IT산업이 더 많은 성장을 위한 준비를 하는 도시공간적 플랫폼을 제공했다. 이 사례는 단지 도시공간적으로만 볼 수 없고, 더 나은 미래와 진보를 위한 강력한 정치적, 사회적 의지를 기반으로 변화하는 사회적 의제에 어떻게 도시개발과 정책이 함께 대응하고 변모하였는지를 잘 보여준다.

모든 성공적인 도시개발과 마찬가지로, 이 사례 또한 현재 진행형이다. 이 사례는 다양한 도시설계 요소들을 통해 오래된 산업지구를 활기찬 비즈니스 및 벤처지구와 매력적인 지역으로 어떻게 변화되었는지를 보여준다. 한편 이 사례를 분석할 때 지속적으로 변화하는 시대적, 정치적, 사회적 맥락의 중요성을 염두에 두고 봐야 한다. 저자는 이 사례의 도시공간적 변화뿐만 아니라, 장소의 선구적인 정신을 유지하고 지속적인 사회와 관련성을 유지하려는 노력이 경제와 산업에 선도적인 역할을 했다는 점에 초점을 두고자 한다.

Introduction

Rebuilding a war-ravaged nation requires a careful yet bold strategic plan that will move the hearts and minds of the public to dream, not just jobs. Over the history of mankind, there are many success stories, but very few cases managed to outlive the very people who experienced such an event. What is remarkable about the G Valley case is that it managed to completely transform itself and yet, it managed to still serve as a place for dreams and opportunities for the young generation.

Such transformation in urban planning and urban design strategies required a formulation of a completely new development and regulatory measures and guidelines. Because of these changes, a set of new standards provided a platform for the next industrial wave that prepared South Korea for more growth. This case owes its success to its national government policy and strong political and social will for a better future and progress. And further transition of power to the local government to cater to the needs of the existing and future occupants. It demonstrates how urban design can serve the changing agenda of a society.

Like all successful urban developments, this case continues to evolve. And this case illustrates how a diverse set of urban design tools have transformed an outdated industrial district into a vibrant business and venture district and an attractive neighborhood. However, with praise for the case, readers must be aware that the times have changed. The context for political and social climate played a significant role in this case which must be carefully understood when analyzing this case. The significance of this case is not only in physical transformation but in that this case managed to maintain its pioneering spirit of the place and stay relevant to society.

국가산업단지의 탄생

한국전쟁 이후 1960년대 중앙정부는 수출 주도 경제발전을 수립하고 이행하기 위해 경제발전을 최우선 과제로 삼았다. 이 계획안은 외화를 벌기 위한 수출품 제조를 위해 인프라뿐만 아니라 인력의 산업화를 강조했다. 1964년 수출산업단지개발법이 제정되어 '한국수출산업단지공사'를 설립했고, 농업부지 개발을 산업단지에 이용 가능한 부지로 조성하는 데 주도적인 역할을 했다. 최초의 '한국수출산업단지'는 농지와 마을로 형성되어 있던 서울지역 외곽에 위치하고 있었다. 1965년 공사가 시작되었을 때, 수출전용 산업체들을 유치하기 위해 계획되었으며, 경쟁력을 갖춘 기업만 입주하면서 최초의 국가산업단지의 기준을 현저하게 높게 설정하는 데 기여했다. 1968년 제1회 한국무역박람회가 산업단지에서 개최되었고, 이는 단지를 홍보하고 근로자를 유치하는 매우 효과적인 행사로 기억된다.

그림 2. 1968년 9월 11일 한국수출산업단지에서 개최된 제 1회 국제무역박람회 (출처: 국가 기록원)
The first Intenational Trade Expo was held at the 'Korea Export Industrial Complex' which opened on the 11th of September, 1968.

Birth of National Industrial Complex

In the 1960s, the central government made economic development a top priority for national policies to establish and implement an export-led economic development. This plan emphasized industrialization of the workforce as well as infrastructure designed to manufacture goods for exports to bring in foreign currency. Under the directives of the national government, Export Industrial Complex Development Act was established in 1964 which allowed an establishment of 'Korea Export Industrial Estate Corporation.' The public corporation spearheaded the development of agricultural plots into sites available for industrial complexes. The first 'Korea Export Industry Complex' was in the outskirts of urbanized area of Seoul defined by farmlands and villages. When the construction began in 1965, it was designed to attract Korean Japanese companies exclusively for exports allowing only companies that were highly competitive setting the standard remarkably high for the first National Industrial Complex. The first Korea Trade Fair opened in 1968 at the first complex to promote the complex's achievements and to attract workers.

The second industrial complex was designated in 1968, and construction was carried out to connect the first and second complexes. The second complex included a multi-story sewing factory with support facilities such as housing and other amenities use. Later, the third industrial complex with 1,190,083m² of land was constructed across the Gyeongbu Line, resulting in the three industrial complexes in proximity of each other. And the surrounding areas quickly filled in with housing and retail uses for the factory workers. In the 70s and 80s, the three complexes were commonly known as Guro Industrial Complex, and to the people of South Korea, it represented the first step to

그림 3. 1957년도 가리봉동 지역 / 1975년도 한국수출산업단지 1차와 2차 단지와 가리봉동 / 2001년도 서울디지털산업단지와 가리봉동 (출처: 서울역사박물관)
Garibongdong Area in 1957 / 1st and 2nd Korea Export Industrial Complex in 1975 / Seoul Digital Industrial Complex in 2001

G 밸리의 탄생

구로산업단지는 전자, 금속, 섬유산업을 대표하는 노동운동의 메카로 불렸다. 3개 단지의 밀도가 높아짐에 따라 노동자들은 더 나은 임금과 근로조건을 요구했다. 80년대 중반 부터 국내 다국적기업과 국내기업, 특히 노동집약적인 산업은 임금 상승과 고령화로 인해 산업단지에서 이주하기 시작했다. 1985년 6월, 구로노동조합은 더 나은 노동조건을 요구하는 파업으로 인해 공장들과 기업들의 이전이 가속화되었다.

90년대에 국가 경제가 빠른 성장에 따라 산업구조 또한 노동집약적 산업에서 기술기반산업 및 서비스산업으로 구조적 변화를 가져왔다. 이러한 변화는 구로산업단지의 오래된 인프라와 노동자의 고령화로 인해 구로산업단지의 경쟁력이 크게 둔화되었다. 그 결과 1995년 구로공단 근로자 수는 구로단지의 초기에 비해 절반 이상 감소하기에 이르렀다.

1990년대 후반, 정부는 노동집약적 산업을 교육받은 노동력을 활용한 부가가치 제품으로 선제적으로 장려하기 위한 새로운 정책 초안을 작성하기 시작했다. 이러한 정책 변화를 반영하기 위해 '한국수출산업단지(KICOX)'는 5개 국공업단지와 합병되어 '한국산업단지공단'을 설립했다. '구로수출공단 선진계획'이라는 계획이 발표되고 구로공단을 공식적으로 개편하는 방안이 시작되었다. 이 계획은 노동집약적인 산업에서 첨단기술 및 지식기반 산업에 이르기까

economic progress for the country. The complex attracted many young workers from all over the country with hopes and dreams for a better future, and they were proud to take part in national policy to build South Korea's economy.

Birth of G Valley

At its peak, Guro Industrial Complex was called the mecca of labor movements representing electronics, metal, and textile industries. As the three complexes densified, laborers demanded better pay and working conditions. Since the mid-80s, domestic multinational and domestic companies, especially the labor-intensive industries, began to relocate out of the industrial complex due to rising wages and ageing laborers. In June of 1985, the Guro Workers Union Alliance Strike demanding better labor conditions which led to increased exodus of relocation of factories and businesses to other newer national industrial complexes in other parts of the country and overseas.

As the country's economy made a significant growth in the 90's, industry sectors had to undergo major structural transformations from labor intensive industries to skill-based industries and service industries, which led to a substantial decrease in competitiveness of the Guro Industrial complex due to outdated infrastructure and ageing laborers. In 1995, the number of workers at the Guro Industrial Complex decreased by more than half compared to the beginning of the complex.

In the late 1990's, the national government began to draft new policies to proactively encourage labor intensive industries to more value-added products taking advantage of educated work forces. To reflect these policy changes, 'Korea Export Industrial Estates

그림 4. 서울디지털산업단지에 위치한 한국산업단지공단 (KICOX) 본사 전경 (출처: 저자)
Headquarter of Korea Industrial Complex Corp. (KICOX) located in Seoul Digital Industry Complex

지 경제발전의 다음 단계를 위한 도시계획 및 도시설계를 공식적으로 준비하기 시작했다.

새로운 장소성의 형성

'구로공단'은 대중에게 매우 잘 알려져 있었기 때문에 그만큼 더 많은 선입견을 주는 곳이다. 이곳은 수출을 통해 외화뿐만 아니라 희망과 꿈을 가져다 준 일자리이자 꿈의 직장으로 많이 알려져 있었으나, 1980년대 말까지 어려웠던 노동환경과 연관된 사건사고로 '구로단지'라는 이름과 함께 편견을 낳게 되었다. 대부분의 사람들은 이 지역에 가본 적이 없더라도, 새로운 투자를 유치하거나 이 곳에서 일하고 싶은 새로운 원동력을 찾는 데 어려움이 되었다. 이러한 편견을 벗어버리기 위한 전략으로 첫 번째 단계는 공식적으로 단지의 이름을 '서울디지털산업단지'로 변경하고 IT 첨단산업단지로 홍보하였다. 그러나 1997년 한국 금융위기 사태로 인해 복합단지의 이름뿐만 아니라 신규 투자 및 첨단기업을 유치하겠다는 계획은 많은

Corporation (KICOX)' was merged with five other national industrial complexes to form 'Korea Industrial Complex Corporation.' And a plan called 'Guro Export Industrial Complex Advanced Plan' was announced and officially revamping Guro Industrial Complex began. This plan officially initiated preparing urban planning and urban design for the next stage of economic development: from labor-intensive industries to high-tech and knowledge-based industries.

Building Place Identity

Because 'Guro Industrial Complex' was well known to the public, the name also carried many preconceptions to the people. It was known to many as the district that brought hopes and dreams as well as foreign currency from exports. However, by the end of 1980's, it also was known as the mecca for labor unions and labor movements; hence, the place carried a stigma and bias along with the name, 'Guro.' Even if people have never been to the area before, the name stood for something distressing and contentious which did not help attract investments into the area. The first step was to rebrand the complex and officially change the name of the complexes to 'Seoul Digital Industrial Complex' and promote the complex as IT high-tech industrial complex. However, the plan to transform not only the name of the complex but attract new investments and high-tech companies to the complex brought to a standstill due to the South Korea economic in 1997.

Later, when the economy picked up and investments restarted in the late 2000's, people started to refer to 'Seoul Digital Industrial Complex' as 'G-Valley' which 'G' stood for initial for three different local districts that started with a letter 'G', Gasan-

그림 5. 서울디지털산업단지 표지판 / 제1단지 중앙도로 경관 (출처: 저자)
Seoul Digital Industrial Complex Signage / Main Street of 1st Complex

부분 실행에 옮겨지지 못하였다.

이후 2000년대 후반 경제가 재개되면서 '서울디지털산업단지'를 'G Valley'라고 부르기 시작했고, 'G'는 가산동, 구로동, 금천동의 첫 영문 이니셜을 표현하고, 미국 실리콘밸리의 '밸리'를 합성한 명칭을 사용하면서 국내 스타트업과 벤처인들을 유치했다. 그리고 구로구와 금천

그림 6. 매년 열리고 있는 G 밸리 위크 홍보포스터 / 넥타이 마라톤 (출처: 서울산업진흥원 SBA)
Annual G Valley Poster / Necktie Marathon

dong, Guro-dong, Geumcheon-dong, and 'Valley' from Silicon Valley hoping to attract high-tech venture companies and startups. And the local governments, Guro-gu and Geuncheon-gu, began to promote G-Valley to start-ups with incentives and job fairs targeted to young professionals. These support programs included educational and networking venues for university graduates as well as young professionals to encourage starting up new businesses in G-Valley. This helped to successfully reinstate the identity of the area to the younger generation and especially for those who are looking to start their ventures and seeking local government's assistance and incentives.

Transforming Factories into Hi-tech Knowledge Centers

As the rebranding of the area helped to get attention from a growing and high value-added industry, transforming the outdated manufacturing factories into a competitive industrial complex proved to be more challenging. The first hurdle was to create zoning amendments to the existing industrial use. Since most parcels were privately owned, the only way to bring new investment was to make each site economically more valuable to rebuild. Much needed amendments to the national urban planning act introduced a concept of mixed use which allowed industrial land use to include not only 'apartment type' factories but support uses, academic, research and development, and other uses required to advance industries.

The physical transformation of the manufacturing factories to knowledge-based industry required a complete overhaul of land use regulations. Also, establishment of a new building typology that suited offices and light manufacturing related to high tech

구는 젊은 전문가를 대상으로 인센티브와 취업박람회를 통해 G 밸리를 스타트업으로 홍보하기 시작했다. 이러한 지원 프로그램에는 G 밸리에서 새로운 사업을 시작하도록 장려하기 위해 대학 졸업생뿐만 아니라 젊은 전문가를 위한 교육 및 네트워킹 장소가 포함되었다. 이는 젊은 세대, 특히 벤처를 시작하고 지방정부의 지원과 인센티브를 추구하는 사람들에게 이 지역의 정체성을 회복하는 데 도움이 되었다.

제조업단지에서 지식산업단지로

이 지역의 새로운 방향에 대한 인지도가 높아지면서 고부가가치 산업단지로서 주목을 받게 되었으나, 오래된 제조산업공장을 경쟁력 있는 지식산업 단지로 탈바꿈 하는 것은 또 다른 문제였다. 첫 번째 장애물은 기존 산업용도를 규제하는 도시계획법의 개정이 필요했다. 대부분의 지역은 사유지이기 때문에 새로운 투자를 가져올 수 있는 유일한 방법은 각 사이트를 경제적으로 재건하여 더 가치가 있게 만드는 것이었다. 국가도시계획법 개정안은 산업토지 이용이 '아파트형' 공장뿐만 아니라 산업발전에 필요한 용도, 학술, 연구개발, 기타 용도를 포함할 수 있도록 혼합사용개념을 도입했다.

제조공장이 지식기반산업으로 물리적으로 전환하려면 토지 이용 규정을 완전히 정비해야 했다. 또한 하이테크 산업, 사무실과 근린생활시설이 모두 함께 같은 건물에 공존하기에는 완전히 새로운 건물 유형이 필요했다. 또한 가장 중요한 것은 기존 공장을 인수하고 재건하기 위한 투자를 유치하기 위해 가능한 개발면적을 늘리고 산업활동을 위한 충분한 영역과 상업 및 소매공간과 같은 임대지원 프로그램을 제공하는 혼합 사용을 허용하는 것이다.

복합토지이용, 복합용도지역, 지식산업구축은 첨단기술관련산업에 대한 수요 증가와 같은 산업구조 변화에 적극 대처하는 산업단지의 특성을 변화시켰다. 사용의 혼합을 허용하는 것은 기존 산업시설에 대한 일회용 구역 및 엄격한 레이아웃에서 벗어나는 것을 의미한다.

industry to coexist. And most importantly, to attract investments to acquire and rebuild the existing factories, increase in FAR as well as allowing mixed use that ensured providing enough areas for industrial activities as well as rentable support programs such as commercial and retail spaces.

Establishment of Mixed Land Use, Mixed-use Area and Knowledge Industry transformed the nature of the industrial complex actively coping with changes in industrial structure like the increased demand for high-tech related industries. Allowing mix of uses meant breaking away from single-use zoning and rigid layout for existing industrial facilities.

The Mixed Land Use (복합용지) Under the 'Industrial Location Act' is a site where industrial facilities, support facilities, public facilities, residential facilities, and commercial facilities can

be in a complex area, but the Mixed-use Area (복합구역) under the 'Industrial Aggregation Act' is a space that can be located around facilities subject to occupancy, universities, research institutes and corporate support facilities in the industrial facility zone. Therefore, there is a difference in the development plan because the complex area allows only limited facilities while the complex land is a comprehensive concept.

In January 2014, the concept of 'complex land' was introduced in 'Article 2 of the Industrial Sites and Development Act', and 'complex area' was designated and managed in December of the same year. After the introduction of the system, most of the newly designated industrial complexes were designated as multi-purpose land, and most of the designated complexes since the introduction of the system in 2014 are on sale or on the rise due to the plan for sale.

As it is described in the Act, 'Knowledge Industry Center' is a multi-story collective building where people can operate light manufacturing, information, and tele-

표1. 업단지 현대화를 위한 제도 제정 및 개편 / Amendments and Enactment to Modernization of Industrial Uses

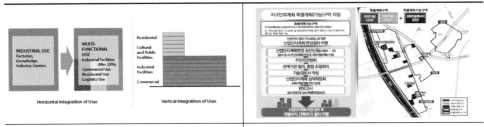

복합용지: 2014년 개정	복합구역: 2011년 신규제정
·산업입지 및 개발에 관한 법률 이 법은 산업입지의 원활한 공급과 산업의 합리적 배치를 통하여 균형 있는 국토개발과 지속적인 산업발전을 촉진함으로써 국민경제의 건전한 발전에 이바지함을 목적으로 한다.	**·산업집적활성화 및 공장설립에 관한 법률** 이 법은 산업의 집적(集積)을 활성화하고 공장의 원활한 설립을 지원하며 산업입지 및 산업단지를 체계적으로 관리함으로써 지속적인 산업발전 및 균형 있는 지역발전을 통하여 국민경제의 건전한 발전에 이바지함을 목적으로 한다.

복합용지는 산업시설, 지원시설, 공공시설, 주거시설, 상업시설이 이용할 수 있는 부지를 말한다. 또한 복합구역은 산업시설구역의 점유시설, 대학, 연구기관, 기업지원시설 등의 시설 주변에 위치할 수 있는 공간이다. 따라서 복합면적은 시설제한만 허용하고 복합단지가 종합적인 개념이기 때문에 개발계획에 차이가 있다.

2014년 1월에는 산업현장개발법에 '복합용지'라는 개념이 도입되었으며, 같은 해 12월에 '복합구역'을 지정·관리했다. 이 제도의 도입 이후 새로 지정된 공단의 대부분은 다목적 토지로 지정되었으며, 2014년 제도 도입 이후 지정 단지의 대부분이 매각 또는 증설로 연결되었다.

또한 '지식산업센터'라는 정의가 처음 소개되면서 도시산업의 새로운 유형을 제시하는 계기가 되었고, 이는 제조업, 정보통신(ICT) 산업을 운영할 수 있는 다층 복합건물로, 현재는 같은 건물에 공존할 수 있다. 토지이용법 개정안을 통해 새로운 구역 및 유형 설정은 산업과 전문가의 건전한 생태계를 육성하는 데 도움이 되는 첨단 산업 클러스터에 대한 증가하는 수요를 해결하는 데 도움이 되었다. 또한 궁극적으로 오늘 날 한국이 4차 산업혁명에 대비하고, 도심에 산업을 유지할 수 있는 데 큰 역할을 했다고 할 수 있다.

communication (ICT) industries and support facilities can now coexist in a building. Through this amendment to the Land Use Act, a new zoning and typology helped to address the growing demand for a high-tech industry cluster which helps to cultivate a healthy ecosystem of industries and professionals. And for the rest of the country, this regulatory adaptation helped to pave the road to prepare for the 4th Industrial Revolution.

Raising Standard of G-Valley's Public Realm

By appealing to the younger generation, G-Valley managed to attract start-ups and venture companies, increasing the demand for office spaces and high-tech light manufacturing spaces. Over a decade, factory sites were quickly redeveloped into massive office buildings and apartment-type factories cladded with glass completely transforming the landscape of factories of 'Guro Industrial Complex'. Along with the new investments in redevelopment of the factories, the need for urban design guidelines for placement of support facilities for workers and businesses became clear because the private developers were only interested in maximizing rentable space on their plots. The existing infrastructure within public right of way produced a lack of adequate sidewalks and green open spaces, not to mention road congestion and lack of parking space. To ensure quality public realm for G-Valley was challenging because the complex was mostly privately owned. Therefore, urban design guidelines were necessary to direct developers to provide open spaces and setbacks for pedestrians and open spaces for workers.

The lack of open space in the G Valley relies on open and full-scale public spaces (공

G 밸리의 공공 영역의 수준을 높이다

G 밸리는 젊은 세대의 관심을 끌며 스타트업과 벤처기업을 유치하여 사무실 공간과 첨단 제조공간에 대한 수요를 증가시켰다. 약 10년 동안 '구로공단'의 지형을 완전히 바꾸어 유리로 덮은 대형 오피스 빌딩과 아파트형 공장으로 빠르게 재개발되었다. 하지만, 공장 재개발에 대한 새로운 투자는 민간 개발업자들로서는 임대공간을 극대화하는 데만 관심이 있었기 때문에 근로자와 기업을 위한 지원시설 배치를 위한 도시설계지침의 필요성이 대두되었다. 공공영역 내의 기존 인프라의 부족은 도로 혼잡과 주차공간의 부족으로 이어졌고, 보행자를 위한 기본적인 보도와 공공녹지 현황은 매우 취약했다. 대부분의 단지는 사유지였기 때문에 양질의 공공영역을 보장하기 위해서는 당시 제도적인 틀로는 불가능하였다. 따라서 도시설계지침을 통해 보행자를 위한 열린 공간과 보행공간을 확보하고, 근로자를 위한 열린 공간과 휴게시설 그리고 지원시설을 확보하기 위해 다양한 제도와 틀을 마련하는 것이 필요했다.

G 밸리의 열린 공간의 부족은 건축양식을 개방적으로 하여 도로변에 공공공간의 확보를

그림 7. 서울디지털산업단지 곳곳에 위치한 옛 산업단지 시설들로 채워진 저이용 필지를 쉽게 찾아 볼 수 있다. (출처: 저자)
Numerous underutilized plots with old factory buildings can be easily found

개공지, 전 면공지) following the transformation of the architectural form. Although Anyang Stream, a large green area, is nearby, it is used passively due to its low accessibility. The existing 'Young Adult Center is scheduled to be newly built as a corporate and worker support facility, and the playground, the only public area in Complex 3, is unable to meet the overall demand of the complex due to the concentration of usage in certain time zones. The open spaces in the G-valley relied on developing a new knowledge industry center planned and approved in a piece-meal fashion. Due to the patchwork of open spaces and linkages, the public realm experience of G Valley can be classified as disconnected pockets of open space and narrow sidewalks.

To rectify the situation, '2030 Comprehensive Development Plan for Subindustrial Areas' was proposed in 2015 after the paradigm of urban management changed from urban development to urban regeneration, and the plan was reorganized with institutional changes, including the 2013 Urban Regulated Urban Regeneration Act. Regarding the Seoul Digital Industrial Complex in the 2030 Comprehensive Development Plan for Sub-industrial Areas, the G Valley Regeneration and Revitalization Plan envisaged the future of 'Creating a Convergence and Complex Base Focused on Creative Industries' and plans to re-emerge as a vibrant complex space through laying the foundation for industrial innovation and physical regeneration.

Considerations for continuous promotion of industrial complex competitiveness through structural upgrading of G-Valley.

- Support function for business workers in areas around industrial complexes, inducing public open space and vitalization of major streets
- Guiding the location of G-valley support functions such as digital transformation in the Garibong-dong promotion zone

요구하였다. 넓은 녹지인 안양천이 근처에 있음에도 접근성이 낮기 때문에 접근성을 해결하는 제안이 필요했다. 기존 '청년센터'는 기업 및 근로자 지원시설로 신축될 예정이며, 3단지 내 유일한 공공지역인 놀이터는 특정 시간대의 사용량으로 인해 복합단지의 전반적인 수요를 충족시킬 수 없었다. 따라서 G 밸리의 공공공간 계획은 개별 개발안에 맞추지 않고, 전체 단지단위의 계획에 맞추어 개별 개발안이 제안되고 승인될 수 있는 제도적 개선이 필요했다.

2015년 G 밸리가 도시관리에서 도시재생으로 도시경영 패러다임이 변경된 후 '2030년 산업지역 종합개발계획'이 제시되었으며, 2013년 도시재생법 등 제도적 변화로 개편되었다. G 밸리 재생 및 활성화계획은 '창조산업에 초점을 맞춘 융합과 복합기지 조성'의 미래를 구상하고, 산업혁신과 물리적 재생의 토대를 마련하여 활기찬 복합공간으로 다시 부상할 계획이다.

G 밸리의 구조적 향상을 통한 산업단지 경쟁력을 위해 다음 항목들을 고려하였다.

- 산업단지 주변 사업장 근로자 지원 기능, 공공개방공간 유도 및 주요 거리 활성화
- 가리봉동 진흥구역의 디지털 전환 등 G 밸리 지원 기능 강화
- 대규모 주거기능 지역에 주거수준 관리시스템 도입 고려
- 도림천 스트림에 녹색 및 보행자 연결 제공
- G 밸리 2단지의 산업지원 기반을 강화하고 기업 근로자 지원환경을 조성하여 산업단지의 경쟁력을 강화
- 독산2동·4동 주거·산업 혼합지역 중소산업 재생 장려
- 대규모 공장, 군사 이전부지 전략사업, G 밸리 연계 지역센터 등 복잡한 개발을 유도한다.

서울디지털산업단지는 2000년대 구조 업그레이드 이후 10,823개 기업과 136,761명의 근로자가 근무하고 있으며, 첨단기술 및 지식산업의 변화로 쾌적한 근무환경과 문화 및 휴게소를 조성하는 것이 시급하게 진행되고 있다. 그리고 이 단지는 최근 몇 년 동안 입주자들의 탈출을 일으킨 판교 테크노밸리, 상암 디지털미디어시티(DMC), 마곡 산업단지 등 서울 을 중심으로 새로 지어진 첨단산업 클러스터와 치열한 경쟁을 겪고 있다. 금천구는 각 건물의 용

- Consider introducing a residential level management system in large residential functional areas

- Provide green and pedestrian connections to Dorimcheon Stream

- Strengthen the competitiveness of industrial complexes by strengthening the industrial support base of G-Valley 2 and 3 complexes and creating an environment for supporting corporate workers.

- Encourage small and medium-sized industrial regeneration in residential and industrial mixed areas of Doksan 2-dong and 4-dong

- Inducing complex development such as large-scale factories, strategic projects of military transfer sites, and regional centers linked to G-Valley.

The Seoul Digital Industrial Complex also saw a surge in tenant companies since the structural upgrading in the 2000s, with 10,823 companies and 136,761 workers currently operating there, and the change to high-tech and knowledge industries has raised the level of workers' awareness, making it urgent to create a pleasant working environment and cultural and rest areas. And the complex is experiencing steep competition from other newly built high-tech industry clusters around Seoul like Pangyo Techno Valley, Sangam Digital Media City (DMC), and Magok Industrial Complex which caused exodus of tenants in recent years. Geumcheon-gu is planning to implement urban landscape management through integrated design of layout, landscape, and exterior in design guidelines for each building's use, and to raise quality of its public realm to compete with the other high-tech industry complexes in Seoul Metropolitan Area.

In recent years, the local government has added key public facilities to strengthen the identity of G Valley by adding layers of its cultural significance to society. The most

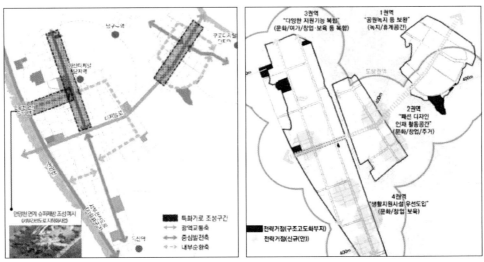

Specialized Street setting   worker-friendly environment

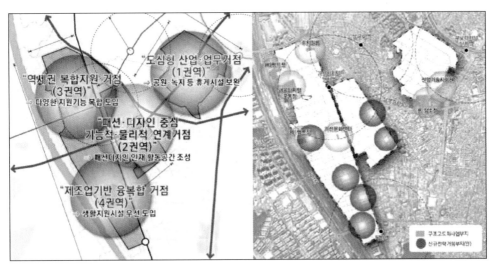

Establish a strategic hub

그림 8. 서울시 2030 산업단지 종합개발계획 (출처: 서울시) / Seoul's 2030 Industrial Areas Comprehensive Plan

도에 대한 설계 가이드라인에서 레이아웃, 경관, 외부의 통합설계를 통해 도시경관 관리를 구현하고, 공공영역의 질을 높이고, 서울시의 다른 첨단산업 단지와 경쟁할 계획이다.

최근 몇 년 동안 지방정부는 사회에 문화적 중요성의 레이어를 추가하여 G 밸리의 정체성

impressive addition is 'Museum G' located on the 2nd floor of a brand-new HQ for Korea's largest mobile game company, Netmarble. The development includes 39 story office tower and a podium with many public amenities including the Museum G. And it provides ample underground parking as well as large public space in the front as well as back side where a large public plaza can hold large public gatherings.

Apart from the obvious benefits of a new development, this real estate development signals a completely new stance on how the underused and underperforming strategic sites will be developed in the future. This development was a project finance venture project which formed a special purpose company between KICOX, Netmarble and other private investors to purchase an obsolete water treatment center within the complex

그림 9. Netmarble 본사 건물 입구 광장 모습 (출처: 저자) / Netmarble HQ Entrance Plaza

을 강화하기 위해 주요 공공시설을 추가했다. 가장 인상적인 것은 국내 최대 모바일 게임 기업인 넷마블의 본사 2층에 위치한 '뮤지엄 G'이다. 개발은 39층 오피스타워와 박물관을 포함한 많은 공공편의시설을 갖춘 연단을 포함한다. 그리고 넓은 지하주차장뿐만 아니라 전면에 큰 공공공간, 큰 공공광장이 대규모 공공집회를 개최할 수 있는 공간을 제공하였다.

새로운 개발에 따른 명백한 이점 외에도, 이 부동산 개발은 미래에 과소평가되거나 실적이 저조한 전략 사이트가 어떻게 개발될 것인지에 대한 완전히 새로운 입장을 신호한다. 이번 개발은 KICOX와 넷마블 등 개인투자자들이 복합단지 내 의외의 수처리센터를 매입하는 특별목적기업을 설립한 프로젝트 금융벤처사업이다. PFV는 부지를 매입하고 가장 인상적인 건물 중 하나를 건설할 뿐만 아니라 인근 근로자와 주민들에게 풍부한 공공편의시설을 제공했다. 이러한 유형의 공공 및 민간 파트너십은 설계지침과 규제 조정 이상을 요구하는 영역을 개편하는 건강한 방법임을 입증한다.

그림 10. 넷마블 본사 2층에 위치한 G 박물관 내부 모습 (출처: 저자)
Musuem G's Exhibition Located inside Netmarble Headquarter Building's 2nd Floor

맺으며

산업단지의 주요 성공요인으로는 기업의 집합, 네트워크 형성, 시장 접근성, 연구시설의 존재, 우수한 인력 확보의 용이성, 쾌적한 생활환경 등이 있다. 그 중에서도 재능 있는 전문가의

1. The PFV purchased the site and constructed one of the most impressive buildings as well as providing a wealth of public amenities to the workers and residents nearby. This type of public and private partnership proves to be a healthy way to revamp the areas that require more than just design guidelines and regulatory adjustments.

Closing Remarks

The key success factors of the industrial complex include the aggregation of enterprises, the formation of networks, market accessibility, the existence of research facilities, the ease of securing excellent workforce, and the pleasant living environment. Among them, the influx of talented professionals is an important success factor for industrial complexes. To attract talent, industrial complexes around the world are focusing on creating a residential environment that can enhance the quality of life (QWL) of workers under the existing concept of simple production-oriented occupancy.

As mentioned earlier, G-valley worked as an industrial city in the past but was renamed Guro Digital Complex and transformed into a mecca for high-tech digital industries centered on knowledge and information industries. In the process, chimney-type factories were removed, and apartment-type factories were built, but due to rapid development, there were underdeveloped places everywhere without considering the convenience and safety of residents. Also, there has been a big problem in the past that there is a lack of consideration between workers and residents and a lack of amenities.

The success of G Valley is still too early to judge; however, if the local governments and KICOX continue to show commitment to make G Valley relevant to the young generations and provide a chance to follow their dreams, the spirit of Guro Industry

유입은 산업단지의 중요한 성공요인이다. 인재 유치를 위해 전 세계 산업단지는 단순한 생산 중심의 점유라는 기존 개념하에 근로자의 삶의 질을 향상시킬 수 있는 주거환경을 조성하는 데 주력하고 있다.

앞서 언급했듯이 G 밸리는 과거에 산업단지에서 '서울디지털단지'로 명칭을 바꾸고 지식산 업과 정보산업을 중심으로 한 첨단디지털산업의 메카로 변모했다. 이 과정에서 굴뚝형 공장 을 철거하고 아파트형 공장이 건설되었지만 급속한 발전으로 인해 주민의 편의와 안전을 고려하지 않았고, 폐공장이 아직 서 있는 사유지가 곳곳에 있다. 또한 각각 사유지 별로 개발이 진행되면서, 주변 근로자와 주민에 대한 배려가 부족하고 편의시설이 부족하다는 큰 문제가 야기되었다. 이렇게 문제가 발견되면 적극적으로 사람 위주의 해결방법을 찾는 것이 매우 인상적이다.

G 밸리의 성공은 아직 판단하기에는 너무 이르다. 그러나 지방자치단체와 KICOX가 G 밸리를 젊은 세대와 관련성 있게 만들고 꿈을 이룰 수 있는 기회를 제공하겠다는 의지를 계속 보여준다면 구로산업단지의 정신은 계속 살아가고 국내에서 가장 유망하고 유망한 산업단지로서 계속 영향력을 발휘할 것이다. 도시 디자인이 세입자와 주민의 미래 요구에 귀를 기울인다면 G 밸리는 서울에서 일하고, 생활하고, 놀 수 있는 건강하고 재미있는 장소가 될 것이다.

참고문헌

1. G 밸리, 서울시 최대 융복합 산업단지로 재탄생…국가산업단지계획, 최초 변경 고시, 서울특별시, https://news.seoul.go.kr/economy/archives/514886, 2021

2. 안치용 외, 구로공단에서 G 밸리로 서울디지털산업단지 50년 50인의 사람들, 한스콘텐츠, 2014

3. 김묵한, 구로공단 그리고/혹은 G 밸리, 서울경제, 2015 한국산업단지공단, www.kicox.or.kr

4. 문석철, G 밸리의 발전과 구조고도화, 생생리포트, 4월호, 서울연구원, 2015

5. https://forum.betacity.center/

Complex will live on and continue to make an impact as the most advanced and most promising industry complex in the country. And if urban design listens to the needs and the future demands of the tenants and residents, G Valley will be a healthy and fun place to work, live and play in Seoul.

그림 11. 오늘날의 '가산디지털단지' 중심가로 전경, 옛 한국수출산업단지의 제3단지 (출처: 저자)
Main Avenue of 'Gasan Digital Industrial Complex' Today, Old 3rd Complex of Korea Export Industrial Complex

References

1. Ryu, Suha & Kim, Saehoon. Investigation of Urban Places in Seoul Digital Industrial Complex (G-Valley). IOP Conference Series: Earth and Environmental Science. 213. 012015. 10.1088/1755-1315/213/1/012015, 2018

2. https://seoulsolution.kr/en/content/urban-planning-news-guro-g-valley-transform-game-industry-center

3. https://www.investkorea.org/ik-en/bbs/i-2486/detail.do?ntt_sn=490772

4. https://forum.betacity.center/

Chapter 3

공장지의 산업재생과 수변의 변화

Transformation of Industrial Area and Waterfront Regeneration

도시가 발전하고 산업이 고도화되면서, 공장과 같은 산업시설이나 하수처리장, 빗물펌프장과 같은 도시기반시설의 도시 이전이 일반화되기 시작하였다. 더불어 산업화의 결과로 오염된 하천과 교통 인프라 스트럭처로 인하여 도시와 격리되었던 수변공간은 시민의 일상을 자연에 가까이 갈 수 있도록 만들어주는 수변공간으로 새롭게 주목받고 있다. 울산 태화강은 오염된 강을 치유하고 생태보호를 강조한 하천 보전 및 관리로의 패러다임 전환의 출발점이며, 시민 사회 중심으로 일어난 환경운동의 일환이었다. 서울 성수동과 마포 문화비축기지, 인천 코스모 40의 사례에서는 기존의 산업시설을 창조적으로 변화시켜 문화활동의 거점공간으로 활용하고 이를 통해서 인접한 지역의 성격이 강화되고 새로운 파급력을 가지는 사례들이다.

As cities developed and industries were advanced, urban relocation of industrial facilities such as factories and urban infrastructure such as sewage treatment plants and rainwater pumping stations began to become common. In addition, as a result of industrialization, rivers are polluted, and the waterfront space, which was isolated from the city due to the transportation infrastructure, is receiving new attention as a waterfront space that makes the daily life of citizens closer to nature. Ulsan Taehwa River was the starting point of a paradigm shift toward river conservation and management that emphasized the healing of polluted rivers and ecological protection, and was a part of the environmental movement centered on civil society. In the cases of Seongsu-dong, Mapo Cultural Oil Tank Culture Park, Incheon Cosmo 40, existing industrial facilities are creatively transformed and used as a base for cultural activities, and through this, the character of the adjacent area is strengthened and a new ripple effect is obtained.

울산 태화강, 공해도시의 변신

이재민 (울산대학교 건축학부 교수)

그림 1. 태화강대공원 / Taewhagang Grand Park

태화강 프로젝트는 중화학공업발전으로 성장한 생태계 오염에 대한 반작용으로 시정부, 시민사회, 전문가 그룹이 민관공동으로 도시의 성장과 생태의 공존을 꾀한 프로젝트이다. 태화강 프로젝트의 목적 및 의의는 크게 다음과 같이 네 가지로 구분할 수 있다. 첫째, 오염된 태화강에 하수종말처리장 증설 등 수질개선 사업을 추진하여 생태계가 복원되었다. 둘째, 오산 십리대밭 벌목에 대한 반작용으로 지역 전문가 그룹은 태화강 홍수통제와 십리대숲 존치에 대해 수리모형 검증 등 연구결과를 중심으로 십리대밭의 보존을 제안하였고 이는 공학 중심의 우리나라 하천정비계획수립방법에서 생태보호를 강조한 하천보전 및 관리에 대한 새로운 패러다임으로의 출발점이었다. 셋째, 환경오염 개선을 위한 '태화강 살기리 운동'과 함께 '태화들 한평 사기 운동' 등 시민사회 중심으로 환경정책의 방향을 결정지었다. 넷째, 시민들이 손쉽게 접근하고 활용

The TaeHwa-gang Transformation, Ulsan

Jaemin Lee (Professor, Ulsan University, Dept of Architecture)

The Tae-Hwa-Gang River Project, led by the city government, civil society, and expert groups, sought the coexistence of urban growth and ecology to control the environmental pollution caused by developing heavy and chemical industries. The purpose and significance of the Tae-Hwa-Gang River Project are: (i) The ecosystem would be restored by a water quality improvement project, including the expansion of sewage treatment plants along the polluted river. (ii) A group of local experts proposed to preserve the Simni Bamboo Forest (Sim-Ni-Dae-Bat) based on research on hydrologic model simulation for Tae-Hwa-Gang River flood control and the preservation of Simni Bamboo Forest, as a reaction to the removal of Simni Bamboo Forest in Osan. This was the starting point of a new paradigm for river conservation and management that emphasized ecological protection. (iii) Civil society-oriented environmental policy was established, such as the 'Tae-Hwa-Gang River Restoration Project' and the 'Buying 1 Pyeong of Tae-Hwa-Gang Field Movement', to improve environmental pollution. (iv) Waterfront parkland was created for citizens to easily access and utilize, to compensate for the insufficient urban park area of Ulsan City.

할 수 있는 수변공원 조성으로 울산시의 부족한 도심공원을 확충하는 계기가 되었다.

들어가며

태화강국가정원은 울산광역시 중심을 가로지르며 역사도심인 중구와 신도심인 남구의 경계로 울산시내로는 태화동, 삼호동, 삼산동, 야음동, 신정동, 중앙동, 학성동, 효문동, 상류로는 울주군의 범서읍, 언양읍 등의 행정구역에 걸쳐있다. 태화강국가정원 혹은 태화강대공원은 중심부 십리대숲이 위치한 태화들뿐 아니라 삼호동대숲 및 철새도래지, 상류의 생태하천구간, 하류의 고수부지구간 등 울산광역시의 주요 퍼블릭 스페이스를 일컫는다.

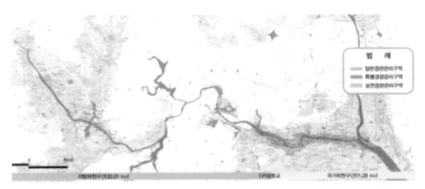

그림 2. 태화강의 위치 / Taewha River Location

시작 배경

산업도시의 탄생

언양에서 울산만, 방어진을 잇는 넓은 논밭의 울산읍은1961년 '국가재건기획위원회'에서 주요 공업지역으로 육성할 것을 결의하면서 대한민국의 대표적인 중화학공업도시로서 발전

Introduction

Tae-Hwa-Gang National Garden, located at the center of Ulsan Metropolitan City, is the boundary between Jung-gu, the city's historical area, and Nam-gu, the new downtown. Tae-Hwa-Gang National Garden refers to the major public spaces of Ulsan Metropolitan City, including Taehwa Field where a 4 km long bamboo forest (Sim-Ni-Dae-Bat), the center of the park, is located, as well as Sam-ho-dong Bamboo Forest and migratory bird sanctuary, the upstream ecological area, and the downstream river terrace area. The park spans Taehwa-dong, Samho-dong, Samsan-dong, Yaeum-dong, Shin-jeong-dong, Jung-ang-dong, Hak-seong-dong, and Hyo-mun-dong in the city, and the river's upstream is bounded by Beom-seo-eup and Eon-yang-eup in Ul-ju-gun.

Background

The Birth of the Industrial City

As the National Reconstruction Planning Committee decided to develop Ulsan-eup as a major industrial city in 1961, the extensive farmland connecting Eon-yang, Ulsan Bay, and Bang-eo-jin was developed into Korea's representative heavy chemical industrial city. Under Korea's First Five-Year Economic Development Plan announced in January 1962, Ulsan-eup was the first national industrial complex. On June 1, the status of the new industrial area was elevated from a township to a city, Ulsan City. In 1964, new factories emerged in the city, starting with Korea National Oil Corporation's Ulsan Refinery Plant, followed by other petrochemical plants, such as the Korea Fertilizer Plant, the world's largest urea fertilizer manufacturer, Dongyang Nylon.

하기 시작한다. 1962년 1월 '제1차 경제개발 5개년계획'에 의거하여 특정공업지구로 결정 공포한 이후 울산읍은 첫번째 국가산업단지로 지정되었고 이에 따라 6월 1일 울산시로 승격된다. 1964년 대한석유공사 울산정유공장을 시작으로 세계 최대의 요소비료공장인 한국비료공장, 동양나일론 등 석유화학계열의 공장이 들어서고 1968년 현대자동차 울산공장, 1974년 현대중공업이 만들어지면서 중화학공업단지를 기반으로 산업도시의 면모를 갖췄다. 이런 산업을 바탕으로 1970년대 27만명 인구의 작은 도시가 1980년대에는 50만명으로 인구가 급증하였고 1990년대는 100만명을 넘으며 대한민국의 대표적인 공업도시로서 성장하였다.

공해백화점

대규모 중화학공업과 급격한 인구 유입으로 인해 울산의 대기환경과 태화강 수질은 1990년대 중반까지 악화되었다. 1991년 공기 중 이산화황(SO_2)은 0.038ppm/년 으로 WHO 권고기준(0.019ppm/년) 보다 2배 높았으며 이산화질소(NO_2)는 0.025ppm/년으로 WHO 권고기준(0.021ppm/년)을 20% 가량 초과하였다. 태화강은 공업용수와 더불어 인구 증가로 인해 유입되는 생활하수와 오폐수로 인해 대한민국 하천 중 최하위 수준의 수질로 생물화학적산소요구량(BOD) 11.3mg/L를 보이며 공업용수는 물론 농업용수로도 사용이 불가능했다.

죽음의 강 : 태화강

봄이면 황어떼, 가을이면 연어떼가 산란을 하는 모래섬과 갈대숲을 자랑하던 태화강은 1970년대까지는 자정작용으로 어느 정도 깨끗한 수질을 유지하였으나 이후 태화강의 환경생태용량(Ecological Carrying Capacity)를 초과하면서 수질오염으로 인해 1990년대부터 '죽음의 강'이라는 별명을 가지게 된다. 갈수기에 하천유지용수와 용존산소 부족으로 1992년에는 다섯 차례의 물고기 떼죽음을 가져왔고 1995년에는 3개월 동안 녹조가 지속되었고 2000년에도 숭어 떼죽음 등 태화강의 오염으로 인해 수생 생태계는 물론 왜가리, 흰뺨검둥오리 등의 철새생태계도 황폐화되었다.

With Hyundai Motor's Ulsan Plant in 1968 and Hyundai Heavy Industries in 1974, it has become an industrial city based on the heavy and chemical industrial complex. Based on these industries, the city's population has rapidly increased from 270,000 in the 1970s to 500,000 in the 1980s. In the 1990s, the number exceeded 1 million, developing into Korea's representative industrial city.

Pollution Variety Pack

The large-scale heavy and chemical industry and rapid population influx worsened air quality in Ulsan and the water quality of the Tae-Hwa-Gang River until the mid-1990s. In 1991, sulfur dioxide (SO_2) content in the air was 0.038 ppm/year, which was twice as high as the WHO recommendation (of 0.019 ppm/year), while nitrogen dioxide (NO_2) was 0.025 ppm/year, which exceeded the WHO recommendation (of 0.021 ppm/year) by approx. 20 %. The biochemical oxygen demand (BOD) of the Tae-Hwa-Gang River was 11.3 mg/L, the lowest among rivers in Korea because of industrial water and the sewage and wastewater flow due to population growth, making it impossible to use for industrial, as well as agricultural purposes.

River of Death : Tae-Hwa-Gang River

Until 1970, Tae-Hwa-Gang River boasted sand islands and reed forests where dace spawned in spring and salmon bred in the fall, as the water quality was maintained by self-purification of the river. However, since the 1990s, water pollution exceeded the ecological carrying capacity of the river, which gave it the nickname of the 'River of Death'. In 1992, lack of water and dissolved oxygen in the dry season caused five massive fish death events; in 1995, algae continued for three months; and in 2000, mullet died en masse. The pollution also destroyed the ecosystem of migratory birds,

그림 3. 물고기 떼죽음(1995년), 중류 부영양화(1996년) / River Pollution (1995, 1996)

1987 태화강 하천정비계획

환경오염으로 인해 태화강은 생태학적 가치를 잃어버렸다. 생태계가 무너진 태화강은 수리 공학자에겐 도시의 홍수를 방지할 수 있는 수로에 불과했고, 이는 1987년 '태화강 하천정비계

그림 4. 태화강과 십리대숲(1960년대) / Taewha River and Simni Bamboo Forest (1960s)

such as heron and spoonbill ducks.

1987 Tae-Hwa-Gang River Refurbishing Plan

Environmental pollution destroyed the ecological value of the Tae-Hwa-Gang River. For hydrologic engineers, the river was only a waterway to prevent flooding in the city. This is clearly shown in the 1987 Tae-Hwa-Gang River Refurbishing Plan. The Ministry of Construction planned to remove the 4 km long Osan Simni Bamboo Forest, which has grown wild since the Silla Dynasty, and build an embankment, based on the river law regulating that 'no perennial trees taller than 1 meter can be placed in the river zone'. Based on this plan, the Busan Land Management Administration excluded the Simni Bamboo Forest and Tae-Hwa-deul areas, where private and state land were mixed at the time, from the river site, and designated them as a residential zone.

Tae-Hwa-Gang River Restoration

To respond to Tae-Hwa-Gang River pollution due to industrialization and the bamboo forest removal plan per the 1987 Tae-Hwa-Gang River Refurbishing Plan, Tae-Hwa-Gang River Conservation Society, a resident volunteer group, was established in 1989. Since 1990, the society held symposia with the theme of 'Let's Save the Tae-Hwa-Gang River', and 'Directions to Save the Tae-Hwa-Gang River', to encourage citizens to participate in the Tae-Hwa-Gang River Restoration Movement.

Flood Management Plan - Removal of Bamboo Forests, Embankment Construction, Dredging of River Basin, Change in Land Use

Despite the opposition of the citizens, the Ministry of Construction and Transportation announced the Tae-Hwa-Gang River Refurbishing Implementation Plan in June 1994

획'에서 잘 나타난다. 건설부는 '하천구역 내에 1m 이상의 다년생 수목은 둘 수 없다'는 하천법 규정을 기반으로 신라시대부터 자생해온 오산 십리대숲을 제거하고 제방을 축조하는 것을 계획했다. 이 계획을 기반으로 부산국토관리청은 당시 사유지와 국유지가 혼재한 십리대숲과 태화들 구역을 하천부지에서 제외하여 주거지로서 개발이 가능하도록 지정하였다.

태화강 살리기

산업화로 인한 태화강 오염, 1987년 태화강 하천 정비계획에 의한 대숲 벌목계획 등에 대응하기 위해 주민자원봉사단체인 태화강 보전회가 1989에 만들어지고 1990년부터 '태화강 살리자', '태화강 살리기의 방향' 등 심포지엄을 개최하여 시민들의 '태화강 살리기운동' 참여를 독려했다.

홍수관리계획 - 대숲 벌목, 제방 축조, 하천 단면확대, 주거지 용도변경

시민들의 반대에도 불구하고 건설교통부는 태화강의 홍수 시 유수 소통 개선을 위해 대숲벌목, 제방축조, 하천 단면확대를 주요 골자로 한 '태화강 하천정비 실시계획'을 1994년 6월 발표하고 울산시는 건교부의 계획에 의해 제방선 안쪽의 태화들 일부분 18만 6,000m²를 자연녹지지역에서 주거지역으로 도시계획 변경했다. 변경된 18만 6,000m²는 대부분 사유지였으며 주거지역으로 고시된 이후 지주들이 조합을 결성하고 아파트 개발에 나서면서 태화들의 개발과 보존에서 시민단체와 갈등하기 시작했다.

그림 5. 십리대숲 산책로 / Walking Trail in Simni Bamboo Forest

to cut down the bamboo forest, construct embankment, and dredge the base of the river, to improve water flow in the event of flooding of the Tae-Hwa-Gang River. The plan changed some 186,000 m² inside the embankment from a natural green zone to a residential zone. The area changing was primarily private land. After the area was announced as a residential zone, the landlords formed a cooperative and started constructing apartments, which deepened the conflict with civic groups.

The Tae-Hwa-Gang River Simni Bamboo Forest Conservation Movement

Along with the Tae-Hwa-Gang River Conservation Society, local media, academia, and civic groups started a more active and systematic 'Tae-Hwa-Gang River Simni Bamboo Forest Conservation Movement'. In November 1994, five months after the 'Tae-Hwa-Gang River Refurbishing Plan' was announced, they held a 'Symposium for the Conservation of the Tae-Hwa-Gang River Bamboo Forest'. At the symposium, experts insisted that even if bamboo forests were preserved, the rise of the Tae-Hwa-Gang River flood level would have negligible effect. Air pollution purification and ecosystem conservation were significantly affected in addition to flood control. By collecting citizen opinion, such as a public campaign, a campaign to sign 70,000 citizens, and a petition to the Ulsan City Council, in October 1995 they urged the Ministry of Construction and Transportation to cancel the Simni Bamboo Forest removal plan. In November of the same year, the Ministry, in response to public opposition, decided to suspend management of the Simni Bamboo Forest. This change preserved the bamboo forest, and led to legislation to allow trees to be planted within a floodplain (UMC 2014: 320).

태화강 십리대숲 보전운동

태화강 보전회와 더불어 지역의 언론계, 학계, 사회단체는 '태화강 살리기'에서 더욱 적극적이고 체계적인 '태화강 십리대숲 보전운동'을 시작했다. '태화강 하천정비실시계획'이 발표되고 5개월 후 1994년 11월에 '태화강 대숲보전을 위한 심포지엄'을 개최했다. 심포지엄에서 전문가들은 대숲을 존치해도 태화강 홍수위 상승에는 미미한 영향을 미치며 홍수통제 외에도 대기오염정화와 생태계 보전에 큰 영향을 미친다고 주장하였다. 이를 바탕으로 대시민 캠페인, 7만명 시민서명운동, 울산시의회 청원 등 전방위적인 시민의견을 취합하여 1995년 10월 건교부에 십리대숲 벌목계획을 폐기할 것을 촉구하였다. 시민의 반대에 건교부는 같은 해 11월 십리대숲 정리를 유보하기로 방침을 정하고 이를 계기로 하천법을 개정하여 하천 둔치에 나무를 심을 수 있게 되었다.

공학중심 홍수통제 하천관리의 제고

건교부의 결정에 부산지방국토관리청은 대나무숲을 존치하도록 결정하였으나 대숲을 유지하면서 홍수위험을 저감시키는 방안이 필요했다. 1999년부터 울산대학교와 울산과학대학의 연구진의 주도로 태화강 수리모형을 제작하고 실험을 통해 제방축조 위치를 변경할 것을 건의했다.

수로굴착계획의 제고

제방위치 변경, 대나무숲 보존, 1994년 용도 변경된 18만 6,000m²의 사유지의 홍수대비라는 새로운 과제에 2003년 부산지방국토관리청은 재정비계획안에서 계획홍수량 확보를 위해 제안된 폭 80m, 깊이 7m, 길이 1.3km의 대규모 수로굴착을 제안한다. 하지만 울산대의 연구진은 수리모형을 기반으로 '하도분담 및 제방방식'의 치수계획은 태화강의 생태경관을 훼손하고 대형수로 설치는 기존 수로의 퇴적현상을 가속시켜 홍수예방효과는 제한적이라는 이유로 태화들 전체를 보전하는 방향으로 태화강 하천정비계획 변경을 주장하였다.

2003년 부산지방국토관리청의 대규모 수로굴착은 1994년에 용도변경된 주거지역을 다시

Improvement of Engineering-oriented Flood Control and River Management

In response to the decision of the Ministry of Construction and Transportation, the Busan Regional Land Management Office determined to preserve the bamboo forest; a measure to reduce the risk of flooding while maintaining the bamboo forest was required. Under the leadership of researchers at the University of Ulsan and Ulsan College in 1990, the Tae-Hwa-Gang River hydrologic model was made. Based on the hydrologic experiments, they proposed changing the embankment construction location.

Improvement of Waterway Excavation Plan

In 2003, the Busan Regional Land Management Office proposed an 80 m wide, 7 m deep, and 1.3 km long large-scale waterway to secure flood prevention capacity. However, based on the hydrologic model, researchers at the University of Ulsan opposed the flood channel in the middle of Tae-Hwa-Deul Field, on the grounds that it was more likely to damage the ecological landscape of the Tae-Hwa-Gang River; also, the flood channel would accelerate the accumulation of soil deposits in the existing river, which might limit the flood prevention effect. They argued for the Tae-Hwa-Gang River Refurbishing Plan to be changed to preserve the whole Tae-Hwa Field.

The large-scale waterway excavation by the Busan Regional Land Management Office in 2003 was a solution created because of complaints and compensation problems that might occur when a residential area whose use was changed in 1994 was returned to a river zone. After 186,000 m² of Taehwa Field was announced as a residential area in 1994, landlords formed a cooperative for apartment construction. There were intensified conflicts with civic groups trying to preserve the waterfront space of the Tae-Hwa-Gang River. Experts and civic groups oppose the construction

하천구역으로 환원할 경우 발생하는 민원 및 보상문제 때문에 만들어진 해결책이었다. 1994년 태화들 18만 6,000m²이 주거지역으로 고시된 이후 지주들은 아파트 개발을 위한 조합을 결성했으며 태화강의 수변공간을 보존하려는 시민단체와의 갈등이 격화됐다. 전문가와 시민단체는 태화들이 주거지역으로 개발될 경우 생태경관 훼손과 새로운 주거지의 홍수 위험이 높다는 이유로 아파트 건설을 반대했다. 이에 더해 시민단체는 태화들 주거지역 용도변경이 석연치 않다고 문제를 제기해 2001년에 이 건에 대해 시민감사권을 발동하고 3년간 개발을 금지하였다.

태화강대공원 조성사업 1단계 (2002~2004)

전문가와 시민단체의 '태화강 살리기운동'이 사유지 개발과 팽팽하게 대척점에서 줄타기를 하고 있는 동안에 태화강을 생태공원으로 변신시키기 위한 노력은 천천히 시작되었다. 먼저 울산시는 1999년 환경부 국비지원을 받아 태화들에 위치한 비닐하우스와 3,500톤의 쓰레기를 제거하고 생태공원으로 조성하는 사업에 지원하여 선정되었다. 울산시는 사업비 38억원으로 태화 및 삼호섬 지구 145,609m²에 십리대숲을 정비하고 자연학습원에 관목류 2만 그루와 초화류 12만 포기를 식재하고 조류 관찰데크, 파고라, 화장실 등 편의시설을 설치했다.

'태화들 한평사기운동' (2004. 5. 27)

태화강대공원 1단계사업 진행과 함께 태화강 보전회 주도로 태화들은 2002년 박맹우 시장이 당선되면서 새로운 국면으로 들어섰다. 건설교통부는 시민의 거센 반발로 태화들 주거지역의 일부만 택지로 개발하는 중재안을 제시했으나 받아들여지지 않았다. 태화강 보전회 등 시민단체와 전문가들은 2004년 태화들 일대를 생태공원으로 보전하기 위해 자발적 성금을 모아 태화들 일대의 사유지를 구입하자는 목적의 '태화들 한평사기운동'을 전개하였다.

of apartments, as developing a residential area in Tae-Hwa Field may damage the ecological landscape and increase the risk of flooding. In addition, a civic group raised the issue that the change of land use was unreasonable. In 2001, according to the citizens' audit rights, land development was banned for three years.

Tae-Hwa-Gang Grand Park Construction Project Phase 1 (2002-2004)

Efforts to transform the Tae-Hwa-Gang River into an ecological park started. The experts and civic groups' Tae-Hwa-Gang River Restoration Movement began a tug-of-war with private land development. First in 1999, Ulsan received financial support from the Ministry of Environment to remove 3,500 tons of garbage and a plastic house located in Taehwa Field, and create an ecological park. Ulsan built the Simni Bamboo Forest on 145,609 m^2 of Tae-Hwa and Sam-ho Island with KRW 3.8 billion, planted 20,000 shrubs and 120,000 plants in the Nature Learning Center, and installed a bird observation deck, pergola, and toilets.

The 'Buying 1 Pyeong of Tae-Hwa-Gang Field' Campaign (May 27, 2004)

With the Tae-Hwa-Gang Grand Park Phase 1 project, the situation changed into a new phase with the election of Mayor Park Maeng Wu in 2002, supported by the Tae-Hwa-Gang River Conservation Society. The Ministry of Construction and Transportation proposed an arbitration plan: developing only a part of the Tae-Hwa Field residential area as residential land, due to strong protests from citizens. The plan was not accepted. In 2004, civic groups and experts, including the Tae-Hwa-Gang River Conservation Society, launched the 'Buying 1 Pyeong of Tae-Hwa-Gang Field' campaign, collecting voluntary donations to preserve the Tae-Hwa Field area as an ecological park.

그림 6. 2006년 태화들 / Taewha Field (2006)

계획과정

'Ecopolis 울산계획' '에코폴리스 울산선언'

울산광역시에서 주요 행정을 담당하던 박맹우 시장은 공해문제 해결이 울산 성장의 발판이라는 판단에 당선 후부터 태화강 수질 개선 등의 사업을 시작했다. 2003년에는 울산발전연구원과 서울대학교 환경생태연구실의 공동연구로 세계의 생태도시계획을 조사하여 'Ecopolis 울산계획'을 출간하며 울산에 맞는 생태도시를 제안했다. 박맹우 시장은 이 계획에서 제시한 110개의 전략계획을 바탕으로 '에코폴리스 울산선언' (2004. 6. 9)을 통해 공업도시에서 생태도시로 발돋움하려는 울산의 미래 비전을 발표했다. 첫 번째 사업으로 태화강의 수질을 개선하고 수변공간을 확충하는 등 2005년에는 태화강 마스터플랜을 수립하였다.

태화강 마스터플랜 (2005)

태화강대공원 1단계 조성사업과 박맹우 시장의 에코폴리스 선언으로 태화강은 본래의 생태하천과 시민공원으로 복원될 수 있는 기회를 찾았다. 울산시는 환경오염을 해결하기 위해 환경보전중기종합계획(1차 1999~2003, 2차 2004~2008)을 수립하고 대기관리 및 수질개

Process of Planning

The 'Ecopolis Ulsan Plan' and the 'Ecopolis Ulsan Declaration'

After election, Park Maeng Wu, the mayor of Ulsan Metropolitan City, started several projects, including water quality improvement in the Tae-Hwa-Gang River. He considered the river to be a steppingstone for Ulsan's growth. In 2003, the mayor published the 'Ecopolis Ulsan Plan', after studying the world's ecocity plan initiated by the Ulsan Development Institute and Environmental Ecology Research Department at Seoul National University. Based on the 110 strategic plans presented in the plan, mayor Park Maeng Wu announced the 'Ecopolis Ulsan Declaration', the future vision of Ulsan to develop from an industrial city to an ecological city (June 6, 2004). For the first project, the Tae-Hwa-Gang River Master Plan was established in 2005 to improve the river's water quality and expand the waterfront space.

Tae-Hwa-Gang River Master Plan (2005)

Due to phase 1 of the Tae-Hwa-Gang Grand Park project and Mayor Park's declaration of Ecopolis, the Tae-Hwa-Gang River presented an opportunity to be restored as an ecological river and citizen park. The government of Ulsan established a mid-term environmental conservation plan (1st, 1999–2003; 2nd, 2004–2008) to address environmental pollution issues. As part of the plan, the water quality of the Tae-Hwa-Gang River has been improved with newly built or expanded sewage treatment facilities to restore natural ecosystems through air management and water quality improvement. The Tae-Hwa-Gang River Master Plan paved the way for restoring the Tae-Hwa-Gang River by proposing to improve water quality, such as ecosystem restoration, and discover and provide historical and cultural resources,

선을 통한 자연생태계 복원을 목표로 하수종말처리시설 신설 및 증설 등을 통해 태화강의 수질환경을 개선해왔다. 태화강 마스터플랜에서는 수질환경 개선뿐 아니라 수질개선을 통한 생태계 복원, 자전거 도로 확보, 수변공간 확충, 태화루 복원 등 역사문화자원을 발굴하는 사업을 제안하여 태화강 살리기의 기틀을 마련했다.

2005년 9월 중앙하천관리위원회는 울산시, 시민사회, 전문가의 적극적인 '태화강 살리기운동'에 동참하여 1994년에 자연녹지지역에서 주거지역으로 용도 변경된 태화들 18만 6,000m²을 다시 주거지역에서 하천구역으로 편입하였다. 이후 추진된 태화강대공원 조성사업 2단계의 주요 사업은 사유지 18만 6,000m²를 하천구역으로 편입하는 후속조치였다. 2004년부터 '태화강 한평사기운동'으로 시민들의 관심은 증대되었지만 시민주도의 성금 모금은 실질적으로 토지를 구입하기에는 부족했고 하천구역 편입을 위한 재정은 중앙정부와 울산시에서 분담하였다. 평당 200만원의 시가로 국비 727억원과 시비 273억원을 조달하여 총 1,000억원을 들여 토지 및 지장물 보상했다.

태화강대공원 조성사업 2단계 (2007~2010)

울산시는 2005년 태화강 마스터플랜을 기초로 2007년부터 2010년까지 태화강대공원 조성사업 2단계를 실행했다. 태화들 주변에 있는 각종 폐기물, 생활쓰레기, 폐건설자재 등 총 3,349톤의 폐기물을 수거하고 새로 하천구역으로 편입된 태화들을 생태공원으로 조성하기 위해 2007년 10월부터 2008년 5월까지 수성엔지니어링 주도로 '태화강 생태공원조성 (2단계) 기본계획 및 기본설계 용역'을 수립했다.

태화강대공원 기본설계

수성엔지니어링은 '태화강의 아름다움', '울산의 상징인 십리대숲', '울산의 균형발전', '시민의 행복추구'의 주요 개념을 가지고 대숲공원을 보존·복원하고 생태공원을 조성하며 태화강의 역사성을 회복할 수 있는 친수공간을 제공하여 지속적인 생태계회복 및 경관관리를 위한 기본설계안을 마련했다. 태화들의 넓은 공간을 생태보전, 생태체험, 역사문화, 문화예술의 테

such as building bicycle trails, expanding waterfront space, and restoring the Tae-Hwa-Ru Pavilion.

In September 2005, civil society, and experts, the Central River Management Committee, reverted the land use designation of 186,000 m^2 of Tae-Hwa Field, which in 1994 had been changed from a natural green zone to a residential zone (Bae, S.H. 2010). The primary content of the second phase of the Tae-Hwa-Gang Grand Park development project was a follow-up measure to incorporate 186,000 m^2 of private land into the river zone. Even though public interest has increased with the 'Buying 1 Pyeong of Tae-Hwa-Gang Field' campaign since 2004, citizen-led fundraising was not sufficient to purchase the land. The central government and Ulsan City paid the expense to change the area to a river zone. At a market price of 2 million won per pyeong, 72.7 billion won from the central government, and 27.3 billion won from the city government, which makes a total of 100 billion won, was paid to compensate for land and obstructions.

Tae-Hwa-Gang Grand Park Construction Project Phase 2 (2007-2010)

Based on the 2005 Tae-Hwa-Gang River Master Plan, Ulsan implemented the 2nd phase of the Tae-Hwa-Gang Grand Park Project from 2007 to 2010. To collect a total of 3,349 tons of waste, including household waste and scrap building materials around Tae-Hwa Field, and develop the newly incorporated Tae-Hwa Field as an ecological park, Soo-sung Engineering & Consulting led the establishment of the Tae-Hwa-Gang River Ecological Park Creation (Phase 2) Basic Plan and Basic Design from October 2007 to May 2008.

그림 7. 태화강대공원 내 조성된 실개천 / A Naturalized Flood Channel in Taewhagang Grand Park

마로 4개 구역으로 나누어 기존의 대숲과 태화강국가정원길까지 생태원, 초화원, 소풍마당, 시민정원, 나비생태원 등의 프로그램을 공원 내에 조성했다. 특히 2003년에 제안되었던 폭 80미터의 대수로 위치에 '실개천'을 조성하여 기존에 제안된 홍수저감 효과는 유지하도록 하고 수로 대신 소프트엣지 등 생태적 요소를 도입하여 주변의 녹지, 정원과 조화를 이룰 수 있는 친수공간을 조성했다.

습지식물원으로서 가치가 뛰어난 태화강대공원의 장점을 바탕으로 계절의 연속성을 가지고 다양한 수종을 도입하여 자연스러운 식재림을 형성하는 것이 식재계획의 주요 요소이다. 봄에

Basic Design of Tae-Hwa-Gang Grand Park

Soo-Sung Engineering & Consulting prepared a master plan to achieve ecosystem recovery and landscape management to provide a waterfront area. The plan aims to achieve several goals, including preserving Simni Bamboo Forest Park, creating an ecological park, and restoring the rich history of the Tae-Hwa-Gang River. The main concepts of the plan were four-fold: 'the beauty of Tae-Hwa-Gang River', 'Simni Bamboo Forest, the symbol of Ulsan', 'balanced development of Ulsan', and 'citizen's happiness'. The park's programs, including an ecology garden, a flower garden, a picnic garden, a citizen's garden, and a butterfly ecology garden, have been created in the park from the existing bamboo forest to the Tae-Hwa-Gang River National Garden trail. The landscape designers divided the Tae-Hwa Field into four zones, with each area fitted-out with the themes of ecological conservation, ecological experience, history and culture, and culture and art. In particular, the landscape architects replaced the 80 meter wide hard-edged flood channel into an ecological stream furnished with a soft edge that can harmonize with the surrounding greenery and gardens.

The Tae-Hwa-Gang Grand Park has great potential to become a wetland botanical garden. Therefore, the landscape plan introduces a variety of species of trees with seasonal features to form a natural plantation forest. The plan intended to plant blady grass, loosestrife, and iris in spring, sagebrush, golden iris, Russian iris, cattail, big bog bulrush, daylily, and rugose rose in summer, and amur silver grass and soft-stem bulrush in autumn.

Tae-Hwa-Gang Grand Park was established to connect with the surrounding city area and the entire Tae-Hwa-Gang River. The existing embankment road was the boundary between Tae-Hwa-Gang Grand Park and Tae-Hwa-dong. A four-lane road was reduced to two lanes, and cherry blossom trees were planted to attract people

는 띠, 부처꽃, 붓꽃, 창포, 여름에는 갯지치, 금불초, 꽃창포, 무늬부들, 송이고랭이, 원추리, 해당화, 가을에는 갯개취미, 물억새, 큰고랭이 등으로 대공원 내 정원의 녹화계획을 수립했다.

태화강대공원은 주변 시가지, 태화강변 전체와 유기적인 연결을 목표로 동선계획을 수립하였다. 기존의 제방도로는 왕복 4차선으로 태화강대공원과 태화동의 경계를 긋는 역할을 수행하였지만 공원으로 진입을 유도하고 공원 경계의 산책로를 조성하기 위해 왕복 2차선으로 축소하고 보행로 확장구간에 벚나무를 식재하여 벚꽃길을 조성하는 등 공원과 보행자 친화공간을 만들었다. 또한 태화강 자전거 계획과 연계하여 남북으로 태화강변의 자전거길, 보행로와 직접 연결하는 도로망을 구성했으며 새로운 십리대밭길을 건설하여 태화강의 반대편 신정동까지 연결하여 보행 네트워크를 조성하였다.

2010년 8월 2단계 태화강대공원 조성사업이 완료되어 현재 국가정원으로 발돋움할 수 있는 기틀을 만들었다.

이해 관계 그룹

'죽음의 강'에서 태화강국가정원으로 변신은 한두 사람의 계획가 혹은 정치인의 리더쉽으로 이루어질 수 있는 것이 아니라 울산시민, 시민단체, 전문가, 정치인의 총체적이 염원과 행동이 바탕이 되었기에 가능한 일이다. 이 섹션에서는 '태화강 살리기', '십리대숲 보존운동', '태화강 한평사기운동', 제방선 변경 및 대수로 굴착안 반대, 태화들 사유지 하천구역 편입 등에 주도적으로 참여한 Stakeholder를 정리하겠다.

환경연합 - 울산 공해추방연합, 울산환경연합

울산은 공업도시로 발전하면서 발생한 경제적 이득과 함께 환경오염에 의한 피해로 시민사회 주도의 환경운동이 자생적으로 발전한 도시이다. 1974년에 온산이 산업기지로 개발되면서 고려아연, 효성알미늄, 한이석유, 온산 동제련 등의 공장이 들어서고 대기오염과 중금속

into the park, which provided a new pedestrian promenade along the park. In addition, in connection with the Tae-Hwa-Gang River bicycle plan, a trail network was formed to directly connect the bicycle trails and pedestrian paths along the Tae-Hwa-Gang River from north to south, and a new Simnidaebat-gil was constructed to connect to Sin-jeong-dong on the opposite side of the Tae-Hwa-Gang River to create a pedestrian network.

In August 2010, phase 2 of the Tae-Hwa-Gang Grand Park Construction Project was completed, and the foundation laid for the current national garden.

Stakeholders

The transformation from the 'River of Death' to the Tae-Hwa-Gang River National Garden could not be accomplished by the leadership of just one or two planners or politicians. It was possible because of the aspirations and actions of Ulsan citizens, civic groups, experts, and politicians, acting as a whole. In this section, the stakeholders who took the lead in the 'Tae-Hwa-Gang River Restoration Project', 'Simni Bamboo Forest Conservation Movement', 'Buying 1 Pyeong of Tae-Hwa-Gang Field Movement', opposing the change of embankment lines and excavation of large waterways, and incorporation of private land Tae-Hwa Field into the river zone were as follow:

Environment Federation - Ulsan Pollution Expulsion Federation and Ulsan Federation for the Environment

Civil society-led environmental movements developed spontaneously in Ulsan due to the economic benefits of development into an industrial city and the damage caused

폐수로 인해 1982년부터 온산 주민 사이의 집단적인 질병 발병에 대해 반공해 등 환경운동이 시작되었다. 이를 계기로 1989년부터 1993년까지 울산공해추방운동연합은 울산공업단지와 온산공업단지에서 배출되는 대기, 수질, 토양오염 문제에 대한 환경운동을 시작했다. 울산 지역문제에 대한 해결만을 관심으로 두었던 공해추방연합은 1993년부터 전국적으로 연대할 필요를 바탕으로 울산환경연합으로 재구성하여 조직적인 환경운동을 진행하기 시작하였다. 울산환경연합의 주요 사업은 공추연과 크게 다르지 않았으며 수질개선, 대기환경 개선, 폐기물 정책, 생태계 보전운동, 환경교육사업, 연구 및 발간사업 등으로 울산환경운동의 구심점으로 활동했다.

시민단체 - 태화강 보전회

1964년부터 대한석유공사 울산공장 등 산업시설이 가동되면서 오염되기 시작한 태화강을 보전하고 생태계를 복원하며 개발을 억제하기 위해 주민자원봉사단체인 '태화강보전회'가 1989년 11월 25일에 구성됐다. 1987년 태화강 하천정비계획에 의한 오산 십리대숲 벌목에 반발하여 태화강보전회는 대대적인 보존운동을 시작했으며 1990년부터 태화강 보존을 주제로 심포지엄을 개최하였다고 공해추방연합 및 환경연합과 함께 1992년부터 태화강 대숲의 원형을 보존하기 위해 '태화강 대숲 지키기 운동'을 전개하였다.

태화강보전회는 2004년 태화들 일대를 생태공원으로 보전하기 위해 자발적 성금을 모아 태화들 일대의 사유지를 구입하자는 목적의 '태화들 한평사기운동'을 비롯한 울산 태화들 내 서널 트러스트 운동을 개진하였으며 자연문화유산 보전 심포지엄을 개최하였다.

학계 전문가 - 울산대학교, 울산과학대학교

울산의 환경운동단체 및 시민단체 외에도 울산의 수리학 전문가의 주도로 태화강의 역사와 생태를 보전하는 데 기여했다. 울산과학대학교 공간디자인학부 이수식 교수는 1994년 홍수 영향으로 십리대숲 벌목을 결정했을 때 'HEC-Ⅱ 프로그램'을 이용해 대숲의 존재가 태화강 홍

by environmental pollution. As Onsan was developed as an industrial base in 1974, factories, such as Korea Zinc, Hyosung Aluminum, Hani Petroleum, and Onsan Copper Smelter, were built in the area. As from 1982, disease due to air pollution and heavy metal wastewater occurred among the residents of Onsan, environmental movements, such as anti-pollution, began. From 1989 to 1993, the Ulsan Pollution Expulsion Federation launched an environmental campaign against air, water, and soil pollution from the Ulsan Industrial Complex and Onsan Industrial Complex. In 1993, the Pollution Expulsion Federation, which was concerned only with solving local problems, was reorganized as the Ulsan Federation for the Environment because of the need for national solidarity, and started an organized environmental movement. The major projects of the Ulsan Federation for the Environment were not much different from those of the Pollution Expulsion Federation, and they played a pivotal role in the Ulsan environment movement, such as water quality improvement, air quality improvement, waste policy, ecosystem conservation movement, environmental education project, and research project.

Civic Groups - the Tae-Hwa-Gang River Conservation Society

The Ulsan Plant of Korea National Oil Corporation began operation in 1964. A group of Ulsan residents established the Tae-Hwa-Gang River Conservation Society on November 25, 1989, to conserve the Tae-Hwa-Gang River, which was becoming polluted by the industrial facilities, to restore the ecosystem and to restrain development. In response to the removal plan of Simni Bamboo Forest in Osan by the Tae-Hwa-Gang River Refurbishing Plan in 1987, the Tae-Hwa-Gang River Conservation Society started a large-scale conservation movement. It began to hold symposia on Tae-Hwa-Gang River conservation from 1990. In 1992, together with the Pollution

수위에 미치는 영향이 미미하다는 결론을 내렸다. 이수식 교수는 2007년부터 태화강보전회의 회장으로 태화강대공원 조성에 앞장섰다.

울산대학교 건설환경공학부 조홍제 교수는 태화강 홍수통제와 십리대숲 존치에 대해 연구하면서 우리나라 하천정비계획 수립과 하천보전 및 관리에 대한 새로운 방향을 제시하였다. 1997년 대숲 존치에 따른 홍수위험 가중에 대한 논문을 시작으로 1999년부터 수리모형을 제작하여 태화강의 제방위치를 변경하면 홍수 위험이 감소한다는 것을 증명해냈다. 또한 2003년 부산지방국토관리청이 홍수통제목적으로 태화들 내 제안한 폭 80m, 깊이 7m, 길이 1.3km의 대규모 수로굴착 사업을 수리모형실험을 바탕으로 반대해 태화강 수변환경과 생태를 보존했다.

정치인 - 울산광역시장 박맹우

태화강처럼 대규모 도시설계 프로젝트가 현실화되는 것은 정책결정자의 과감한 결단 없이는 불가능하다. 민선 제3기 박맹우 시장은 1997년부터 울산광역시에서 기획조정실장, 내무국장, 건설교통국장 등 도시 전반에 대해 이해하고 있었으며 태화강 살리기에 대한 시민의 여망에 공감했다. 이에 2002년 당선 후 태화강 수질환경 개선을 위한 사업을 계획하고 실천했

그림 8. 태화강 국가정원 / Taewhagang National Garden

Removal Expulsion Federation and the Ulsan Federation for the Environment, the 'Tae-Hwa-Gang River Bamboo Forest Protection Movement' was launched to preserve the original Tae-Hwa-Gang River bamboo forest.

In 2004, the Tae-Hwa-Gang River Conservation Society launched the Ulsan Taehwa Field National Trust Movement, including the 'Buying 1 Pyeong of Tae-Hwa-Gang Field Movement' campaign to collect voluntary donations and purchase private land in the Taehwa Field area for the purpose of preservation of the Taehwa Field area as an ecological park. They also held symposia on heritage conservation.

Academic Experts - the Role of the University of Ulsan and Ulsan College

Along with Ulsan's environmental activists and civic groups, Ulsan's hydrologic experts contributed to preserving the history and ecology of the Tae-Hwa-Gang River. Based on the simulation using HEC-II program, Professor Lee Susik of the Department of Spatial Design at Ulsan College concluded that the bamboo forest had little effect on the floodplain of the Tae-Hwa-Gang River (Jeong, J.R. 2009). Based on his contribution, Professor Lee has served as the Tae-Hwa-Gang River Conservation Society president since 2007, and contributed to transforming Tae-Hwa-Gang Grand Park.

Professor Cho Hong Je of the Department of Civil and Environmental Engineering at the University of Ulsan researched Tae-Hwa-Gang River flood control and the preservation of Simni Bamboo Forest. Based on that research, he suggested a new direction for establishing Korea's river refurbishing plans and river conservation and management. Starting with a dissertation on the aggravation of flood risk due to the preservation of bamboo forests in 1997, he has since 1999 built hydrologic models to prove that changing the location of the Tae-Hwa-Gang River embankment reduces the flood risk. Also, in 2003, based on a hydrologic model experiment, he opposed the

다. 대표적으로 하수종말처리장을 중·개축하여 처리용량을 증대시키고, 태화강으로 연결되는 지천 개선 사업, 태화강 퇴적오니 준설사업을 통해 태화강 생태를 복원했다. 또한 태화들 사유지 하천구역 편입 시 소요되는 재정을 당시 국회의원 정갑윤과 함께 국비로 확보하는 등 역량을 발휘했다.

설계안/디자인 이슈

태화강대공원에서 제기된 설계 및 디자인 이슈는 다음과 같다.

수질오염 개선

공업도시로 발전하는 동안 축적된 태화강의 수질오염으로 기존의 생태계는 파괴되었고 이를 복원하기 위해 울산시는 다각도로 공해물질 감소를 위한 계획을 발표하고 사업을 진행했다. 환경보전중기종합계획(1차 1999~2003, 2차 2004~2008)과 2005년 태화강마스터플랜을 기반으로 태화강 상하류의 수질을 관리하는 사업을 추진했다. 하수처리시설을 확충하고 우수와 오수를 구분하여 하수처리량을 줄이며 축산분뇨 및 생활오수, 비료, 농약, 공업폐수 등에 의한 오염원을 확인하고 처리하는 환경정책을 펼쳤다.

생태복원

오염물질 유입방지 및 처리와 더불어 환경오염으로 인해 파괴된 생태계를 복원하는 사업도 태화강을 복원하는데 중요했다. 인공소재의 하천변을 자연소재로 재조성하여 동·식물의 서식처를 복원하고 생태연결통로를 설치하는 등 생태계 회복과 종다양화를 도모하는 사업을 진행했다. 오산 십리대숲은 생태공원으로 조성하고 콘크리트 호안을 자연형 호안으로 정비했으며 삼호대숲은 철새도래지로 조성하였다. 이런 각고의 노력으로 '죽음의 강'으로 인식된 태화강에 40년만인 2003년에 연어가 돌아왔고 은어, 황어 등 회귀어류의 서식지가 복원되었다. 어

Busan Regional Land Management Office's plan to build the large-scale waterway of 80 m width, 7 m depth, and 1.3 km length in Taehwa Field for flood control, and thus preserved the waterfront environment and ecology of the Tae-Hwa-Gang River.

Politician - Mayor of Ulsan, Park Maeng Wu

A large-scale urban design project like Tae-Hwa-Gang River does not happen without bold decisions by policy makers. Mayor Park Maeng Wu once worked as the head of the planning and coordination office, the head of the interior affairs bureau, and the construction and transportation bureau director, and understood the city as a whole. He empathized with the aspirations of citizens to revitalize the Tae-Hwa-Gang River. Accordingly, after he was elected in 2002, he planned and implemented a project to improve the water quality of the Tae-Hwa-Gang River. The treatment capacity was increased by extending and remodeling the sewage treatment plant. The ecology of the Tae-Hwa-Gang River was restored through the stream improvement project connected to the Tae-Hwa-Gang River and the Tae-Hwa-Gang River sedimentary sludge dredging project. He also secured government funds with the then-member of the National Assembly, Jeong Gap Yun, to finance the transfer of the private land of Taehwa Field to the river zone.

Draft Design/Design Issues

The design issues of Tae-Hwa-Gang Grand Park are as follows.

류뿐 아니라 태화강철새공원에 8,000여 마리의 백로가 서식하며 노랑부리백로, 고니, 큰기러기, 물수리, 솔개 등 멸종위기종의 서식이 확인되었으며 바지락, 성게, 해삼류의 서식지가 태화강 안쪽으로 확장되고 수달, 삵, 너구리 등 포유류의 모습도 관찰되어 생태계가 복원되었다.

공학중심 홍수통제 vs. 생태적 하천관리

태화강대공원의 서막은 1987년 십리대숲 벌목계획이라고 할 수 있다. 환경오염으로 생태학적 가치를 잃어버린 태화강은 수리공학자에겐 홍수예방을 위한 수로로 인식되었다. 기존의 하천법으로는 하천 구역 내에 1m 이상의 다년생 수목은 둘 수 없었고 오산 십리대숲은 홍수 시 하천의 흐름을 막는 장애물이었다. 따라서 십리대숲을 벌목하여 홍수를 통제하는 것은 이성적인 선택이었다.

하지만 울산의 환경단체, 시민사회, 전문가집단은 공학중심의 홍수통제를 위한 수로보다는 도시의 역사와 문화가 내재된 태화강을 살리기로 마음을 먹고 사회, 정치, 과학에 걸쳐 전방위로 십리대숲 보전운동을 펼쳤다. 특히 수리공학 전문가가 기존 우리나라 하천정비기본계획수립방안을 과학적인 실험을 통해 비판하고 하천보전 및 관리에 대한 새로운 패러다임을 제시하는 계기가 되었다는 점에서 큰 의미를 찾을 수 있다.

수변공간활용

태화강 마스터플랜의 실행은 태화강의 생태계를 복원하고 강변을 보존하며 부수적으로 얻어지는 수변공간을 활용하여 울산에서 부족한 도심공원을 확충하고 태화강을 중심으로 자전거, 보행자 네트워크를 조성하는 기회가 되었다.

Improvement in Water Quality

Pollution from industrialization destroyed the ecosystem of the Tae-Hwa-Gang River. Ulsan Metropolitan City implemented a project to reduce the pollutants in various aspects. Based on the Environmental Conservation Medium-Term Comprehensive Plan (1st 1999–2003, 2nd 2004–2008) and the 2005 Tae-Hwa-Gang River Master Plan, the Tae-Hwa-Gang River upstream and downstream water quality management project was promoted. Sewage treatment facilities were expanded, and sewage treatment volume was reduced by separating rainwater and sewage. In addition, environmental policies were implemented to identify and treat pollutants, such as livestock manure, domestic wastewater, fertilizers, pesticides, and industrial wastewater.

Ecological Restoration

Along with the pollutant reduction and treatment, restoring the ecosystem was also crucial in restoring the Tae-Hwa-Gang River. The project promoted ecosystem recovery and species diversification by recreating artificial riversides with natural materials to restore habitats for animals and plants, and provide ecological connection pathways. Osan Simni Bamboo Forest was created as an ecological park. The concrete shore was converted into a natural shore. The Samho Bamboo Forest was refurbished as a migratory bird habitat. Because of such efforts, in 2003, salmon returned to Tae-Hwa-Gang River. After 40 years, the 'river of death' once again became the habitat of salmon, sweetfish, dace, and many more. Also, more than 8,000 egrets now inhabit the Tae-Hwa-Gang River Migratory Bird Park, and endangered species, such as yellow-billed egret, cygnus, geese, osprey, and kite, have been observed. The habitats of clams, sea urchins, and sea cucumbers expanded inside the Tae-Hwa-Gang River, and wild mammals, such as otters, wildcats, and raccoons, were also observed.

맺으며

　태화강은 공업도시 울산의 환경공해문제를 시민과 정부 주도로 해결하는 발자취를 보여준 성공사례이다. '죽음의 강'에서 '생명의 강'으로 탈바꿈하는 것은 어느 한 사람이나 단체로 가능하지 않다. 시민의 공통된 염원과 행동으로 공업도시 울산이 생태도시로서의 개벽이 가능

그림 9. 태화강 십리대숲 수변공간 처리 / Riverfront Promenade along the Taewha River and Simni Bamboo Forest

Engineering-based Flood Control vs. Ecological River Management

The beginning of the Tae-Hwa-Gang Grand Park was the 1987 Simni Bamboo Forest logging plan. For hydrologic engineers, the Tae-Hwa-Gang River, whose ecosystem had collapsed, was only a waterway to prevent flooding in the city. According to the existing river law, 'no perennial trees larger than 1 m can be placed in the river zone,' and Osan Simni Bamboo Forest was an 'obstacle' that blocked the flow of the river in the event of a flood. To control flooding, logging Simni Bamboo Forest was a sensible option.

However, Ulsan's environmental groups, civil society, and expert groups decided to save the Tae-Hwa-Gang River, which contains much of the city's history and culture, rather than using it as an engineering-oriented waterway for flood control. They carried out Simni Bamboo Forest conservation campaigns across society, politics, and academia. In particular, it is meaningful, as it served as an opportunity for a hydrologic engineering expert to criticize the existing method of river maintenance in Korea through scientific experiments, and present a new ecological paradigm for river conservation and management.

Utilization of the Waterfront Space

Implementation of the Tae-Hwa-Gang River Master Plan became an opportunity to restore the ecosystem of the Tae-Hwa-Gang River. The plan preserves the riverside, expands the urban park that was lacking in Ulsan by utilizing the waterfront space, and creates a recreational bicycle and pedestrian network around the Tae-Hwa-Gang River.

했다. 태화강의 민관공동주도의 도시변혁의 사례로서 국내외에서 우수사례로 다양한 분야에서 수상하였고 여러 도시, 학술, 환경단체에서 벤치마킹되고 있다. 생태하천으로의 복원, 시민사회의 염원, 도시공간으로서의 활용성을 높게 평가해 2019년 7월 12일 산림청은 태화강대공원을 두 번째 국가정원으로 지정했다.

Closing Remarks

Tae-Hwa-Gang River is a successful case led by citizens and the government to solve the environmental pollution of the industrial city, Ulsan. Tae-Hwa-Gang's transformation from a 'river of death' to a river of life is a collective effort that was and is driven by the citizens, experts, and leaders. The citizens' common aspirations and actions enabled the transformation of the industrial city into an ecological city. Tae-Hwa-Gang River, an example of a public–private joint initiative for urban transformation, has been recognized in various fields as a case of excellence by awards at both the domestic and international level, and is being benchmarked by diverse cities, academics, and environmental groups. On July 12, 2019, Tae-Hwa-Gang Grand Park was designated by the Korea Forest Service as the 2nd National Garden, thanks to its high value for restoration as an ecological river, aspirations of civil society, and usability as an urban space.

참고문헌

1. 강정원. "시민과 함께하는 '태화강 살리기' 지금도 진행중" 2013.7.23. 울산신문, 2013 https://www.ulsanpress.net/news/articleView.html?idxno=153854

2. 김석택, 이상현, 이상윤, 김영혜. "울산광역시 제2차 환경보전중기종합계획" 울산발전연구원, 2002

3. 김석택 et al., "Ecopolis 울산계획", 울산발전연구원, 2004

4. 김석택 et al., "태화강 화보", 울산광역시, 2014

5. 박맹우, "죽음의 강에서 생명의 강으로!" 하천과 문화 vol.8 no.2, 2012

6. 배석희, "민관 함께 만든 '태화강대공원' 개장", Landscape Times. (2010.05.28) http://www.latimes.kr/news/articleView.html?idxno=6139, 2010

7. 수성엔지니어링 주도로 '태화강 생태공원조성 (2단계) 기본계획 및 기본설계 용역'

8. 울산광역시. "태화강국가정원 백서" 울산광역시, 2020

9. 울산광역시. "태화강백서" 울산광역시, 2014

10. 울산신문. "태화강 생태공원 2단계 조성부지 태화들 대대적 환경정비사업 추진" 2007.5.15. 울산신문 https://www.ulsanpress.net/news/articleView.html?idxno=12915, 2007

11. 조홍제, 바람직한 하천정비기본계획 수립방안 (태화강 사례를 중심으로)' 물과 미래, vol.39, no.8, 2006

12. 정재락, "하천공학자다운 정책대안 낼 것'. 동아일보, 2009.10.07. https://www.donga.com/news/Society/article/all/20061215/8385513/1, 2009

13. 한희동 et al., '태화강 생태공원조성 (2단계) 실시계획 용역', 수성엔지니어링, 2009

14. 한상진, 정우규, 구도완, '울산의 환경문제와 환경운동' 울산학연구센터, 울산발전연구원, 2008

태화강국가정원 홈페이지

https://www.ulsan.go.kr/s/garden/main.ulsan

References

1. Bae Seok-hee, "Opening of 'Taehwagang Grand Park' created jointly by the public and private sector", Landscape Times. (2010.05.28), 2010

2. Cho Hong Je, "A method for establishing a desirable basic river maintenance plan (focused on the Taehwagang River case)", Water and the future, vol.39, no.8, 2006

3. Han Hee-dong et al., 'Taehwagang River Ecological Park Creation (2nd Phase) Implementation Plan Service', Soosung Engineering & Consulting, 2009

4. Han Sang Jin, Jung Woo Gyu, Koo Do Wan, 'Environmental Issues and Environmental Movements in Ulsan' Ulsan Research Center, Ulsan Development Institute, 2008

5. Jeong Jae-Rak, "Propose a policy alternative like a river engineer". Dong-A Ilbo, Oct. 7, 2009, 2009

6. Kang Jung won. "Reviving the Taehwagang River together with citizens is still ongoing" 2013.7.23. Ulsan Press, 2013

7. Kim Seok Taek, Lee Sang Hyun, Lee Sang Yun, Kim Young Hye. "The 2nd Medium-Term Comprehensive Plan for Environmental Conservation of Ulsan Metropolitan City" Ulsan Development Institute, 2002

8. Kim Seok-taek et al., "Taehwagang River Pictorial", Ulsan Metropolitan City, 2014

9. Park Maeng Wu, "From the River of Death to the River of Life!" Rivers and Culture vol.8 no.2 p.13, 2012

10. Soosung Engineering & Consulting, The Taehwagang River Ecological Park Creation (Phase 2) Basic Plan and Basic Design.

11. Ulsan Metropolitan City, "Taehwagang River National Park White Paper" Ulsan Metropolitan City, 2020

12. Ulsan Metropolitan City (UMC), "Taehwagang River White Paper" Ulsan Metropolitan City, 2014

13. Ulsan Press, "Taehwagang River Ecological Park Phase 2 Development Site Taehwa Field Large-scale Environmental Renovation Project" 2007.5.15. Ulsan Press, 2007

서울 성수동, 준공업 지역의 부활

김호정 (단국대학교 건축학부 교수)

들어가며

성수동은 서울 중심에 위치한 보기 드문 준공업지역이다. 과거에는 의류, 수제화, 가방, 자동차 등 다양한 업종의 공장들이 성수동 곳곳에 자리하고 있었다. 그러나 지난 20년간 준공업의 쇠퇴로 많은 공장들이 문을 닫고, 지금은 수제화 산업 및 자동차 정비업소 위주로 그 명맥을 이어오고 있다. 2000년대 초기부터는 일부 공장부지가 공동주택으로 개발되면서 아파트와 공장의 불편한 공존이 이루어지기 시작하였다.

2004년에 조성된 서울숲은 인접한 성수동을 새롭게 인식하게 되는 계기가 되었으며, 주변의 초고층 개발은 한강의 스카이라인을 바꾸고 있다. 도심 및 강남과 인접한 접근성, 상대적으로 저렴한 임대료, 준공업지역의 높은 개발밀도, 상대적으로 큰 필지규모와 같은 여러 조건들은 성수동을 매력적인 지역으로 인식하게 되는 출발점이 되었다.

2009년부터 서울시는 준공업지역 종합발전계획을 수립하고 성수동을 우선정비대상구역으로 지정하여 오래된 공장과 주거가 혼재되어 노후화되고 있는 지역을 정비하고 있다. 공공에서는 IT 중심으로의 산업구조 개편 및 수제화 산업의 보존 및 육성을 중점적으로 지역정비를 진행하고 있으며, 민간에서는 대림창고를 시작으로 버려진 공장과 창고가 카페와 예술공

Seoul Seongsu-dong, Resurrection of a Semi-industrial Area

Hojeong Kim (Professor, Dankook University, Dept of Architecture)

Introduction

Seongsu-dong is a rare semi-industrial area located in central Seoul. In the past, factories in various industries, such as clothing, handmade shoes, bags, and cars, were located throughout Seongsu-dong. However, the decline in heavy industry over the past two decades has resulted in many factories closing down, while the area has remained famous mainly for the handmade shoes industry, and auto repair shops. From the early 2000s, some factory sites were developed into multi-family housing, and an inconvenient juxtaposition of apartments and factories developed.

Seoul Forest was created in 2004; being adjacent, it led to a new recognition of Seongsu-dong; and the development of high-rise buildings around the area changed the skyline of the Han River. Various advantages, such as access to the adjacent city center and Gangnam, relatively low rents, high development density in semi-industrial areas, and relatively large land size, have combined to encourage Seongsu-dong to be recognized as an attractive area.

Since 2009, the Seoul Metropolitan Government has established a comprehensive

그림 1. 성수동의 다양한 모습(출처: 김호정) / Various scenes in Seongsudong, Seoul

간으로 변모하는 변화가 이루어지고 있다. 창업/창작플랫폼 언더스탠드 에비뉴는 기업, 지자체, 비영리단체가 공동으로 운영하고 있으며, 성수역 인근에는 다수의 지식산업센터가 들어서면서 젊은 층이 유입되고 유동인구가 늘어나고 있다. 이제 성수동은 기존 산업과 공장, 주거와 신산업, 예술과 문화가 공존하는, 독특한 지역으로 새로운 가능성을 열어가고 있다.

redevelopment plan for quasi-industrial areas, and designated Seongsu-dong as a priority maintenance zone to repair areas where old factories and houses are aging. The public is reorganizing the industrial structure centered on IT, and preserving and fostering the handmade shoe industry, while the private sector is changing the abandoned factories and warehouses into cafes and art spaces. Understand Avenue, a start-up and creative platform, is created and operated by companies, local governments, and non-profit organizations, and as a number of knowledge industry centers are built near Seongsu Station, young people are flowing in, and the ambient population is increasing. Seongsu-dong is opening up new possibilities as a unique area, where existing industries and factories, housing, new industy, and art and culture coexist.

그림 2. 성수동 도시재생 및 지구단위 계획 지역(출처: 김호정)
Urban Regeneration and District Planning Area in Seongsudong, Seoul

조성과정

성수동은 수자원이 풍부하고 도심으로의 접근성이 높아서 1940년대부터 영등포에서 이전한 공장들이 자리 잡기 시작한 공업지역이다. 1968년 토지구획정리사업에 의하여 현재와 같은 격자형의 가로체계가 형성되었으며, 수제화산업과 자동차정비산업 관련 업체가 오랜 시간 자리잡아 지역의 특화산업으로 자리잡았다. 해당 공장에 자재나 부품을 납부하는 업체나 일부 공정을 제공하는 업체까지 공존하면서 산업생태계를 이루었다.

그러나 1987년 수도권정비계획법 수립 이후 일부 중대형 공장이 이전하여 제조업체 수가 감소하기 시작하였다. IMF 이후 정부는 성수공단 재정비를 위하여 노력을 하였으나 지역은 계속 쇠퇴하였고 대규모 공단은 이전하고 소규모 공장들은 영세화되어, 주거기능이 강화되기 시작하였다. 준공업지역으로서의 성수동은 쇠퇴의 시기를 맞았으나 성수동 서쪽 서울경마장 부지가 서울숲으로 변모되어 강북을 대표하는 녹지공간이 되었다. 서울숲 인근 및 한강변을 따라서는 도심 재개발, 재건축사업이 전개되어 한강의 스카이라인을 변화시키는 고층주거 건축물이 들어섬에 따라서 지역의 전반적인 가치가 재인식되는 계기가 되었다.

지난 10년간 준공업지역 정비계획을 통해 성수동이 새롭게 변화하기 위한 틀이 만들어졌다. 서울시는 기존의 전면 재개발 대신에 지역산업과 특성을 최대한 고려하고 일자리를 창출하면서 민간과 공공이 함께하는 방식으로 지역이 변화할 수 있는 방향으로 지역정비를 유도하고 있다. 성수동 준공업지역의 가장 중심인 2호선 남단은 성수동 2가 도시계획정비구역으로 지정되었고, 성수동 2가 정비구역 서쪽으로는 성수 도시재생시범사업지구, 동쪽으로는 성수 IT 산업개발진흥지구가 지정되어 지역 전체를 새롭게 개편하고 있다. 2000년대 이후 성수동은 수제화, 인쇄, 자동차정비 중심의 생활밀착형 산업의 집적지로서 존재함과 동시에, 문화예술복합공간, 소셜벤처 플러스터로 변신 중이다.

Creation Process

Seongsu-dong is an industrial area where factories began to relocate from Yeongdeungpo in the 1940s, due to the abundant water resources and high accessibility to the city center. In 1968, the current grid-shaped street system was formed by the land compartmentalization project; the handmade shoes industry and automobile maintenance industry were established for a long time, and became specialized industries in the region. An industrial ecosystem has developed as companies that pay for materials or parts from factories coexist with companies that provide various services.

However, following the establishment of the Seoul Metropolitan Area Readjustment Planning Act in 1987, the number of manufacturers began to decrease, due to the relocation of some medium and large factories. At that time, Seongsu-dong, a semi-industrial area, was in decline, but the site of the Seoul Racecourse on the west side of Seongsu-dong was transformed into Seoul Forest, and became a representative green space of the district north of the Han River. Urban redevelopment and reconstruction projects near Seoul Forest and along the Han River have been carried out, which has led to the renewed recognition of the overall value of the region, as high-rise residential buildings that are changing the skyline of the Han River are being built.

Through the semi-industrial area maintenance plan over the past decade, a framework has been created for the renewed change of Seongsu-dong. Instead of abrupt comprehensive redevelopment, the Seoul Metropolitan Government is encouraging regional maintenance in a way that can evolve the region so that the private and public sectors can come together, while considering local industries and characteristics as much as possible, and creating jobs. The southern end of Line 2

성수동 2가 일대 정비계획과 수제화산업

성수동 2가 일대는 성수동 변화의 가장 중심에 있다. 2011년 도시환경정비계획 수립 및 구역지정 용역이 발주되고 자문회의 및 주민설명회를 거쳐 2015년 정비계획이 확정되었다. 지자체는 공공지원 및 자율적 갱신을 바탕으로 한 공공지원 수복형 정비계획을 수립하여 대상지가 가지고 있는 기존 산업보호 및 지역 활성화를 추진하고 있다. 해당 지역은 영세한 구두 제조업체와 관련 부자재 공급업체가 밀집해 있는 상황으로 전면철거방식이 아닌 기존 대지 및 도로형태를 유지하는 방식으로 정비계획이 이루어졌다. 대신 전체 지역을 세분화하여 산업활성화지구, 공동개발지구, 주거밀집지구, 산업관리지구 등 자율적인 소단위 정비를 유도하는 방식으로 접근하고 있다.

성수동에서는 기존 수제화산업의 보호 및 육성사업을 위해서 다양한 형식의 지원이 이루어지고 있다. 가격경쟁력 약화 문제와 숙련된 기술자의 고령화, 임대료 상승문제 등으로 인하여 성수동 수제화산업은 과거에 비해서 많이 위축되어 있는 상황이지만, 성수동 수제화거리에는 국내 수제화 제조업체의 70%가 밀집되어 500여 곳의 수제화사업체와 부자재, 판매장이 모여 있다. 특히 성수동의 수제화거리는 1970년대 이후부터 염천교에서 명동으로, 다시 성수동으로 모인 수제화 관련 사업장의 밀집지역이다. 다만 현재는 젠트리피케이션 방지를 위한 상생협약에도 불구하고 임대료의 상승으로 말미암아 수제화거리의 상인은 과거에 비해서 1/3 정도로 줄었다.

공공에서는 성수동의 수제화산업 알리기와 수제화거리 홍보, 수제화 및 가죽산업 창업인력 지원, 지역 소상공인을 위한 시설장비지원과 같은 사업을 전개하고 있다. 2011년 서울성동제화협회는 수제화 판로 확보를 위해 수제화 공동판매장인 성수수제화타운을 열었으며, 설립 7개월 만에 매출 5억원을 달성하며 행정안전부의 우수마을기업에 선정되기도 하였다. 2013년에는 구두테마역(슈스팟 성수), 수제화 공동매장(from SS)이 개관하였는데, 성수역을 구두테마역으로 지정하고 수제화 공동매장을 오픈하여 수제화 특화거리의 경쟁력을 확보하기 위한 노력을 이어갔다. 2017년 성수 수제화 희망플랫폼은 성수동 수제화를 한눈에 볼 수 있는 체

of the Seoul Metro, the center of the Seongsu-dong semi-industrial area, has been designated as an urban planning and maintenance zone for Seongsu-dong 2-ga, with the Seongsu IT industry development promotion zone is concentrated to the east.

Maintenance Plan and Handmade Shoes Industry in Seongsu-dong 2-ga

Seongsu-dong 2-ga area is at the center of the change in Seongsu-dong. In 2011, the urban environment readjustment plan was established, zoning services were ordered, and after consultation meetings and resident briefing sessions, the maintenance plan was finalized in 2015. Local governments are promoting the protection of existing industries and regional revitalization of the target sites by establishing a public support recovery plan based on public support and autonomous renewal. As the area is crowded with small shoe manufacturers and related suppliers of subsidiary materials, the maintenance plan was made by maintaining existing land and road types, rather than removing them altogether. The approach that is being taken is one of segmenting the entire area to induce autonomous small-unit maintenance, such as industrial revitalization zones, co-development zones, dense residential zones, and industrial management zones.

In Seongsu-dong, various forms of support are provided for the protection and fostering of the existing handmade shoes industry. Due to the weakening price competitiveness, aging of skilled engineers, and rising rents, the Seongsu-dong handmade shoe industry is shrinking compared to the past, but 70 % of the domestic handmade shoe manufacturers are concentrated in Seongsu-dong handmade shoe

그림 3. 구두테마역인 성수역 하부 수제화 공동매장(출처: 김호정) / Handmade shoes store at Seongsu Station

그림 4. 성수동 수제화 거리(출처: 김호정) / Seongsudong Handmade Shoes Street

street, attracting more than 500 handmade shoe companies, subsidiary materials, and sales. In particular, Seongsu-dong's handmade shoe street is a dense area of handmade shoe-related businesses that migrated there from Yeomcheon Bridge to Myeong-dong and Seongsu-dong since the 1970s. However, despite the current win–win agreement to prevent gentrification, the increase in rent has reduced the merchants of handmade shoe streets by one-third, compared to the past.

The public is carrying out projects, such as informing the handmade shoe industry in Seongsu-dong, promoting the handmade shoe street, supporting the start-up personnel in the handmade shoe and leather industry, and supporting facility equipment for local small business owners. In 2011, the Seoul Sungdong Shoe Association opened Sungsoo Shoe Town, a joint market for handmade shoes, which was selected as an excellent village company by the Ministry of Public Administration and Security, achieving 500 million won in sales within seven months of its establishment. In 2013, Shoe Theme Station (Shoespot Seongsu and 'from SS') and Sujehwa Joint Store (from SS) opened. They designated Seongsu Station as a shoe theme station, and opened a Sujehwa Joint Store to secure the competitiveness of Sujehwa Specialized Street. In 2017, Seongsu Handmade Shoes Hope Platform is an experience-type showroom where the extensive range of Seongsu-dong Handmade Shoes can be seen at a glance, and it is operated as a handmade shoe exhibition hall and an experience workshop to foster handmade shoes craftsmanship, and expand the market.

In 2016, the Seoul Handmade Shoe Academy opened a two-year course, and it operates a handmade shoe manufacturing process, a designer and MD course, while the Seoul Metropolitan Government supports the entire education fee. In addition, the Seoul Metropolitan Government and the Seoul Business Agency (SBA) opened the

험형 쇼룸으로 수제화 장인을 육성하고 판로를 넓히기 위한 공간으로 수제화 전시장과 체험
공방으로 운영되고 있다.

2016년에는 서울수제화 아카데미가 2년 과정으로 개설되어 수제화 제조과정과 디자
이너·MD 과정을 운영하고 있고 교육비는 전액 서울시가 지원하고 있다. 더불어 서울시와
SBA(서울산업진흥원)는 업무공간, 시제품 제작공간 및 공동장비실을 갖춘 성수 수제화 제작
소를 2017년 개관하여, 다품종 소량생산의 유연생산체계 지원체계를 구축하고 있다. 성수 수
제화 제작소는 서울수제화 아카데미를 마친 교육생은 물론 서울 소재의 디자이너와 프리랜서
들을 위한 샘플 제작을 지원해 주고 있으며, 창업 초기 6개월 간 공간을 무상 지원하고 있다. 제
작소의 공용장비터는 지역소상공인을 위하여 시설장비 무료사용 지원이 이루어지기도 한다.
이외에도 성동지역경제혁신센터 내 수제화공방 교육시설, 청년창업공방 등 수제화나 가죽
관련 산업에 관심을 가진 인력들을 위한 지원시설을 제공하고 있다. 공공은 산업 전반의 숙련
된 기술의 고령화에 대비하고 젊은 인재들의 육성 및 새로운 디자인과 신기술을 접목한 수제
화 산업을 변화를 핵심가치로 수제화 산업의 변화를 유도하고 있다.

민간에 의한 공장과 창고건물의 재생

수제화거리와 함께 성수동이 젊은이들에게 널리 알려진 계기는 대림창고가 그 시작이라고
할 수 있다. 대림창고는 수제화 거리 인근에 위치하여 한 때는 수제화를 보관하던 곳이었으며
지금은 복합문화공간으로 공연을 겸한 카페로 활용되고 있다. 대림창고는 서울의 다른 지역
에서 흔히 볼 수 없는 공장건물이 가지는 높은 층고, 철제 트러스와 박공지붕, 낡고 오래된 벽
돌과 미장마감 등 성수동이 지닌 역사를 감성적으로 드러내 보이는 공간으로 자라잡고 있다.
대림창고 이외에도 인쇄공장에서 카페로 변신한 자그마치나 할아버지 공장과 같은 카페들은
과거 공장에서 사용하던 소품을 그대로 사용하기도 하고, 과거의 모습과 현대의 감성을 적절

Seongsu Handmade Shoes Production Center in 2017 to promote the influx of young and competent human resources into the handmade shoes and leather industries. This initiative aims to support a flexible production system of multi-breed small-volume production, which is the core of the fourth industrial revolution, as it is equipped with a start-up space, a prototype production space, and a joint equipment room. Seongsu Handmade Shoes Factory supports sample production for Seoul-based designers and freelancers, as well as trainees who have completed the Seoul Handmade Shoes Academy, and provides free space for the first six months of its foundation. Free use of facility equipment may be provided for local small business owners at the public equipment site of the production center.

In addition, it provides support facilities for people interested in handmade shoes and leather-related industries, such as handmade shoe workshops and youth start-up workshops in the Seongdong Regional Economic Innovation Center. The public is preparing for the aging of skilled technologies across the industry, fostering young talent, and encouraging the change of the handmade shoes industry with new design and new technology as its core values.

Regeneration of Factories and Warehouses by the Private Sector

Along with the handmade shoe street, Seongsu-dong is most widely known to young people because of Daelim Warehouse. Daelim Warehouse is located near the handmade shoes street, and it was once a place where handmade shoes were stored, while now it is used as a cafe for performances as a complex cultural space. Daelim

성수동 골목길 탐방 코스

뚝섬역
'블루보틀'
아틀리에길
성수동 구두 테마 공원 수제화갤러리 '카페 수다'
수제화 투어
'두리 컴퍼니'
성수 수제화 희망플랫폼
성수동 갈비골목   '빵선생'
서울숲
성수 수제화 타운
브루클린 투어
'어반소스'
성수역사 내
'어니언'·'카페봇'
언더스탠드애비뉴
'로우키'
수제화 전시장
연무장길
'오르에르'
벽화골목
'자그마치'
카페 투어
'베란다
인더스트리얼'
'성수연방'
수도박물관
경찰기마대
'카페 할아버지 공장'

그림 5. 성수동 골목길 탐방 코스(출처: 박구원) / Seongsu-dong Alleyway Tour Course

그림 6. 성수동 문화허브의 출발점이 된 대림창고(출처: 김호정) / Multi-cultural space Daelim Warehouse

Warehouse consists of factory buildings of high story height, which are uncommon in other parts of Seoul, and it is a place that shows the history of Seongsu-dong, such as iron trusses, thin roofs, old brick, and beautiful finishes. In addition to Daelim Warehouse, cafes such as Jagmachi and Grandpa's Factory, which were transformed into cafes in print factories, are being recognized as new attractions by properly

그림 7. 대림창고 내부 모습(출처: 김호정) / Interior space of Daelim Warehouse

하게 조화시키면서 새로운 명소로 인식되고 있다. 대림창고 인근 성수 연방도 과거 화학공장을 복합문화공간으로 바꾼 대표적인 사례로 카페, 음식점, 편집숍, 소품숍 등이 입점하여 있다. 건축가는 기존에 있던 2개의 건축물 사이의 외부공간에 디자인을 집중하면서 이 지역에서 보기 힘든 의미 있는 녹색공간을 만들었고 최상층부에 증축된 온실 같은 느낌의 카페공간도 젊은 층에게 널리 알려져 있다.

성수역의 북쪽에 위치한, 자동차정비업소가 많은 지역에 있는 어니언 카페도 성수동을 알리는 데 큰 역할을 하였다. 이곳은 소규모이지만 50년이라는 시간동안 주택, 슈퍼, 정비소, 공장 등 여러 용도로 사용된 건물의 흔적을 적나라하게 드러내는 방식으로 색다른 감성을 자극하는 공간이다. 지난 10여 년간 레트로나 빈티지 감성이 주목을 받으면서, 소셜미디어와 언론에서 성수동이 젊은 층들이 찾는 인기지역으로 알려지기 시작하고 여러 크고 작은 카페와 식당이 들어서게 되었다. 대기업에서도 이러한 트렌드를 인식하고 아모레퍼시픽에서는 오래된 자동차정비소 건물을 개조하여 아모레 성수라는 이름의 뷰티라운지를 만들었다. 상품을 판매하는 곳이 아닌 오로지 제품을 경험하고 즐기는 공간이라는 개념으로 만들어진 이곳은 우리나라에서 자생하는 식물들로만 이루어진 정원이 꾸며져 있다.

성수동의 문화공간, 맛집과 카페들은 지금도 여기저기에 흩어져서 생겨나고 있으며, 여전히 작동되고 있는 크고 작은 공장들과 묘한 대조를 이루면서 공존하고 있다. 성동구와 문화컨설팅 회사는 '성수 브루클린 투어'를 통해서 성수동의 과거와 현재가 공존하는 장소들을 2~3시간 걸으면서 둘러보는 경험을 제공한다. 뚝섬역에서 출발하여 20년 넘게 성수동 주민과 공장 노동자들이 찾는 갈비골목과 아틀리에길에 있는 카페와 편집숍, 수제화 매장과 공방을 구경하고, 서울숲과 인근에 있는 창업지원공간 언더스탠드 에비뉴에서 여정을 마친다. 이외에도 성수도 카페 투어, 수제화 투어를 통해서 이곳을 체험할 수 있는 프로그램이 제공된다.

harmonizing past and modern sensibilities.

Seongsu Yeonbang, near Daelim Warehouse, is a representative example of converting a chemical factory into a complex cultural space, with cafes, restaurants, select shops, and accessory shops. The architect created a meaningful green space that is hard to find in this area, while concentrating on the design of the exterior space between the two existing buildings; the cafe space that feels like a condensed greenhouse on the top floor is also widely known among young people.

Onion cafon located in the north of Seongsu Station in an area with many auto repair shops, has also played a significant role in promoting Seongsu-dong. Although this place is small, it is a space that stimulates different emotions, by revealing the traces of buildings used for various purposes for 50 years, such as houses, supermarkets, repair shops, and factories. As Seongsu-dong began to become known as a popular area for young people in social media and the mass media, many large and small cafes and restaurants were built; recently, Amore Pacific renovated an old auto repair shop building to create a beauty lounge named Amore Seongsu. Created not as a place to sell goods and services, but as a place to experience and enjoy products, this place has a garden composed only of plants native to Korea.

The cultural spaces, restaurants, and cafes in Seongsu-dong are still scattered all over the place, coexisting in strange but creative contrast to the large and small factories that are still operating. Seongdong-gu and the cultural consulting company provide a 2 to 3 hour walk, the 'Sungsu Brooklyn Tour', around places where the past and present coexist in Seongsu-dong . Departing from Ttukseom Station, tourists visit cafes, editing shops, handmade shoe stores and workshops in Galbi Alley and Atelier-gil, where residents and factory workers have visited for more than 20 years, and finish their journey at Seoul Forest and the nearby Understand Avenue.

그림 8. 성수연방의 외부공간(출처: 건축사사무소 푸하하하 프렌즈) / Outdoor spaces of Seongsu Yeonbang

창업지원공간 언더스탠드 에비뉴

언더스탠드 에비뉴는 청년창업가들의 창작과 창업을 도와주는 공간이다. 입점 교육부터 시장 테스트, 금융 투자 및 지원에 판매까지 가능하도록 기업이 도와주는 곳으로 성동구 일자리센터와 작은 상점들이 입점해 있다. 이 프로젝트는 신한은행그룹의 사회공헌사업의 일환으로 시작되었으며, 일자리사업과 문화예술활동을 결합한 청년취업 및 창업플랫폼이다. 신한은행이 디지털 인프라 등 금융 인프라를 활용한 취업플랫폼을 제공하고, 비영리 청년 스타트업 아르콘이 스타트업 교육 및 인큐베이션, 공간콘텐츠 기획 및 운영을 준비하며, 성동구는 일자리

그림 9. 성수동에서 열리는 아트 마켓 안내(출처: 김호정) / Posters for art market held in Seongsu

그림 10. 성수동의 기존 건축물을 이용한 카페와 문화공간(출처: 김호정)
Cafe and cultural spaces using existing buildings in Seongsu

그림 11. 창업지원공간 언더스탠드 에비뉴(출처: 건축사사무소 메타) / Understand Avenue, start-up support space

창출과 관련된 행정지원 서비스를 제공하고 있다.

언더스탠드 에비뉴에서는 창업교육, 취업지원, 인큐베이터센터, 마켓과 페스티벌, 자영업 지원과 같은 취·창업 지원이 이루어지며, 성동구가 주최하는 많은 행사들이 여기에서 이루어지는 관계로 지역 커뮤니티 센터의 역할을 하기도 한다.

언더스탠드 에비뉴는 컨테이너 박스로 만들어진 가설 건축물로 서울숲역과 서울숲 사이에 위치하여 도시와 공원을 연결하는 연결통로로써 작동한다. 가운데 위치한 외부공간을 중심으로 모두 120여 개의 컨테이너가 양 옆으로 길게 늘어선 형식으로 배치되어 있다. 일부 박스들은 2, 3개 층으로 적층되기도 하고 일부 컨테이너 상부는 외부 테라스 공간으로 계획되어 여러 색깔의 다채로운 도시경관을 만들어내고 있다. 이 컨테이너들은 기성 공업용 컨테이너는 아니고 메타건축의 설계 의도에 따라서 특수 제작된 조립식 건축물이다. 서로 마주보고 있는 2열의 컨테이너 사이의 외부공간은 봄부터 가을까지 청년 스타트업 또는 창작자가 셀러가 되어 가능성 있는 제품을 선보이는 영 크리에이터 마켓인 마주치장으로 사용된다. 공개모집을 통해서 청년 셀러를 선정하고 다양한 이벤트를 함께 기획하기 때문에 이 장소는 창업공간뿐만이 아니라 문화공간, 상업공간으로도 많이 알려져 있다. 아트스탠드에서는 연극공연도 상시로 이루어진다.

Start-up Support Space, Understand Avenue

Understand Avenue is a space that helps young entrepreneurs create and start their own businesses. Seongdong-gu Job Center and small shops are located to help companies sell from entry education to market testing to financial investment and support. The project began as part of Shinhan Bank Group's social contribution project, and is a youth employment and start-up platform that combines job projects with cultural and artistic activities. Shinhan Bank provides a job platform using financial infrastructure, such as digital infrastructure; non-profit youth startup Arcon prepares startup education and incubation, and space content planning and operation; and Seongdong-gu District Office provides administrative support services related to job creation.

Understand Avenue provides job start-up support, such as start-up education, job support, incubator centers, markets and festivals, and self-employed businesses, and many events hosted by Seongdong-gu are held here, which also serves as a local community center.

Understand Avenue is a hypothetical building made of container boxes, located between Seoul Forest Station and Seoul Forest, and operates as a link between the city and the park. A total of 120 containers are arranged in a long-stacked format on both sides, centering on the outer space located in the middle. Some boxes are stacked in two or three layers, with some containers being planned as outer terrace spaces, creating colorful urban landscapes of different colors. These containers are not ready-made industrial containers, but are prefabricated structures specifically designed for the purpose of meta-architecture. The outer space between the two-row containers that stand facing each other is used as an encounter market, a young creator market

성수 IT 산업개발 진흥지구와 소셜벤처 산업지원

　수제화산업, 봉제산업, 자동차정비산업과 같이 전통적인 산업 이외에 성수동에서는 IT 산업으로의 산업구조 변화도 이루어지고 있다. 이는 정부의 정책적인 개입으로 이루지고 있으며, 서울시는 2009년 신산업 뉴타운프로젝트를 수립하고 2011년 성수2가 3동 일원을 IT 산

그림 12. 성수 IT 산업개발 진흥지구 내 신축 건축물(출처: 김호정) / New construction in Seongsu IT Industry Development Area

where young startups or creators become sellers from Spring to Fall, and showcase potential products. Because young sellers are selected through open recruitment, and various events are planned together, this place is known not only as a start-up space, but also as a cultural space and a commercial space. At the Art Stand, theatrical performances are also held at regular times.

Seongsu IT Industry Development Promotion Zone

In addition to traditional industries, such as the handmade shoes industry, sewing industry, and automobile maintenance industry, industrial structure changes from Seongsu-dong to IT industry are taking place. This is a policy intervention by the government, whereby the Seoul Metropolitan Government established the New Industrial New Town Project in 2009, and designated Seongsu 2-ga 3-dong as an IT industry development promotion zone in 2011. The purpose of the designation was to develop the Seongsu IT district into a convergence model of high-tech IT technology with traditional industries, such as handmade shoes, printing, and automobile maintenance, and to create a corporate ecosystem through network revitalization. As at September 2011, a total of 271 IT/BT/R&D companies were concentrated in the Seongsu IT district, accounting for 17.1 % of all companies in the district. Three specialized industrial zones, consisting of the Seongsu IT district, Jongno Precious Metal District, and Mapo Design Publishing District, were included in the plan to strengthen the foundation of small and medium-sized manufacturing industries, and foster next-generation new growth engines.

In 2011, Seongdong-gu opened Seongdong Arc Valley, a Seongsu IT anchor facility

업개발 진흥지구로 지정하였다. 성수 IT 지구를 수제화, 인쇄, 자동차정비업 등 기존 전통산업과 첨단 IT기술 융합모델로 발전시키고, 유기적인 기업지원시스템을 구축해 네트워크 활성화를 통한 기업생태계를 조성하는 것이 지구 지정의 목적이었다. 2011년 9월 당시 성수 IT 지구에는 총 271개의 IT/BT/R&D기업이 집적해 있었으며 이는 지구 내 전체 기업의 17.1%에 해당된다. 성수 IT 지구 이외에도 종로 귀금속 지구, 마포 디자인 출판지구의 3개 특화산업지구가 서울의 중소제조업 기반 견고화 및 차세대 신성장동력산업을 육성하기 위한 계획에 포함되었다.

2011년 성동구는 첨단 IT 산업을 육성·선도하기 위해 성수 IT 산업개발진흥지구 내에 총 면적 9909m², 지하 1층, 지상 6층 규모의 성수 IT 앵커시설인 성동 아크밸리를 개관하였다. 앵커시설은 일종의 거점 지원시설로, 성동 아크밸리에는 IT 산업분야 경영지원과 교류협력, 연구개발, 판매지원 등을 담당할 종합컨설팅 시설과 영세 제조업체의 창업보육을 담당할 임대시설 등이 다수 입주해 있다. 이 시설을 통해서 성수동 준공업지역에 산재되어 있는 중소 IT 업체에 대한 기술·경영·홍보 등에 대한 지원이 일원화되었다. 더불어 권장업종시설 건축 시 용적률, 건폐율, 높이제한 완화 등의 지원을 위한 1종 지구단위계획을 수립하였다. 권장업종 입주업체에게는 취·등록세 및 재산세를 5년간 감면하고, 기업들에게 자금융자을 제공하여 해당 산업을 집중 육성하고자 하였다.

성수동은 IT 산업개발진흥지구로 지정되어 있는 서울 중심권의 유일한 준공업지역이며, 지자체의 전폭적인 지원 아래 IT 관련 업종의 수가 증가하고 지식산업센터의 신축이 늘어나고 있다. 지자체는 성수동 내 IT/BT/R&D 분야 중소기업의 경영컨설팅, 기술사업화, 판로개척, 홍보 등 다양한 사업을 통한 기업성장을 지원한다. 이 지역 기업들이 가장 큰 어려움을 겪고 있는 자금조달과 전문인력을 양성하기 위해 투자유치 설명회를 개최하고, 성수 IT CEO 아카데미 등 전문인력 양성 프로그램도 운영하고 있다.

2020년에는 서울창업허브 성수가 문을 열었다. '서울창업허브 성수'는 성수지역 산업환경 정비 및 핵심산업 육성을 위하여 2011년도에 개관한 중소기업 종합지원센터가 리모델링을 거쳐 새롭게 조성된 공간이며, 감염병, 건강, 안전, 환경 등 도시문제 해결을 위한 기술기업을

with a total area of 9,909 m² with one basement floor and six ground floors, in order to foster and lead the high-tech IT industry. Anchor facilities are a kind of base support facility, and Seongdong Arc Valley houses a number of comprehensive consulting facilities in charge of management support, exchange cooperation, research and development, and sales support in the IT industry. Through this facility, support for technology, management, and public relations for small and medium-sized IT companies scattered in the semi-industrial area of Seongsu-dong was unified. In addition, a first-class district unit plan was established to support the reduction of floor space rate, building rate, and height limitations for the construction of recommended industrial facilities. It was intended to reduce employment and registration taxes and property taxes for five years for tenants in recommended industries, and to focus on the industry by providing financial loans to companies.

Seongsu-dong is the only semi-industrial area in central Seoul that is designated as an IT industry development promotion zone, and with the full support of local governments, the number of IT-related industries is increasing, and the number of new knowledge industry centers is increasing. Local governments support corporate growth through various projects, such as management consulting, technology commercialization, market development, and the promotion of small and medium-sized enterprises in the IT/BT/R&D sector in Seongsu-dong. Companies in the region hold investment attraction briefing sessions to raise funds and train professionals who are suffering the most, and also run professional training programs, such as the Seongsu IT CEO Academy.

In 2020, Seongsu of the Seoul Startup Hub opened. Seongsu of the Seoul Startup Hub is a newly created space after remodeling the Small and Medium Business Comprehensive Support Center, which opened in 2011 to improve the industrial

그림 13. 서울창업허브 성수(출처: 서울시) / Seoul Start-up Hub Seougsu

육성하는 소셜벤처 육성공간으로 마련되었다. '서울창업허브 성수'는 창업 7년 미만의 중소기업을 위한 보육공간과 T-스페이스 등의 시설과 더불어 모바일 테스트가 가능한 앱 테스트베드가 제공된다.

업무지구로서의 성수동은 강남 테헤란로 일대나 구로-가산디지털단지와 비교하면 여전히 부족한 수준이지만 사회적기업 및 소규모 소셜벤처를 중심으로 커지고 있다. 카셰어링업체 쏘카, 인공지능(AI) 스타트업 스캐터랩, 에듀테크 기업 에누마코리아, 그리고 큐브엔터테인먼트가 성수동에 위치하며 특히 2017년 사회적기업을 위한 공유오피스 '헤이그라운드'가 문을 연 이후 소셜벤처가 몰려들기 시작하여, 현재 성동구에 기반을 둔 사회적경제기업이 300여 개가 넘고 있다. 더불어 성수동은 작은 단위 소셜벤처가 밀집되면서 공유 오피스 격전지로도 주목받는 분위기이다. 국내 1위 공유 오피스 기업인 '스트파이브'가 성수동에 2개 지점을 운영하고 있고, '헤이그라운드', '상상플래닛', '에스팩토리'도 성수동에 위치한다.

environment and foster core industries in the Seongsu area. It has been prepared as a social venture fostering space to foster technology companies to solve urban problems such as infectious diseases, health, safety, and the environment. Seoul Startup Hub Sungsoo will be provided with a mobile test-bed along with facilities such as childcare space and T-space for small and medium-sized enterprises less than seven years old.

Seongsu-dong as a business district is still insufficient compared to the area of Teheran-ro in Gangnam or Guro-Gasan Digital Complex. However, Seongsu-dong is growing around social enterprises and small social ventures. Car-sharing company Socar, artificial intelligence (AI) startup Scatter Lab, edutech company Enuma Korea, and Cube Entertainment are located in Seongsu-dong. In particular, since the opening of the shared office Heyground for social enterprises in 2017, social ventures have begun to flock, and there are currently more than 300 social and economic companies based in Seongdong-gu.

In addition, Seongsu-dong is drawing attention as a shared office battleground as small social ventures are concentrated. 'St Five', the nation's No. 1 shared office company, operates two branches in Seongsu-dong, while 'Heyground', 'Sang Sang Planet' and 'Espactory' are also located in Seongsu-dong.

Seongsu Urban Regeneration Pilot Project Area

The Seoul Metropolitan Government selected the Seongsu-dong area as the target of this project after a public contest for the Seoul-type urban regeneration pilot project in 2014, and opened the Urban Regeneration Support Center in 2015. This was followed by recruiting residents' participation groups, operating an urban regeneration academy,

성수 도시재생 시범사업지구

　서울시는 2014년 서울형 도시재생시범사업 공모를 거쳐 성수동 일대를 사업대상지로 선정한 바 있으며, 2015년 도시재생지원센터를 개소하였다. 이어서 주민참여단 모집, 도시재생아카데미 운영, 도시재생소식지 발행 등이 이루어졌으며, 2016년 주민협의체 구성을 거쳐 2017년에는 성수동 도시재생시범사업인 도시재생활성화계획이 수립되었다. 성수동 도시재생활성화지역은 사업 대상지(88만 6,560m²)의 80% 가량이 준공업지역으로 서울숲 북쪽지역과 서울숲 동쪽지역으로 성수 1가 및 2가 지역을 모두 포함한다. 주거와 산업이 혼재되어 있어 주민들의 정주환경 개선에 대한 열망이 높았으며, 토착산업 쇠퇴현상도 문제였다.

　도시재생시범사업은 마중물 사업비 100억이 투입되어 일터재생, 삶터재생, 쉼터재생, 공동체재생의 4개 분야의 사업이 계획되었으며, 산업혁신공간 조성, 각 하부공간 개선, 우리동네 안심길 조성, 생활자전거 순환길 조성, 지역문화 특화가로 조성, 성장지원센터 건립, 나눔공유센터 건립 등과 같은 사업이 전개되었다. 마중물 사업 외에 연계사업으로 젠트리피케이션 방지를 위한 공공임대점포 취득, 사회적경제 패션 클러스터 조성, 성수동 도시경관사업, 태조이성계 축제, 사회적경제 지원센터 건립 등 23개 사업(443억원)을 서울시와 자치구, 민간에서 진행하였다.

서울숲 북쪽 붉은벽돌 지원사업

　서울시와 성동구는 성동구 내 붉은벽돌로 된 건축물이 68%가 밀집해 있는 성수동 서울숲 북쪽 일대를 '붉은벽돌 마을' 시범사업 대상지로 정해 지역 건축자산으로 보전하고 마을을 명소화하는 계획을 수립하였다. 개별적인 가옥뿐만 아니라 붉은벽돌 형태의 공장, 창고 등 산업유산 건축물도 보전해 지역정체성을 강화하고자 하고 있다. 이와 관련해 2017년 서울시는 뚝섬주변지역 지구단위계획을 재정비하면서 특별계획구역(4·5구역)이 해제된 저층주거지를

and publishing an urban regeneration newsletter. After forming a residents' council in 2016, a plan to revitalize urban regeneration project in Seongsu-dong was established in 2017. The Seongsu-dong urban regeneration revitalization area is about 80 % of the project target area (886,560 m²), which includes both Seongsu 1-ga and 2-ga areas to the north and east of Seoul Forest. Residents' desire to improve the settlement environment was high due to the mixed housing and industry, and the decline of the indigenous industry was also a problem.

In the urban regeneration pilot project, 10 billion won was invested in projects in four areas: workplace regeneration, living, shelter regeneration, and community

그림 14. 성수도시재생지원센터 네이버 블로그(출처: 성수도시재생지원센터)
Seongsu Urban Regerneration Center Naver Blog

그림 15. 붉은벽돌 지원사업 지원 건축물 (성수동 1가 685-259)(출처: 성동구 홈페이지)
Construction supported by Red Brick Construction Support Project (Seongsu-dong 1-ga 685-259)

regeneration. Projects included creating an industrial innovation space, improving the upper and lower space, creating a life cycle route, building a growth support center, and building a sharing center. In addition to the priming water project, 23 projects (44.3 billion won) were carried out in Seoul, autonomous districts, and private sectors, including the acquisition of public rental stores to prevent gentrification, the creation of social and economic fashion clusters, the Seongsu-dong Urban Landscape Project, and the Social Economy Support Center.

Red Brick Cladding Support Project in the Area North of Seoul Forest

The Seoul Metropolitan Government and Seongdong-gu established a plan to preserve the area as a local architectural asset and make the town a tourist attraction by selecting the area north of Seoul Forest in Seongsu-dong, where 68% of the red brick buildings in Seongdong-gu are concentrated. Not only individual houses, but also industrial heritage buildings such as factories and warehouses in the form of red bricks are being preserved to strengthen the regional identity. In this regard, in 2017, while reorganizing the district unit plan for the area around Ttukseom, the city of Seoul raised the use area for low-rise residential areas where special planning zones (zones 4 and 5) were lifted (type 1 and 2 residential area → type 2 general residential area) region). Accordingly, when building with red bricks, a reduced floor area ratio of 10.8 to 36% is applied.

In addition, standards have been established to support up to 20 million won within 1/2 of the construction cost for conversion to a red brick building including new and

대상으로 용도지역을 상향(1종 및 제2종 일반주거지역 → 2종 일반주거지역)했다. 이에 따라 붉은벽돌로 건축할 경우 최대 10.8~36%까지 완화된 용적률을 적용하고 있다.

더불어 신·증축을 포함해 붉은벽돌 건축물로 전환 시 공사비용 1/2범위 내 최고 2천만원까지, 대수선·리모델링 시 공사비용 1/2범위 내 최고 1천만원까지 지원하는 기준을 마련하였다. 붉은벽돌 건축물 보전정책으로 옥외광고물 특화사업, 전신주 디자인 개선, 도로포장 정비, 붉은벽돌 마을 안내시설, 붉은벽돌 플랜트박스 조성 등 마을환경 개선을 위한 기반시설을 설치하고 정비하고 있다. 성수동 나눔공유센터 내 붉은벽돌 지원센터를 설치하고 마을건축가와 코디네이터를 운영하고, 벽돌건축물의 구조적 안정성을 위한 대책으로 리모델링·신축 시 건축·구조분야 전문가 검토를 실시하고 있다.

맺으며

성수동은 주거와 산업이 혼재된 준공업지역이다. 수제화, 봉제, 인쇄산업 등 과거 지역의 특화산업이 시대적인 상황에 따라서 쇠퇴하게 되고 지역 노후화가 가속화되는 상황 속에서, 지자체는 수제화산업의 보존 및 육성과 지역경제의 신동력으로서 IT 산업 지원이라는 두 가지 방향성을 가지고 이 지역을 탈바꿈시키고 있다. 이를 위해서 지난 10여 년간 공공에서는 도시계획정비구역 지정, 도시재생시범사업지구 지정, 성수 IT 산업개발진흥지구 지정과 같은 도시계획 차원에서의 큰 틀을 마련하여 계획을 실행하고 있다.

대표적으로는 수제화거리의 활성화와 창업지원시설을 통해서 지역의 전통적인 특화산업을 지원하고 있다. 성수동이 새롭게 변모하고 수제화거리에 사람들이 모이면서 민간에 의해서 과거 버려진 공장과 창고는 문화공간, 카페, 식당으로 변모하였다. 성수 수제화거리는 서울시 도시재생 프로젝트의 중심에 있으며, 대림창고, 성수연방, 성수 구두테마공원, 어니언, 자그마치 카페, 수피, 사진창고, 오르에르 등 오래된 공장이나 창고가 문화예술공간으로 재탄생하여 지역에 활력을 불어넣고 있다.

extension, and up to 10 million won within 1/2 of the construction cost for major repairs and remodeling. As a red brick building conservation policy, The Seoul Metropolitan Government and Seongdong-gu are installing and maintaining infrastructure for the improvement of the village environment, such as outdoor advertising specialization projects, power pole design improvement, road pavement maintenance, red brick village information facilities, and red brick plant boxes. The Red Brick Support Center is set up in the Seongsu-dong Sharing Center, and a coordinator is operated with the village architects, and as a measure for structural stability of brick buildings, experts in the field of architecture and structural engineering are reviewing the remodeling and new construction.

Closing Remarks

Seongsu-dong is a semi-industrial area mixed with housing and industry. In the past, specialized industries, such as handmade shoes, sewing, and printing industries, have declined according to the times, and local governments are transforming the region in two directions: preservation and fostering of the handmade shoes industry, and supporting the IT industry as a new engine for the local economy. To this end, the public has been implementing plans in the past decade by establishing a large framework for urban planning, such as designating urban planning maintenance zones, urban regeneration pilot projects, and high-demand IT industry development zones.

Typically, the area supports traditional local specialized industries through the revitalization of handmade shoe streets and support facilities for start-ups. As Seongsu-dong was newly transformed and people gathered on the handmade shoes

지역개발 측면에서는 IT 업종 관련 시설 신축 시의 규제 완화를 통한 인센티브로 신축을 유도하여 업무공간을 확보, 새로운 지역 클러스터 산업을 만들고 있다. 성수동이 힙한 장소로 떠오르고 미디어의 주목을 받게 되자 성수동의 입지적 중요성 또한 재인식되고 지역 내 재건축사업의 활성화와 건축물의 고층화가 이루어지고 있다. 공공에서는 앵커시설을 통한 중소 IT 업체에 대한 기술·경영·홍보 지원이 이루어지고 있으며, 무엇보다도 젊은 직장인의 유입으로 성수동이 다시 살아나고 있다.

　서울의 다른 구도심지역이 상업화되고 인지도가 높아짐에 따라서 생겼던 젠트리피케이션의 피해를 사전에 예방하고자, 성수동에서는 임대로 상생협약 및 기존 지역이 가지고 있는 상권을 살리면서 신산업을 육성하여 지역을 재생시키고자 하였다. 성수동은 오래된 공장과 아파트, 신사옥과 고층아파트, 저층주거와 상업시설이 혼재되어 도시의 여러 단편이 모인 독특한 지역으로 변모하고 있다. 성수동이 낡고 낙후된 회색공장지역의 이미지를 벗고 '성수-스러움'을 만들어가고 있는 것은 사실이나, 가로 및 보행환경의 문제, 주차문제, 근린 공원의 부족과 같은 오래된 도심의 문제는 여전히 존재하는 것이 현실이다.

　도심 산업 및 도시형 제조업 지원은 어떻게 이루어져야 하는지, 비균질적 산업 집적공간과 주거공간은 어떻게 공존할 수 있을지, 복합용도지역으로의 성수동의 가능성은 어떻게 극대화될 수 있을지, 이 지역이 계속 풀어가야 할 과제가 쉽지 않다는 점에서 지난 10여 년간의 변화와 앞으로의 미래가 더욱 흥미롭다.

street, factories and warehouses abandoned in the past by the private sector were transformed into cultural spaces, cafes, and restaurants. Seongsu Handmade Shoe Street is at the center of Seoul's urban regeneration project, and old factories or warehouses, such as Daelim Warehouse, Seongsu Shoes Theme Park, Onion Cafe, Jagmachi Café, Sufi, Photo Warehouse, and Orre, are reborn as cultural and artistic spaces, revitalizing Seongsu Handmade Shoe Street.

In terms of regional development, new regional cluster industries are created by encouraging new construction with incentives through deregulation when new facilities related to the IT industry are built. Seongsu-dong has emerged as a hip place and attracted media attention, while the location importance of Seongsu-dong is undergoing renewed recognition, while reconstruction projects in the region are being revitalized, and high-rise buildings are being developed. In the public realm, technical, management, and public relations support is being provided to small and medium-sized IT companies through anchor facilities; and most of all, Seongsu-dong is experiencing a revival, due to the influx of young workers.

To prevent damage to gentrification caused by commercialization and awareness of other old downtown areas in Seoul, Seongsu-dong planned to foster new industries and regenerate the region by saving the win–win agreement and commercial districts of existing areas. As a result, Seongsu-dong is transforming into a unique area with a long history of mixed old factories, apartments, new buildings and old low-rise commercial facilities and various snapshots of the city. It is true that Seongsu-dong is breaking away from the image of an old and underdeveloped gray factory area and creating an local identity. However, the reality is that old urban problems such as street and walking environments, parking, and lack of neighborhood parks still exist.

The changes over the past decade and the future are more interesting in how to

참고문헌

1. 성수동 2가 일대 도시환경정비구역지정 및 정비계획(안), 2015

2. 서울형 도시재생 시범사업 성수동 도시재생활성화계획(안), 2017

3. 성수동 준공업지역 공장건축물의 건축행위 특성에 관한 연구, 양유상, 박소현, 대한건축학회 논문집, 2018. 12

4. 도시재생의 관점에서 본 뚝섬지역 도시공간 변천에 관한 연구, 김수현, 백진, 한국건축역사학회 추계학술대회 논문집, 2014.11

5. 성수 도시재생 : '장인의 마을'은 가능한 것인가?, 한국도시설계학회 웹진. 2017.7

6. 오늘의 도시 설계 : 성수동, 한국도시설계학회 웹진. 2019. 3

7. 2030 준공업지역 종합발전계획, 서울시, 2015

8. 서울시 성수동 도시재생사업에 대한 산업지원 및 도시재생 통합적 관점의 특성 분석, 심소희, 구자훈, 서울도시연구 18(1), 2017.3

9. 서울생활문화자료조사 성수동, 서울역사박물관, 2014

10. 성수 도시재생지원센터 블로그 https://blog.naver.com/sd_seongsu

11. 성동구 홈페이지 https://www.sd.go.kr/main/contents.do?key=4344&

support urban industry and urban manufacturing, how non-homogeneous industrial integration and residential spaces can coexist, how the possibility of Seongsu-dong as a complex area can be maximized.

References

1. Designation and maintenance plan for urban environment improvement for Seongsu 2-ga area, 2015

2. Urban renewal pilot project, urban renewal plan for Seongsu, 2017

3. A Study on the Characteristics of Construction Behavior of Factory Buildings in Quasi-industrial Areas in Seongsu, Yusang Yang and Sohyun Park, Proceedings of the Korean Institute of Architects and Engineers, 2018. 12

4. A Study on the Change of Urban Space in the Ttukseom Area from the Perspective of Urban Regeneration, Kim Soo-hyeon and Baek Jin, Proceedings of the Fall Conference of the Korean Society of Architectural History, 2014.11

5. Seongsu Urban Regeneration: Is a 'Craftsman's Village' Possible?, Webzine of the Korean Society of Urban Design. 2017.7

6. Today's Urban Design: Seongsu, Webzine of the Korean Society of Urban Design. 2019. 3

7. 2030 Comprehensive Development Plan for Quasi -Industrial Area, Seoul, 2015

8. Analysis of the characteristics of industrial support for urban regeneration projects in Seongsu, Seoul and an integrated perspective of urban regeneration, Sohee Shim and Jahoon Koo, Seoul Urban Studies 18(1), March 2017

9. Seoul Living Culture Data Survey - Seongsu, Seoul Museum of History, 2014

10. Seongsu Urban Regeneration Support Center Blog https://blog.naver.com/sd_seongsu

11. Seongdong-gu Office website https://www.sd.go.kr/main/contents.do?key=4344

인천 코스모 40, 산업자원을 통한 공간재생

김환용 (한양대학교 에리카 건축학부 교수)

그림 1. 코스모40 전경 / CoSMo40 Facade

Incheon COSMO 40, Regeneration through Industrial Resources

Hwanyong Kim (Professor, Hanyang University ERICA, Dept of Architecture)

Changes in Chemical Factories

Founded in 1968, Hankook Titan industry Co., Ltd. changed its name to Hankook Titanium Industry Co., Ltd. Three years later, the firm constructed a factory in Gajwa-dong, Incheon, and in 2003, changed its name to Cosmo Chemical. In 2006, Cosmo moved to Ulsan City, and sold its 76,000 m² site in Incheon. At the time, the 76,000 m² site had 45 buildings. Most of the site was sold in units of around 1,000 m². Only one building that was used as a plant to refine and reuse the sulfuric acid used in the chemical factory remained undemolished, and this building number 40 was named 'CoSMo40'.

Through the process of planning and remodeling centered on the private sector, the existing chemical factory was transformed into a new cultural space, and showed Incheon's unique charm as an industrial city from a new perspective. CoSMo40 is an old industrial facility that has been remodeled as a cultural complex, which project resembles other industrial building regeneration projects. However, there is a clear difference, as the private sector is the key for the planning and operation of this project.

화학공장 변화의 의미

1968년에 설립된 ㈜한국지탄공업은 3년 후 ㈜한국티타늄공업으로 사명을 바꾸어 인천 가좌동에 공장을 준공하였고, 그후 2003년 코스모화학으로 다시 사명을 바꾸었다. 이후 2016년 울산시로 이전하면서 76,000m² 규모의 인천 부지를 매각하였다. 코스모화학 공장 이전으로 생겨난 76,000m²의 부지는 매각 당시 모두 45동의 건물이 있었다. 부지는 분할되어 대부분 1,000m² 내외로 나뉘어 재매각되었고, 그 중 화학공장에서 사용된 황산을 정제하여 재사용하는 플랜트로 사용했던 건물만이 철거되지 않은 채 남게 되었는데, 연번으로 40번에 해당했던 이 건물은 'CoSMo40'(코스모40) 이라는 이름을 얻게 되었다.

민간이 중심이 되어 1년 10개월간의 기획과 리모델링 과정을 통해 기존의 화학공장이 새로운 문화공간으로 탈바꿈하게 되었고, 결과적으로는 산업도시로 각인된 인천 고유의 매력을 새로운 시각에서 보여주기에 이르렀다. 연면적 3,508m², 지상4층, 지하1층으로 구성된 코스모40은 산업시설을 재생한 복합문화공간으로, 산업시설 재생을 활용하는 다른 프로젝트에서 보여지는 재생의 전략과 그 모습은 상당히 유사하다. 그러나 민간영역이 주축이 되어 기획되고 운영된 시설이라는 점에서 다른 사례와는 차별성을 가진다. 도시재생의 관점에서 이루어지는 많은 행위들이 공공의 영역에서 기획되고 실행된다는 점을 감안할 때, 코스모40이 구축된 과정, 그리고 그 과정에서 이루어진 많은 의사결정들이 민간 중심이었다는 점은 적지않은 의미를 가진다.

장소의 기억

코스모40이 위치한 지역은 인천 가좌동이다. 가좌동은 예로부터 가재울이라는 지역명에서 유래하듯 가재가 나오는 개울이라는 의미를 갖는다. 1970년대부터 공업단지가 들어서면서 1차 산업 중심의 다수의 제조업 관련 업체들이 입주하게 되었다. 아울러 우리나라 최초의 고

Considering that many urban regeneration projects are planned and implemented in the public domain, the fact that the processes of CoSMo40 remodeling and its many decisions were made mainly by the private sector suggests a considerable significance.

그림 2. 코스모40 내부 / Inside of CoSMo40

The Memory of the Place

CoSMo40 is located in Gajwa-dong, Incheon. The name 'Gajwa' comes from the old local name Gajae-ul, which is a stream where there are crayfish. When the industrial complex was established in the 1970s, primary industry became the major business, and various manufacturing-related companies moved in. Additionally, Gyeongin expressway, which is Korea's first highway, opened in 1967 from Gajwa-dong to Yeongdeungpo, Seoul. Gajwa-dong has built unrivaled hegemony in Incheon.

Gajwa-dong and its area, which experienced hegemony, are now regarded as past glory, while asking for another change in history. Industrial complexes, which seemed to remain unchanged, were shaken by various internal and external factors, such as

그림 3. 가좌동 전경 (출처: http://www.kyeongin.com) / Birdeye view of Gajwa area

속도로인 경인고속도로가 1967년 가좌동 톨게이트에서 서울 영등포 구간으로 개통되면서 가좌동은 인천 내에서도 독보적인 헤게모니를 구축하게 된다. 소금을 만들어내는 염전에서 출발하여 공업단지의 조성, 그리고 경인고속도로 개통을 통해 가좌동은 고유의 독특한 정체성을 확보한 것이다.

이렇게 부흥기를 경험한 가좌동 일대는 역사의 흐름 속에 또다른 변화를 요구 받게 되고 자연스럽게 지역의 번성은 과거의 영광으로 치부되게 된다. 변치 않을 것 같았던 공업단지는 산업구조의 개편, 인구변화, 경제성장 등 다양한 내·외부적 요인으로 인해 지역 정체성이 흔들리고 도시공간적 헤게모니는 주도권을 점차 잃어가게 되었다. 이렇기에 누군가는 가좌동의 정체성이 다 사라지기 전에 남기고자 노력하였으며 기억해야 하는 의미를 찾아 끊임없이 고민하였다. 이러한 노력이 코스모40이라는 복합문화공간이 탄생하게 된 배경이다. 가좌동이 갖는 의미에 대한 고민, 버려진 산업자원의 활용을 통한 정체성 확립이 코스모40의 가장 큰 의미라 볼 수 있다.

industrial restructuring, population change, and economic growth, and its urban spatial hegemony gradually lost its initiative. Therefore, it was necessary to retain Gajwa-dong's identity before it all disappeared, and was constantly agonized over at the time. This effort has led to the birth of the complex cultural space 'CoSMo40'. Concerns about the identity of Gajwa-dong and its founding character by reusing abandoned industrial assets are the key elements of CoSMo40.

Coexistence of the Past and the Present

It is never easy to add a modern interpretation to a building full of traces of the past built by the passage of time. Remodeling was needed to maintain the original appearance as much as possible, but to accomodate the necessary new functions, a design was required to represent the position of the artist, the main user of the space. CEO Shim Ki-bo and CEO Sung Hoon-sik of Bin Brothers, who have lived for generations in Gajwa-dong, Incheon, spent more than two years on carrying out the project. The two planned the CoSMo40 project through regeneration for purposes that could contribute to the local community. To remodel the existing building, considering the position of creators, artists, and sellers who will become the new users, the first stage renovation was conducted in 2018 with Yang Soo-in as architect, and in 2020, the second stage was directed to revitalize the interior space with Lim Seung-mo, CEO of the SML Architects office.

Under the current law, fireproof treatment is essential for the main structure of remodeled buildings. Usually, special paint or concrete is used, which eventually distorts the memory of the place through building regeneration, as these treatments

과거와 현재의 공존

　군이 무언가를 더하지 않아도 시간의 흐름 자체로 구축된 과거의 흔적 가득한 건물에 현대적 해석을 덧붙인다는 건 결코 쉽지 않은 일이다. 가능한 본래의 모습을 유지하되 필요한 새로운 기능을 잘 소화할 수 있는 리모델링이 필요했고, 공간의 주된 사용자인 예술가의 입장을 대변하는 디자인이 요구되었다. 인천 가좌동에 대를 이어 거주하고 있는 심기보 대표와 빈브라더스의 성훈식 대표는 2년 여의 시간을 들여, 이 건물을 통해 지역사회에 기여할 수 있는 용도로의 재생을 통한 코스모40 프로젝트를 기획하였다. 기존 건물의 모습은 유지하면서 새로운 사용자가 될 창작자나 예술가 그리고 판매자의 입장을 고려한 리모델링을 위해 양수인 건축가와 2018년 1단계 리노베이션을 진행하였고, 2020년에는 2단계로 SML 건축사사무소 대표 임승모 건축가와 내부공간 활성화를 위한 디렉팅을 진행하여 현재의 모습을 갖추게 되었다.

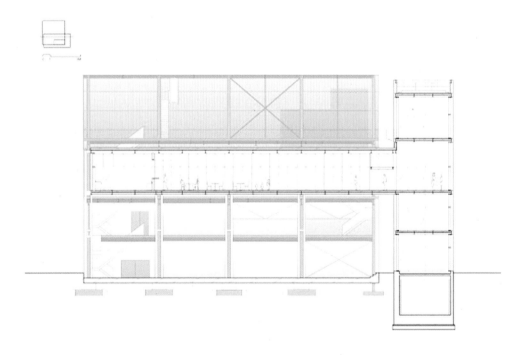

그림 4. 코스모40 단면 (출처: http://lifethings.in/gajwa/) / CoSMo40 section

erase the traces of the old building. So, the newly constructed space and the pre-existing space were designed separately to maintain CoSMo40's unique beauty. The operation method was also differentiated. Art is a program that frequently appears in urban regeneration facilities, but there are not yet many people in Korea who consume art-related content. Cultural programs are mostly private projects, so it is not easy to ensure sustainability. In the early stage of planning, CoSMo40 was participated in by Bean Brothers, who are famous for coffee, and used coffee as a mediator of space utilization. Cafés with low entry barriers can act as a starting point of interest in the building itself, and even the entire region.

The interior space is largely divided into the Main Hall and Hoist Hall, and the Main

그림 5. 코스모40 내부 전시공간 / CoSMo40 exhibition space

현행 법규상 오래된 건축물을 재생하기 위해서는 주 구조물에 내화처리가 필수이다. 이를 위해 특수 페인트나 콘크리트 등의 새로운 재료를 사용하면 옛 건물의 흔적은 사라지고 이는 결국 건축물 재생을 통해 장소의 기억을 왜곡하게 된다. 이를 위해 신축되는 공간과 기존 존재하는 과거의 공간이 맞닿지 않게 설계하였고 코스모40 특유의 폐허미를 유지할 수 있도록 하였다. 운영방식 또한 차별화되어 있다. 예술은 도시재생시설에 자주 등장하는 프로그램이지만, 아직 한국에서 예술 관련 콘텐츠를 소비하는 층은 많지 않다. 때문에 문화 프로그램은 민간사업으로서 지속 가능성을 담보하기 쉽지 않다. 코스모40은 기획 당시부터 커피로 유명한 빈브라더스가 참여하여 공간활용의 매개로서 커피를 사용하였다. 진입장벽이 낮은 카페는 작게는 건물 자체, 크게는 그 지역에 대한 관심의 시작점으로서 작용할 수 있다.

내부공간은 크게 Main Hall과 Hoist Hall로 나누어져 있으며 Main Hall은 다양한 공연과 이벤트를 수용할 수 있는 10m 높이의 공간이고 Hoist Hall은 옛 공장 3층을 사용하는 14m 높이의 대공간이다. 14m의 대공간 여러 곳엔 과거의 흔적이 남아 있다. 4층엔 과거 기계 작동을 위해 부분적으로 설치되었던 그레이팅 바닥이 남아 있으며, 굴토작업 중에 드러난 기존 공장의 기초부분이 위치한 곳에는 증축을 위한 구조적 장치만 추가한 채로 Bunker 공간으로 활용하였다. 옛 공장부분은 관입된 현대적 신관의 폴딩 도어를 통해 공간의 연결과 분절을 동시에 보장하는 장치가 되어준다. 2개의 홀은 여러 가지 프로그램으로 운영되지만 지역주민들에게 보다 더 다가갈 수 있도록 주·월 단위의 정기적인 음악공연, 명상, 요가, 워크숍 등과 같은 지역밀착형 프로그램을 시도하는 점은 코스모40이 추구하는 운영철학으로 볼 수 있다.

지역공간으로의 코스모40

우리나라는 짧은 산업화 기간에 비하여 미관상의 이유 또는 시설에 대한 거부감 등의 이유로 많은 수의 산업시설들을 큰 고민없이 철거해 버리는 경우가 많았다. 최근 증가하는 도시재

Hall is a 10 meter high space that can accommodate various performances and events, while the Hoist Hall is a 14 meter high large space using the third floor of the old factory. There are traces of the past in various large 14 m long spaces. The fourth floor was partially installed for machine operation in the past, and used as a bunker space with only structural devices for expansion. The old factory part serves as a device that simultaneously guarantees the connection and segmentation of space through the folding door of the new building. The two halls operate various programs, such as weekly and monthly regular music performances, meditation, yoga, and workshops; all of these can be seen as the operating philosophy pursued by CoSMo40.

CoSMo40 as a Local Space

Compared to its short industrialization period, Korea often demolished a large number of industrial facilities without much consideration, due to aesthetic reasons or reluctance to use the facilities. Interest in urban regeneration, which has recently increased, is improving these consuming efforts in a more positive direction under the name of regeneration architecture and regeneration space. The number of cases of old developments being reborn as unique commercial and cultural spaces by retaining all or part of the old space has increased significantly.

However, most of these attempts are focused on one-time events or buildings themselves, and are not often accepted as part of efforts to preserve harmony or locality with their surrounding areas. In urban and spatial regeneration, only the subject of space is often highlighted. Creating a spatial foundation that can contribute to the community through various exchanges and experiments with the area will play

생에 대한 관심은 이러한 소모적 노력을 그나마 재생건축, 재생공간이라는 이름으로 보다 긍정적인 방향으로 개선시키고 있다. 옛 공간의 전체 또는 부분을 남겨 독특한 상업 및 문화공간으로 재탄생시킨 사례가 많이 증가하였다.

다만, 이러한 시도들이 다분히 일회성 이벤트 또는 건축물 자체에 집중되어, 주변 지역과의 조화나 지역성을 살리기 위한 노력의 일환으로 받아들여지는 경우는 많지 않다. 도시 및 공간재생에 있어 주로 그 대상이 되는 공간에 대한 주제만 부각되는 경우가 많은데, 진정한 의미의 공간재생을 위해서 그 공간이 위치한 지역과의 다양한 교류 및 실험 등을 통해 시행착오를 겪으면서 커뮤니티에 기여할 수 있는 공간적 기반을 만드는 것이 도시재생의 근원적 차원에서 중요한 역할을 할 것이다.

도시라는 유기적 생명체에서 새로운 지역이 생겨나는 것만큼이나 낙후된 지역들이 발생하는 것은 매우 필연적인 현상이다. 이러한 도시의 자연스러운 변화과정에서 낙후된 지역들을 다시 살리기 위한 노력을 통해 새로운 정체성을 부여받을 수 있는 기회를 제공하고, 더 나아가 새로이 해석된 공간이 주변 지역과 조화될 수 있도록 만들어 나가는 것이 진정한 의미의 공간재생, 도시재생의 목적이 될 것이다.

그림 6. 코스모40 전시공간과 전시물 / Exhibition and art works inside CoSMo40

an important role in the fundamental level of urban regeneration.

The occurrence of underdeveloped areas in cities is as inevitable as the creation of new areas. In the process of the natural change of a city, efforts to revive underdeveloped areas will provide an opportunity for the city to be given a new identity. Furthermore, creating a newly interpreted space in harmony with the surrounding area will be the real purpose of spatial regeneration and urban regeneration.

CoSMo40, which was almost demolished, is in its second heyday, in which its meaning is very different from the past. The value pursued by CoSMo40 to establish a new identity is reinterpreted and re-granted in the modern era. These efforts can be found in establishing a master plan for the surrounding area at the same time as the establishment of CoSMo40. The win–win value of CoSMo40, which is aimed to pursue with the region through a masterplan, is centered on the old house of the Shim family, who has lived in Gajwa-dong for 400 years, and that value is sufficient to provide a new perspective on space regeneration.

Over three years, Shim's old house 'Gwanhaegak' in the surrounding Cheongsong district was newly transformed into a hanok cafe, the CoSMo40 area was selected for the '2020 Village Art Project', and various murals and public sculptures were installed throughout the surrounding area. What is noteworthy about these changes is that various artists and businessmen from or living in Incheon are gathering to operate showrooms, studios, and surf shops around CoSMo40. In other words, if the beginning of CoSMo40 was to create a spatial re-creation and cultural space, CoSMo40 will now establish itself as an anchor facility that can grow and move forward with the surrounding area.

철거될 뻔했던 코스모40은 제2의 전성기를 맞았다. 두 번째 전성기에서 코스모40이 갖는 의미는 과거의 모습과는 많이 다르다. 현대에 재해석되어 다시 부여된 새로운 정체성 확립을 위해 코스모40이 추구한 가치는 지역과의 상생이다. 이러한 노력은 코스모40이 구축되면서 주변지역에 대한 마스터플랜을 동시에 수립하는 시도에서 찾아볼 수 있다. 가좌동 일대에서 400년 터전을 일구며 살아온 심씨 가문의 고택을 중심으로 주변 지역을 아우르는 마스터플랜을 통한 지역적 배려 및 지역과 함께 추구하고자 하는 코스모40의 상생적 가치는 최근 단발성 또는 이벤트성으로 치부되는 공간재생에 대한 새로운 시각을 제공하기에 충분하다.

3년이 넘는 시간 동안 주변 청송 심씨 고택 관해각은 한옥 카페로 새로이 변화하였고 코스모40 일대가 '2020 마을미술 프로젝트 사업'에 선정되어 다양한 벽화, 공공조형물이 주변 곳곳에 설치되었다. 이러한 변화에서 괄목할 부분은 인천 출신 또는 거주 중인 다양한 예술가 및 사업가들이 모여 코스모40 주변에 쇼룸과 스튜디오, 서프샵 등을 운영한다는 사실이다. 즉, 코스모40의 시작이 폐허미를 활용한 공간적 재창출과 문화공간 조성이었다면 이제 코스모40은 주변 지역과 함께 성장하고 나아갈 수 있는 하나의 앵커 시설로 자리매김한다.

맺으며 : 시너지를 위한 노력

인천은 수도권에 위치한 도시로서 서울과는 다른 도시적 경관을 갖고 있다. 오랜 기간 방치된 낡은 공장들, 산업단지, 공업단지들과 함께 성장한 도시공간은 인천에게 산업화시대의 정체성을 수립해준 고마운 장치들이다. 현대에 이르러 쇠퇴하는 지역의 전형으로 여겨지는 이러한 공간들은 사실상 근대 건축물 또는 근대 도시경관의 유산으로 가치가 있을 뿐 아니라 코스모40과 같은 공간재생을 통해 얼마든지 새로운 생명력을 부여할 수 있다. 인천에 존재하는 수많은 산업화와 공업화의 흔적은 구시대적 유물이 아닌 인천만이 갖는 독특한 산업경관으로서 활용될 수 있으며 이러한 경관자원으로서 폐공장의 현대적 해석을 통한 지역성 구축은 도시재생사업의 새로운 모델이 될 수 있을 것이다.

Closing Remarks : Efforts to Create Synergy

Incheon is a city located in the metropolitan area, and has a different urban landscape from Seoul. Urban spaces that have grown with old factories and industrial complexes that have been neglected for a long time are grateful devices that have allowed Incheon to establish the identity of the industrial era. These spaces, considered a decline in modern times, are both valuable as modern buildings or the heritage of modern urban scenery, and can give new vitality through spatial regeneration, such as CoSMo40. Many traces of industrialization in Incheon can be used as a unique industrial landscape only in Incheon, not as an outdated relic. The establishment of locality

그림 7. 코스모40 주변 신규시설들: 심씨 가문 사택을 리모델링한 카페, 서핑샵, 주민시설 등 다양한 소규모 건축들이 코스모40 과 함께 지역성을 구축한다 / Surrounding areas of CoSMo40

코스모40은 민간의 영역이 먼저 나서 공간재생을 위한 노력을 보여준 대표적인 예이다. 그러나 앵커시설이 주변의 쇼룸, 스튜디오 등과 연결되기 위해서는 기본적으로 공공의 역할이 중요하다. 보행자를 위한 배려, 앵커시설을 위한 대중성의 확보, 지역성 확립을 위한 주민들과의 적극적인 교류 등은 민간의 활동만으론 명확한 한계를 보인다. 개인의 관심과 노력에서 시작된 이러한 의도가 도시를 재생하고자 하는 제도적 실천에 궁극적으로 적용되기 위해선 보다 적극적인 공공과의 접점을 만드는 노력이 필요할 것이다. 코스모40을 통한 민관이 함께하는 노력이 향후 변화하는 도시공간에 새로운 도시재생, 공간재생으로의 이정표를 만들 수 있을 것으로 기대한다.

참고문헌

1. [2020 어반 오디세이]8 사라진 공장과 남아있는 40번 건물: 인천 코스모40, 방승환, 국토연구원, 2020.8

2. 공장의 재탄생: 코스모40 기획이야기, SPACE, 2019.3

3. 인천 가좌동의 디자인 허브, 코스모40, 월간디자인, 2021.3

4. 코스모40 홈페이지: https://www.cosmo40.com/

5. 낡은 화학공장을 문화공간으로, 가좌동 '코스모 40', 인천도시공사, 2018.11

6. 혐오시설에서 힙한 문화단지 꽃피운 인천 '코스모40', 시사저널, 2020.3

7. 가좌동 코스모40에서 펼쳐진 복합문화예술, 인천인, 2018.12

through the modern interpretation of abandoned factories as landscape resources can be a new model for urban regeneration projects.

CoSMo40 is a representative example of the private sector's efforts to regenerate space first. However, for anchor facilities to be connected to nearby showrooms and studios, the public role is of basic importance. Consideration for pedestrians, securing popularity for anchor facilities, and active exchanges with residents to establish locality have distinct limitations only with private activities. For these intentions, which began with individual interest and efforts, to be ultimately applied to institutional practices to regenerate the city, efforts for more active contact with the public are needed. It is expected that the public–private efforts through CoSMo40 will create a milestone for new urban regeneration and spatial regeneration in the future.

References

1. [2020 Urban Odyssey] 8 The disappeared factory and remaining building No. 40: Incheon Cosmo 40, Bang Seunghwan, Korea Research Institute, 2020.8
2. Factory Rebirth: CoSMo40 Planning Story, SPACE, 2019.3
3. Design hub in Gajwa-dong, Incheon, Cosmo 40, Monthly Design, 2021.3
4. CoSMo40 homepage: https://www.cosmo40.com/
5. An old chemical factory as a cultural space, "Cosmo 40" in Gajwa-dong, Incheon City Corporation, November 2018
6. Incheon CoSMo40, which blossomed a hip cultural complex in a hate facility, Current affairs Journal, 2020.3
7. Complex Culture and Arts at CoSMo40 in Gajwa-dong, Incheonin, 2018.12

서울 마포 문화비축기지,
산업 유휴시설의 재탄생

김호정 (단국대학교 건축학부 교수)

들어가며

문화비축기지는 1970년대부터 사용되었던 석유탱크와 인접한 외부공간을 새롭게 리노베이션하여 다목적 복합문화공간으로 재생시킨 사례이며, 서울 상암월드컵경기장 서쪽 마포구 중산로 87에 위치한다. 기존 5개의 석유탱크는 T1에서부터 T5까지 각기 다른 프로그램을 가진 문화공간으로 변화되었고, T6는 커뮤니티 및 정보교류센터로 새롭게 조성되어 2017년 개장하였다.

문화비축기지는 조성과정에서 장소와 시설의 활용방안에 대해 시민의 적극적인 논의를 바탕으로 한 공론화 과정으로 만들어졌다는 점에서 기존의 관주도의 도시재생 사례와는 차이가 있다. 근대산업시설의 재생이라는 측면에서 오스트리아 빈의 Gasometer City, London의 Tate Modern Gallery, Truman Brewery와 유사한 사례라고 할 수 있으며, 쓰임을 다한 유휴시설을 재생하여 새로운 모습으로 재탄생시킴으로써 과거를 기억함과 동시에 새로운 활력을 만들어나간다는 데 그 의미가 있다.

서울시는 문화를 하나의 자원으로 인식하고 '탱크에 문화를 채우다'라는 의미를 새기고자 이 프로젝트를 문화비축기지로 명명하였다. 40년 이상 보안시설로 존재를 알 수 없던 시설이 시민들에게 개방되었다는 상징성과 더불어 석유비축탱크가 가지는 건축 유형의 특이성으로

Seoul Mapo Oil Tank Culture Park, Rebirth of Industrial Unused Equipment

Hojeong Kim (Professor, Dankook University, Dept of Architecture)

Introduction

The Oil Tank Culture Park is an example of regeneration as a multi-purpose complex cultural space. The oil tank used since the 1970s was newly renovated along with the adjacent external space, and is located at 87 Jungsan-ro, Mapo-gu, west of the Sangam World Cup Stadium in Seoul. The existing five oil tanks T1 to T5 have been transformed into cultural spaces with different programs, and T6 was newly created as an information exchange center with the community, and opened in 2017.

The Oil Tank Culture Park is different from the institution-led urban regeneration case, in that it was created based on the process of publicization based on the active discussion of citizens about the utilization plan of places and facilities in the composition process. From the perspective of regenerating modern industrial facilities, it can be considered similar to Gasometer City in Vienna, Austria, the Tate Modern Gallery in London, and Truman Brewery. It is meaningful in that it creates new vitality while remembering the past, by regenerating the used and unused equipment into a new shape.

그림 1. 문화비축기지와 상암월드컵 경기장(출처: 서울시 공원 홈페이지)
Oil Tank Culture Park and Sangam World Cup Stadium

인하여 조성 초기부터 많은 관심을 받았으며, 현재는 시민문화단체가 중심이 되어 다양한 문화프로그램을 제공하고 있다.

조성과정

1973년 중동전쟁으로 인한 석유파동 이후, 서울시는 1976~78년에 걸쳐 비상상황에 대비하여 매봉산 자락에 민수용 유류 저장시설을 건설하였다. 당시 서울 시민의 한 달 석유 사용량인 40만 배럴(6,908만 리터)의 비축을 위해 5개의 유류 저장탱크를 설치하고, 유류 출하기 1대, 사무실, 창고, 기계실, 관사와 같은 부대시설을 조성하였다. 석유비축기지는 일반인의 접근과 이용이 통제된 1급 보안시설로 분류되었으며, 40여 년간 시민들은 그 존재를 알 수 없이 철저히 격리된 공간이었다.

이후 2000년 석유비축기지 동쪽에 2002년 월드컵 개최를 위한 상암동월드컵경기장이 조

The Seoul Metropolitan Government has named this project 'The Oil Tank Culture Park' to recognize culture as a resource, and to inscribe the meaning of 'Filling the tank with culture'. In addition to the symbolism that the facility, whose existence could not be known to the public for security reasons for more than 40 years, was opened to the citizens, the peculiarity of the building type of the oil storage tank has received a lot of interest from the early stage of construction. Currently, civic cultural organizations play a central role in offering various cultural programs.

그림 2. 문화비축기지 항공 사진(출처: World Architecture) / The Aerial Photo of Oil Tank Culture Park

성되었는데, 경기장과 너무 근거리인 관계로 안전상의 우려가 제기되었다. 당시 월드컵 개최 시까지는 너무 촉박하여 석유비축기지는 일단 폐쇄된 후 이전이 결정되었으며, 석유비축기 지 전면의 외부공간은 월드컵 행사를 위한 임시주차장 용도로 쓰이게 되었다. 이후 10여 년간 시설은 방치되었고 지역주민들과 자치구청은 시설의 공원화를 위해서 서울시에 건의를 하기 에 이른다. 이에 서울시는 석유비축기지에 대한 활용방안을 모색하기 시작하여 2013년 일반 시민대상 아이디어 공모전을 실시하고, 2014년 전문인을 대상으로 '마포 석유비축기지 재생 을 위한 공원화사업'을 위한 국제설계경기(Creating a Cultural Depot from an Oil Tank Depot)를 시행하였다. 그 결과 RoA 건축사사무소 허서구 건축가의 '땅으로부터 읽어낸 시 간'이 당선작으로 선정되었다.

그당시 서울시는 서울형 도시재생이라는 명칭으로 다수의 도시재생사업을 실행하였는데, 도시공간의 역사를 그릇 삼아 지역시민들이 향유할 수 있는 장소 만들기를 목표로 하였다. 서 울역 인근 고가도로를 보행자 전용 스카이 워크로 변신시킨 '서울로 7017', 쇠락해 가는 세운 상가를 철거하는 대신 주변과 연결하고 도심창의산업과 연계한 거점공간으로 조성하는 것을 골자로 하는 '다시세운 프로젝트'가 마포 문화비축기지와 비슷한 시기에 조성되었다.

특히 서울시는 문화비축기지 조성 초기부터 이 프로젝트를 시민주도형의 문화재생사업으 로 진행하고자 하는 의지가 있었다. 기존의 관 주도방식인 시설완공 이후 운영방식의 결정 및 운영자 선정을 진행하는 관례에서 벗어나고자 하였으며, 시민 참여로 기획·운영방안을 마련 하고 그에 최적화된 시설을 설계·시공하는 '新도시재생 프로세스'를 적용하였다. 이를 위해서 기본설계(2014.10~2015.4) 과정에서 실제 기획·연출·운영 실무 전문가로 구성된 '워킹그룹' 이 제안한 의견을 반영해 최종 설계안을 확정했다. 이 과정에서 당초 계획범위에 들어가 있지 않았던 주차장부지가 포함되어 '문화비축기지' 조성 대상지는 총 14만m² 규모로 유류저장탱 크(10만1,510m²) 주차장부지(3만5,212m²), 산책로(3,300m²)로 확장되게 된다. 더불어 전 문가 그룹과는 별개로 시민기획단 '탐험단'를 만들어 지역주민 및 민간 예술가 집단과 활성화 방안을 논의하였다. 최종적으로는 2014년부터 3년간 전문가그룹을 통한 설계자문 24회, 실 무회의 41회를 통해 문화비축기지의 운영 및 이용 활성화 방안을 마련하였다.

Creation Process

After the oil crisis caused by the Middle East War in 1973, the city of Seoul built a private oil storage facility at the foot of Maebong Mountain from 1976 to 1978, in preparation for an emergency. At the time, five oil storage tanks were installed to stock up to 400,000 barrels (69.08 million liters) of oil to be used by Seoul citizens per month, and auxiliary facilities, such as an oil loading arm, offices, warehouses, machine rooms, and government offices, were built. The Oil Tank was classified as a first-class security facility with controlled access and use by the general public, and it was a completely isolated space, where for more than 40 years, citizens could not know of its existence.

Since then, the Sangam-dong World Cup Stadium was created to the east of the Oil Tank in 2000 to hold the 2002 World Cup, raising safety concerns, due to its proximity to the stadium. At that time, until the World Cup was held, the oil tank was closed and then relocated, and the space in front of the oil stockpile was used as a temporary parking lot for the World Cup events. The facility was abandoned for the next 10 years, and local residents and the autonomous region began to proposition the Seoul Metropolitan Government to turn the facility into a park. Seoul Metropolitan Government began exploring ways to utilize oil storage bases, held an idea competition for the general public in 2013, and targeted experts in 2014 for the Mapo Oil Tank park project, 'Creating a Cultural Depot from an Oil Tank Depot' that was carried out. As a result, 'Time Read from the Land' by Seo-gu Heo of the RoA Architects Office was selected as the winning work.

This time, the Seoul Metropolitan Government has been running a large number of urban development policies under the name of Seoul urban regeneration. They aimed to create a place where local citizens could enjoy the history of urban spaces. Seoullo

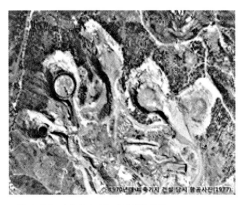

그림 3. 석유 비축기지 건설 당시 항공사진
The aerial photo of the construction of the oil storage

총 사업비는 470억이 소요되었고 공사 기간은 2015년 12월~2017년 8월이며, 2017년 9월 1일 개원하여 10월 14일 개원 기념시민축제가 있었다. 현재 마포 문화비축기지는 서울시 푸른도시국과 문화비축기지협치위원회가 공동으로 민관 협치로 운영되고 있으며 연간 운영비는 50억이다.

건축적 접근방식

기존 석유탱크는 매봉산 자락의 암반을 폭발시킨 다음 후면 옹벽을 설치하고 다시 그 내부에 석유탱크를 설치하여 전면 차폐벽을 조성하는 형식으로 조성되었으며, 모두 5개의 탱크가 있었다. 건축가는 '문화비축기지 구축과정은 과거 석유비축기지 구축과정의 역순으로 진행된다'라는 개념으로 프로젝트를 시작하였다. 프로젝트를 진행하기 위해서는 과거에 석유탱크가 조성되었던 구축과정을 다시 발굴할 필요가 있음을 이야기 하면서, 묻혀 있던 구축 과정의 발굴을 통하여 새로이 들어서야 할 계획의 방향을 정당화하였다.

건축가는 각각의 석유탱크에 대해서 활용방법과 존치형식을 각각 다르게 제안하면서, 과거 되메워진 차폐지형을 걷어내고 작업로의 암반지형을 노출하며 전면 차폐벽의 개폐와 변형 여부를 각 탱크마다 서로 다른 방식으로 결정하였다. 암반 절개지, 콘크리트 옹벽, 석유 탱크라는 세 가지의 핵심 디자인 요소가 설정되고 각 요소들 간의 관계 재설정을 통해 디자인의 전개가 이루어졌다.

건축가는 프로그램상 각각의 탱크에 덧붙여지는 매스는 최소하고 전체적으로 절제된 언어로 형상을 구현하였으며, 기존의 석유탱크에 대해서는 탱크 자체를 보강하거나 구조물로 사용하지 않는 것을 원칙으로 하고 시간이 지남에 따라서 자연스럽게 부식되도록 하였다. 석유

7017 transformed an overpass near Seoul Station into a pedestrian-only sky walk; the Dasi Sewoon Project of Seoul's Sewoon Arcade, whose main goal is to create a base space that connects with the surroundings and connect with the city's creative industry instead of demolishing the fading Sewoon Plaza, was built around the same time as the Mapo Oil Tank Culture Park.

In particular, Seoul Metropolitan Government was willing to proceed with this project as a citizen-led cultural revitalization project from the early days of the formation of the Mapo Oil Tank Culture Park. They deviated from the habit of deciding the operation method and selecting the operator after completion of the facility, which is the existing government-led method, and devised a plan and operation plan with citizen participation, and applied the 'New Urban Regeneration Process' to design and construct an optimized facility. Therefore, in the process of basic design (October 2014 — April 2015), the final design proposal was finalized by reflecting the proposed opinion of the 'working group' composed of experts in actual planning, production, and operation. In this process, the site of the parking lot, which was not within the scope of the original plan, was included. The target site will be expanded to a total of 140,000 m^2 of oil storage tank (101,510 m^2) parking lot (35,212 m^2), and walking trail (3,300 m^2). In addition, apart from the expert group, a civic project 'Exploration Team' was formed to discuss revitalization plans with local residents and a group of private artists. Finally, for three years from 2014, we prepared a plan to revitalize the operation and use of the cultural storage base through 24 design consulting sessions and 41 business meetings.

The total project cost was 4 million dollars, and the construction period was from December 2015 to August 2017. A citizen festival commemorating the opening of the park was held on September 1, 2017, and again on October 14, 2017. Currently, the Mapo Oil Tank Culture Park is jointly operated by the policy green city of Seoul and the

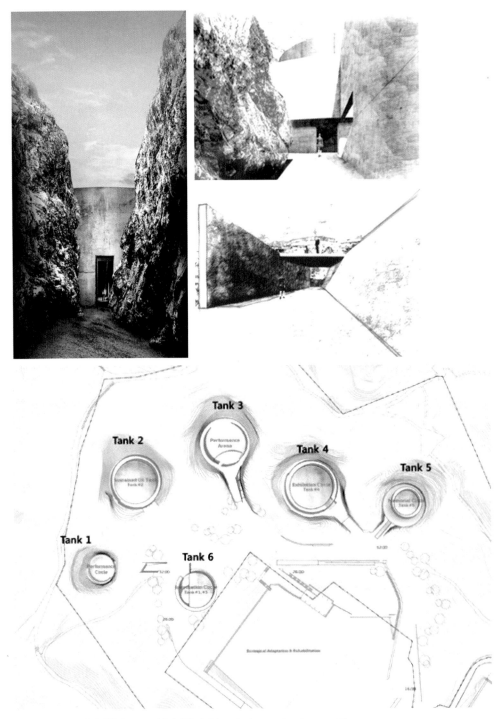

그림 4. 현상설계 당선안(출처: RoA 건축사사무소) / The winning works of competition

Mapo Oil Tank Culture Cooperation Committee under a public-private partnership, with an annual operating cost of 5 million dollars.

Architectural Approach

Existing oil tanks were constructed in the form of exploding the bedrock at the foot of Maebongsan Mountain, then installing the rear retaining wall, installing an oil tank inside it, and creating a front shielding wall, there being five tanks in total. The architect started the project with the concept of 'the construction process of the Mapo Oil Tank Culture Park proceeds in the reverse order of the construction process of the past oil tank'. To proceed with the project, it was said that it was necessary to re-excavate the construction process where oil tanks were built in the past, and justify the direction of the new plan through excavation of the construction process that had been buried.

The architects propose different usage methods and types of maintenance for each oil tank, removal of the refilled shielding terrain, exposure of the rocky terrain of the work furnace, and determination of whether the front shielding wall is opened or deformed in a different way for each tank. Three core design elements were established: rock incision, concrete retaining wall, and oil tank; and the development of the design was accomplished by reestablishing the relationship between the elements.

The architect's program was to minimize the mass added to each tank, and implement it in a completely conservative language. Existing oil tanks, in principle, were not reinforced or used as structures, and naturally corroded over time. In some cases, the rear retaining wall for oil tank protection was reinforced, and finally the rock wall of the access road was devised.

탱크 보호를 위한 후면 옹벽은 경우에 따라서 암벽을 보강하고 최종적으로는 진입로 암벽을 구상하는 방식으로 작업을 진행하였다.

건축적 변형과 디자인

기존의 5개의 탱크 높이는 모두 15m이고 가장 작은 것은 1번 탱크로 지름이 15m, 2번과 4번 탱크가 가장 큰 탱크로 지름이 33.7m에 이른다. 건축가는 각 탱크들이 각기 다른 방식으로 변형되어 재발견되는 형식으로 차별화하여 디자인을 전개하였으며, 각 탱크는 산책로와 연결되어 있다. 5개의 탱크 중 첫 번째 탱크 T0는 기존 탱크와 동일한 형상의 유리 파빌리온으로 대체되었고, 세 번째 탱크 T3는 원형으로 보존 존치하여 후대에 과거의 모습을 그대로

그림 5. T0 문화마당(출처: 서울시 공원 홈페이지) / T0 Culture Field

Architectural Transformation and Design

The five existing tanks are all 15 m high, and the smallest one is the 1st tank, with a diameter of 15 m, while the 2nd and 4th tanks are the largest tanks, with a diameter of 33.7 m. The architects develop their designs in a way that each tank is transformed and rediscovered in a different way, and each tank is connected to a sidewalk. The first of the five tanks, T0, was replaced with a glass pavilion that had the same shape as the existing tank; while the third tank, T3, was retained in its original shape to convey its past appearance to future generations. The new tank, T6, was placed closest to the Cultural Square, which spreads upon entry, so that visitors would recognize it first. This tank directly showed that the existing tanks 1 and 2 could be reconstructed with dismantled materials to reconstruct the legacy of the past and regenerate it.

T0 is a cultural area created by minimal transformation of the existing parking lot site so that large-scale outdoor events can be held. Early in the design, the architects tried to design a suitable structure for this space to shield the tank at the back, which was apparent at a glance because the area was overly open. However, the civic cultural groups associated with the project preferred a changeable open outdoor space, and the design was finished in its current state. Except in winter, the Cultural Plaza holds multiple events every weekend and is positioned as a space that is full of tourist attractions and attractions for the citizens.

For T1, an existing tank was dismantled to create a glass pavilion with the same shape. The oil tank at T1 was the smallest in size, 15 m in diameter. It is reported that the tank where the existing 3,019 L of gasoline was stored was highly volatile, and presented great difficulty for management. Inside the pavilion, the environment surrounding the tank from which the display is made can be seen, with the rocks and

전할 수 있도록 하였다. 새로운 탱크인 T6를 진입과 동시에 펼쳐지는 문화광장에서 가장 가까운 거리에 위치시켜 방문자가 가장 먼저 인지할 수 있도록 하였다. 이 탱크는 기존 탱크 1, 2가 해체된 재료를 가지고 재구성하여 과거의 유산이 재구성되어 새롭게 재생될 수 있음을 직접적으로 보여 주었다.

T0는 기존 주차장 부지를 최소한으로 변형하여 야외 대규모 행사가 가능하도록 조성한 문화마당이다. 디자인 초기에 건축가는 해당 공간이 지나치게 개방형이어서 한눈에 드러나고 후면부에 있는 탱크의 차폐를 위하여 이 공간에 적절한 구조물을 설계하고자 하였으나, 프로젝트와 관련된 시민문화단체에서는 다목적의 열린 야외공간을 선호하여 현재의 상태로 디자인이 마무리되었다. 문화광장은 겨울철을 제외하고 주말마다 여러 이벤트가 열려서 시민을 위한 볼거리, 즐길거리가 가득한 공간으로 자리매김 하고 있다.

T1은 기존 탱크를 해체하고 같은 형상으로 유리 파빌리온을 만든 것이다. T1에 있던 석

그림 6. T1 유리 파빌리온에서 옹벽을 바라본 모습
(출처: World Archtiecture)
A view of the retaining wall from the T1 glass pavilion

그림 7. T4 탱크와 옹벽 사이 공간의 재구성
(출처: World Archtiecture)
Reconstruction of the space between the T4 tank and the retaining wall

retaining walls on the outside. The dramatic atmosphere of the bedrock and buildings seen beyond the retaining wall is created, and the contrast is extreme, especially at night. This glass pavilion has an additional exhibition space called the Cultural Passage, which was also created. The total area is 554 m², it accommodates 120 people inside and outside, and can be used for exhibitions and performances. The original dismantled tank was used for the interior finish of the T6, and anchoring was installed to stabilize the existing rock.

그림 8. T1 유리 파빌리온 (다목적 공간)과 문화통로(출처: 서울시 공원 홈페이지)
T1 Glass Pavilion : Multipurpose Space and Culture Passage

유탱크는 사이즈가 가장 작아서 지름이 15m이다. 기존에 휘발유 3,019리터가 보관되어 있던 탱크로 휘발성이 커서 관리에 큰 어려움이 있었다고 전해진다. 파빌리온 내부에서는 외부에 있는 암반과 옹벽이 그대로 보여 원래의 탱크를 둘러싼 환경을 볼 수 있다. 옹벽 너머로 보여지는 암반과 건축물이 극적인 분위기를 연출되는데, 특히 밤에 그 대조가 극을 이룬다. 이 유리 파빌리온에는 덧붙여진 문화통로라 불리는 별도의 전시공간도 있으며, 전체 면적은 554m²로 120명 내외의 인원을 수용하며 전시와 공연이 가능하다. 해체된 원래 탱크는 T6의 내부 마감에 사용되었고, 기존 바위를 안정시키기 위해서 앵커링을 하였다.

그림 9. T2 공연장(출처: 서울시 공원 홈페이지) / T2 Concert hall

T2는 기존의 탱크와 옹벽 일부를 절단하여 상부는 야외공연장을 만들고 하부는 200석 규모의 지하공연장을 만들었다. 하부공연장 면적은 2579m²로 400명 내외 인원이 이용 가능하며, 상부 야외공연장은 입구로부터 자연스럽게 경사를 따라 올라가면 만나게 되는 암벽과 차

For T2, a part of the conventional tank and retaining wall was cut to create an outdoor performance hall at the top, and an underground venue with 200 seats at the bottom. The lower venue area is 2,579 m² and can accommodate around 400 people, while the upper outdoor performance hall is a cozy exterior surrounded by rocks and shield walls that is met when climbing naturally along the slope from the entrance. At the bottom of this exterior space, cubic stone cushions are geometrically arranged and used as an exterior space, so that the public can use it even when there are no performances. This is the most impressive exterior space in Mapo Oil Tank Culture Park.

그림 10. T2공연장 상부 야외 공연장(출처: World Architecture)
Outdoor performance hall on the upper level of T2 concert hall

As mentioned earlier, tank T3 is preserved in its original form, so that viewers can see the original appearance of the oil tank. For T3, the original tank was preserved as it

그림 11. T3 탱크 원형으로의 접근로(출처: World Architecture)
Access to the T3 tank

그림 12. T3 탱크 원형(출처: 서울시 공원 홈페이지)
T3 tank in original shape

폐벽에 의해 둘러싸인 위요감이 있는 외부공간이다. 이 외부공간의 바닥에는 정육면체의 돌방석이 기하학적으로 배치되어 공연이 없을 때도 머무를 수 있는 외부공간으로 활용되고 있다. 이 문화비축기지에서 가장 인상적인 외부공간 이기도 하다. T3는 앞서 언급한 바와 같이 원형 그대로 보존하여 관람자는 석유탱크 본연의 외관모습을 볼 수 있다. T3는 기존의 탱크를 원형 그대로 보존하여 후대에도 원래의 모습을 알 수 있도록 하였다. T3 내부로는 접근이 불가하고 외부 옹벽과 탱크 사이의 공간에 접근하여 과거의 모습을 그대로 볼 수 있다.

T4는 탱크의 독특한 내부형태를 활용한 복합문화공간으로 탱크 내부를 원형 모습 그대로 체험할 수 있다. 면적은 984m²로 130명 내외의 인원이 이용 가능하다. 탱크 안으로 쏟아져 들어오는 빛과 세장한 철제 원형 기둥으로 이루어진 원뿔 지붕형상의 공간은 지금까지 체험해보지 못한 새로운 느낌을 준다. 대체적으로 어두운 공간으로 미디어 아트와 같은 형식의 전시가 어울린다.

was, which made it possible for posterity to know its original appearance. The interior of the T3 is not accessible, while the space between the outer retaining wall and the tank can be accessed to see the past.

For T4, a multicultural space was created that utilizes the unique internal shape of the tank, and the inside of the tank can be experienced in its original form. The area is 984 m², and can be used by 130 people inside and outside. The conical roof-shaped space consisting of the light shining into the tank and the elongated iron cylinder give a new feeling that has not been experienced before. As it is generally a dark space, an exhibition in the same format as media art is suitable.

T5 offers a story hall and a space for exhibits on the history of Mapo Oil Tank Culture Park. The tank has mass and ramps inserted to increase the use of the exhibition space, and a protective retaining wall is used outside the oil tank. The tank is used as a 360 degree video screening space, with an area of 890 m², and can be used by up to 100 people. For T5, the public are able to experience both the exterior and exterior of the tank, and the partially enclosing concrete retaining wall, rock and incision; and a staircase was set up between the exterior of the tank and the retaining wall.

T6 offers a newly built building that used as an interior / exterior material that reassembled the iron plates dismantled from the other two tanks inside and outside of the community center. It provides convenient facilities, such as an operation office, conference rooms, lecture rooms, and cafeteria to communicate with citizens. T6 is located in the foreground against the backdrop of the five existing tanks, and there are some criticisms of its position in the foreground. However, the recycling of rust-written iron plates can make the building look as if it had existed in the past, or can act as a device to show the history of the place.

그림 13. T4의 내부 전시 공간(출처: World Architecture) / Internal exhibition space in T4

T5는 이야기관으로 문화비축기지의 역사에 대한 전시가 이루어지는 공간이다. 일부 매스와 램프가 삽입되고 석유탱크 바깥 보호 옹벽을 사용하여 전시공간으로써의 활용도를 높였다. 탱크 내부는 360도 영상 상영공간으로 활용되며, 면적은 890m²으로 최대 100여 명이 이용가능하다. T5에서는 탱크 내 외부, 콘크리트 옹벽, 암반 및 절개지를 모두 경험할 수 있으며, 탱크 외부와 옹벽 사이에 계단이 설치되었다.

T6는 커뮤니티 센터로 다른 두 탱크에서 해체된 철판을 내외부에 재조립하여 내외장재로 사용한 신축건물이다. 운영사무실, 시민과 소통하는 회의실, 강의실, 카페테리아와 같은 편의시설을 제공한다. T6는 기존에 있던 5개의 탱크를 배경으로 가장 앞쪽에 위치하여 그 위치가 전면에 있다는 것에 대해서 비판하는 일부의 시각도 존재한다. 그러나 녹 쓴 철판의 재활용은 마치 이 건물이 과거로부터 존재하고 있었던 건물인 것처럼 보이게 만들기도 하고 이 장소의 역사를 보여주는 장치로서 작동하기도 한다.

그림 14. T4 복합문화공간(출처: 서울시 공원 홈페이지) / T4 Complex Cultural Space

그림 15. T5 이야기관의 외관(출처: World Architecture) / The exterior of the T5 historical story tank

그림 16. T6 커뮤니티센터 내부 공간(출처: 서울시 공원 홈페이지) / The interior space of the T6 community center

프로그램과 활동

기존의 석유탱크가 전문문화공간으로 쓰임새를 다하기에는 물리적인 한계가 있었기 때문에, 문화비축기지는 다목적 문화공간을 표방한 시민주도형 공간으로 계획되어 그 단점을 극복하고자 하였다. 복합문화공간은 그 정체성에 의문을 가질 수가 있는데, 서울시는 이 장소가 시민공동체의 중심지가 되기를 원했고 문화, 사람, 자연으로 채워지는 이상적 공간이길 바랬다. 이를 위해 초기부터 시민 '워킹 그룹'을 형성하여 공간사용방식과 프로그램에 대한 탐구가 함께 이루어졌다.

서울시는 공원 조성단계에서 워킹그룹 및 탐험단을 운영하면서 시민참여를 위한 운영 기본계획을 수립하였다. 이후 문화비축기지 협치위원회를 통해 시민주도적 공간운영전략 수립하고 개원기념 시민축제 기획을 하면서 지속 가능한 운영모델을 시험하였다. 개원 이후에는 모

그림 17. T5 이야기관의 전시공간(출처: 서울시 공원 홈페이지) / The interior space of the T5 historical story tank

그림 18. T6 커뮤니티 센터(출처: 서울시 공원 홈페이지) / T6 Community Center

그림 19. 2019년 서울 프린지 페스티벌 (출처: https://www.artinsight.co.kr/news/view.php?no=49200)
Seoul Fringe Festival 2019

그림 20. 2019년 서울 프린지 페스티벌 (출처 :https://www.artinsight.co.kr/news/view.php?no=49200)
Seoul Fringe Festival 2019

그림 21. T6 커뮤니티센터 외관(출처: World Architecture) / Exterior of T6 Community Center

Programs and Activities

Since there was a physical limit to the many uses of the existing oil tanks as a specialized cultural space, the Mapo Oil Tank Culture Park will be planned as a citizen-led space advocating a multipurpose cultural space, and its shortcomings overcome. The complex cultural space can be used to question its identity, and the city of Seoul hoped that it would be the center of civil society, and an ideal space that was full of culture, people, and nature. For this reason, from the beginning, a citizen working group was formed to explore space utilization methods and programs together.

The Seoul Metropolitan Government established a basic management plan for citizen participation while running a working group and expedition team at the park development stage. Afterwards, a citizen-led space management strategy

그림 22. 2019년 서울 프린지 페스티벌 (출처: https://www.artinsight.co.kr/news/view.php?no=49200)
Seoul Fringe Festival 2019

니터링, 가드닝, 해설활동 중심으로 시민이 직접 콘텐츠를 생산하여 만드는 커뮤니티 공간을 만들어 가고 있으며, 서울시는 운영주체로서 시민 커뮤니티 형성, 이용자를 넘어서는 시민주체 형성에 목표를 두고 있다. 일련의 과정을 통해서 공원의 환경을 스스로 만들어가는 과정을 체험교육방식으로 진행하여 협동과 사회적 공감대를 형성하였으며 지역공동체 활성화에도 이바지하였다. 이러한 경험을 계속 지속시키기 위해서 기지의 공간활용과 쓰임새를 탐구하기 위한 조사, 연구, 기획 프로그램이 공간전환학교라는 이름으로 매년 개설되고 있다.

석유비축기지에 대한 역사에 대한 전시는 이야기관인 T5에서 상설전시가 이루어지고 있으며, T0, T1, T2, T4의 모든 공간이 시민에게 다목적 용도로 개방된다. 이에 상설 전시 이외의 모든 프로그램은 특별 프로그램으로 운영된다. 특별 프로그램은 시민문화기획팀에 의해서 초기기획이 이루어지고 공모를 통해서 운영팀을 결정하는 방식으로 운영되고 있다. 대표적으로 시민 스스로 가꿀 수 있는 작은 텃밭에서 재배까지 경험하는 시민자율 운영 공동체인 소소텃밭 프로그램, 생태 및 생활문화 프로그램으로 야외놀이 프로그램, 손도구놀이 프로그램, 에코생활 프로그램 등이 운영되고 있다.

was established through the Mapo Oil Tank Culture Park Coordination Committee, and a sustainable operation model was tested while planning a citizen festival to commemorate the opening of the park. After the opening, the citizens have created their own content to create a community space created by the citizens, focusing on monitoring, gardening, and commentary activities, and Seoul Metropolitan Government aims to form a citizen community as the operating entity, and to form a citizen entity beyond users. Through a series of processes, the process of self-creating the environment of the park was implemented as a method of hands-on education, forming cooperation and social consensus, and contributing to the revitalization of the local community. To continue such experiences, research, planning, and programs for exploring the space utilization and usage of the base are established every year under the name of the Space Switching School.

The history of the oil storage base is permanently exhibited by T5, and all the spaces of T0, T1, T2, and T4 are open to the public for multipurpose use. All programs other than this permanent exhibition will be operated as special programs. The special program is run by a method in which the initial planning is done by the citizen culture planning team, and the management team is decided through open recruitment. Representatively, there is a small garden program, a self-operated community of citizens who experience small gardens that citizens can self-cultivate. Outdoor play programs, hand tools play programs, and eco-life programs are operated as ecological and life culture programs.

In addition, various transformation projects are operated to demonstrate Seoul as a 'transition city', which has been declared to practice the work of confronting climate change and environmental pollution on a city-by-city basis. Everyone's market, dishwashing zone and compost bin, zero plastics project, and upcycling production

또한 기후변화와 환경오염에 맞서는 일을 도시단위로 실천하고자 선언한 '전환도시' 서울을 실천하기 위한 다양한 전환 프로젝트를 운영하고 있다. 모두의 시장, 설거지존과 퇴비함, 플라스틱 제로 프로젝트, 업사이클링 제작 프로그램 등이 연 중 운영되고 있다.

이외에도 밤도깨비 야시장, 에코투어, 거리예술마켓, 숲-생태, 텃밭, 적정기술 워크숍, 가구 만들기 워크숍 등 다양한 축제 및 문화 이벤트가 상시 열리고 있다. 이처럼 문화비축기지가 도시재생의 랜드마크인 만큼 공원은 석유 시대 이후 시민의 여가와 시민자율의 생태 및 생활 문화 프로그램, 환경의 지속 가능성과 미래에 대한 모색 프로그램을 중심으로 운영되어 시민 문화플랫폼, 협동경제 공원으로 자리매김하고 있다.

그림 23. 문화비축기지 리플렛 / Leaflet of the Oil Tank Culture Park

환경문제에 대한 접근

문화비축기지 조성 초기의 우려사항은 환경오염에 대한 의문이 있었다. 지표조사를 통하여 해당 지역의 토양 오염과 탱크 내 공기 오염에 대한 조사가 실시되었으며, 실제로는 우려와는

program are in operation throughout the year.

In addition, various festivals and cultural events, such as the Bamdokkaebi Night Market, Eco Tour, Street Art Market, Forest-Ecology, Garden, Appropriate Technology Workshop, and Furniture Making Workshop are held at all times. As such, Mapo Oil Tank Culture Park is a landmark of urban regeneration, and since the Age of Oil, the park is operated around citizen's leisure, citizen's autonomous ecology and lifestyle programs, and programs to seek environmental sustainability and future. It has established itself as a park of cooperative economy.

Approach to Environmental Issues

The concern at the beginning of the construction of the Mapo Oil Tank Culture Park was the problem of environmental pollution. Through a survey of indicators, we investigated soil pollution in the area and air pollution in the tanks, and despite the concerns, there were no problems with oil spills or air pollution.

The architects sought to regenerate modern oil tanks as a new cultural space while minimizing their impact on the environment through an environment-friendly approach. The interior of the tank was painted to prevent further corrosion. All heat sources for heating and cooling use geothermal heat, and pipes are embedded 205 m underground to apply the bottom pump method. 30 ton of heavy water processing facility is used as toilet water, and 300 t of rainwater storage is used for landscaping and management. Mapo Oil Tank Culture Park received the highest energy efficiency rating from the Korea Institute of Building Energy Technology, and the Green Building Certification (G-SEED) of the Korea Industry Technology Certificate, and has

달리 기름 유출이나 대기 오염의 문제는 없었다.

　건축가는 근대의 석유탱크를 새로운 문화공간으로 재생시키면서 친환경적인 접근을 통해서 환경에 대한 영향을 최소화하고자 하였다. 탱크의 내부는 더 이상의 부식의 방지하기 위하여 페인트가 칠해졌다. 냉난방은 모든 열원 자체가 지열을 이용하며 지하 205m에 파이프가 매립되어 히프 펌프 방식을 적용하고 있다. 중수처리시설 30톤은 화장실 용수로 활용되고 빗물저장 300톤은 조경 및 관리용수로 활용된다. 문화비축기지는 한국건물에너지기술원에서 수용하는 최우수 에너지 효율 등급을 획득하고 한국산업기술인증원의 녹색건축인증을 받아 사회적 지속 가능성 뿐만이 아니라 건축환경 측면에서의 지속 가능성 확보에도 노력을 기울였다.

맺으며

　녹슬어 가는 탱크와 암벽, 높은 옹벽과 좁은 통로, 오래된 콘크리트 매스의 낯설음… 문화비축기지는 40여 년간 시민들에게 감추어졌던 석유탱크가 재생됨으로써 반전의 역사가 시작되는 곳이다. 마포구 성산동 매봉산 자락에 위치한 석유기반 경제를 상징하는 석유비축기지가 친환경 문화복합공간으로 변모하였다. 산업화시대 유사시 상황에 대비하여 석유를 비축하던 5개 탱크는 각각 다목적 파빌리온, 실내외 공연장, 기획·전시공간, 정보교류센터 등 '문화비축기지'로 다시 태어나 시민들을 만나고 있으며, 전면의 외부공간은 다양한 문화활동을 위한 큰 마당으로 작동하고 있다.

　문화비축기지는 관람형 문화예술, 창작자-생산자가 분리된 문화공간을 지양한다. 대신 시민이 직접 생산에 참여하고 지역자원을 키우며, 지역 예술인 중심으로 역량을 강화하고, 시민과의 상호작용에 가치를 두는 민관 협치 모델을 기반으로 운영되고 있다.

　마포 문화비축기지가 완성됨에 따라서 인근지역은 쓰레기처리장이었던 난지도를 매립한 노들공원, 체육공원인 월드컵 공원과 함께 도심의 중요한 녹지, 문화, 체육 공원이 완결되어 중요한 지역의 공공 인프라가 완성되었다.

endeavored to secure not only social sustainability, but also sustainability in terms of the built environment.

Closing Remarks

Rusty tanks and rocks, high retaining walls and narrow passages, the strangeness of old concrete masses... Mapo Oil Tank Culture Park is the place where the history of the war begins with the regeneration of oil tanks that have been hidden from citizens for over 40 years. The Oil Tank, which symbolizes the oil-based economy, located at the foot of Maebongsan Mountain in Seongsan-dong, Mapo-gu, has been transformed into an eco-friendly cultural complex. In preparation for the emergency oil supply situation during the industrialization era, the five tanks that secured oil will be reborn as the 'Oil Tank Culture Park', with a multipurpose pavilion, indoor and outdoor venues, a planning–exhibition space, and information exchange center. The exterior spaces in front of the tanks will be used as a large garden for diverse cultural activities.

Mapo Oil Tank Culture Park avoids spectator-type culture and art, and a cultural space where creators and producers are separated. Instead, it is operated based on a public–private partnership model, in which citizens directly participate in production, grow local resources, strengthen competence centered on local artists, and value interaction with citizens.

With the completion of the Mapo Oil Tank Culture Park, the area's public infrastructure was finally completed, along with Nodeul Park where Nanjido was reclaimed, which used to be a waste disposal facility, and World Cup Park, an athletic park, together with important green areas, cultural and sports parks in the city center.

참고문헌

1. 문화비축기지, 땅으로부터 읽어낸 시간, 허서구, 월간 환경과 조경, 2017년 12월호

2. 문화비축기지 조성 완료 및 개원 현장 설명회, 서울시, 2017. https://www.youtube.com/watch?v=SboOBq75zH4

3. Heritage Value through Regeneration Strategy in Mapo Cultural Oil Depot, Seoul, Jiae Han and Soomi Kim, Sustainability, 2018, 10, 3340.

4. 문화비축기지 홈페이지 http://parks.seoul.go.kr/template/sub/culturetank.do

5. https://www.artinsight.co.kr/news/view.php?no=49200

6. https://worldarchitecture.org/architecture-projects/hccev/oil_tank_culture_park_project-pages.html

References

1. Mapo Oil Tank Culture Park, Time reading from the Ground, Seo-gu Heo, Environment and Landscape, 2017.12.

2. Mapo Oil Tank Culture Park Opening Briefing Session, Seoul Metropolitan Government, 2017. https://www.youtube.com/watch?v=SboOBq75zH4

3. Heritage Value through Regeneration Strategy in Mapo Cultural Oil Depot, Seoul, Jiae Han and Soomi Kim, Sustainability, 2018, 10, 3340.

4. http://parks.seoul.go.kr/template/sub/culturetank.do

5. https://www.artinsight.co.kr/news/view.php?no=49200

6. https://worldarchitecture.org/architecture-projects/hccev/oil_tank_culture_park-project-pages.html

Chapter 4

도시교통체계와 환승거점개발

Urban Transportation System and Transit-oriented Development

교통운송기능은 도시의 성장형태를 결정하는 핵심 체계이다. 특히 도시를 관통하는 철도와 도시철도는 도시민의 출퇴근동선을 결정하고 환승거점의 거대 유동인구를 통해 주변의 상업 및 업무구역의 개발을 결정해 왔다. 이 장은 서울의 진입부인 남산 예장동의 지하에 도심 버스주차공간을 조성하고 지상부에 조성된 예장공원, 대구의 고속철도역인 동대구역과 인접해 고속(시외)버스터미널과 쇼핑센터로 개발된 동대구 복합환승센터, 경주의 철도체계의 변화와 이에 따른 도시확장과 역사보존의 상충되는 도시성장관리의 노력, 그리고 서울의 대표적인 폐선부지였던 경의선의 선형공원화와 그 지하에 조성된 일군의 도시교통 환승거점의 사례들을 살펴본다.

The transportation is the core functional system that guides the growth pattern of cities. Railroads and urban mass transits that pass through the city center have determined the commuting route of city residents and the movement of the floating population whose activities, particularly at the transfer base, often goes hand in hand with the development of commercial and business uses. A set of four cases will be examined in this chapter: Seoul Namsan Bus Parking Facility on the basement of Yejang Park on Namsan area; Daegu Rail-Bus Transfer Complex with shopping centers next to the high-speed railroad station; the historical changes in the railway system of Gyeongju that have generates the historical conflicts between suburban development and historic area preservation; a group of mass transit transfer stations underneath of the Gyeongui Railroad Park, a linear urban park which was recently converted from abandoned railroad.

서울 남산 예장공원

장항준 (전 시아플랜건축 디자인 총괄, 현 MDA건축사사무소 대표)

남산 예장자락 공원화사업은 서울시가 남산 전체를 문화와 예술이 연결된 특화공간으로 디자인해 관광명소로 육성하기 위한 '남산 르네상스' 사업의 일환으로 추진되었다. 남산 르네상스사업은 남산 주변을 장충, 예장, 회현, 한남, N타워 등 5개 지구로 나눠 각각 갤러리파크, 미디어아트, 콘서트, 생태, 전망존(Zone)으로 특화하고, 예술인마을이나 관광숙박촌, 악기전문상가 등을 지원해 배후시설로 조성하는 사업이다. 남산공원을 중심으로 숭례문에서 지하철 명동역, 충무로역을 거쳐 동대문입구역까지 도심 동-서 구간과 남쪽의 한강진역 주변을 연결하는 역삼각형 형태의 남산 일대 90만m²에 대해 '도심 재창조사업'과 연계해 남산의 스토리텔링 전개 방향과 물길 및 수공간 조성방향 등을 통해 남산의 정체성과 비전을 제시하는 마스터플랜 사업이다. 이렇게 서울시가 추진한 남산 르네상스 사업지구의 하나인 남산 예장자락사업은 도심에서 자락이 갖고 있는 도시재생의 방안과 의미를 모색하고, 향후 사람, 자연, 도시가 연계하여 지속적으로 치환되는 터의 공간을 제안하는 것이다.

도시재생방식으로 출발한 본 계획안은 과정적 재생의 의미를 해석하는 방식에서 시작되었다. 재생이라는 명분은 새로운 가치를 부여하여 다시 사용한다는 의미이다. 이미 존재하는 것들에 대한 존중과 기존의 도시 인프라를 자연과 사람이 공존하는 공간으로 치환하는 것은 매우 의미 있는 작업이었다. 삽입되는 새로운 콘텐츠와 원래 가지고 있었던 역사의 기억과 충돌하면서 또 다른 의미를 만들어낼 것으로 기대한 것이다.

Seoul Namsan Yejang Park

Hangjoon Jang (Former Design General Manager, SIAPLAN Architecture/
CEO, MDA Architecture Office)

The Namsan Yejang Jarak Park Project was promoted as part of the Namsan Renaissance project to design the entire Namsan Mountain as a specialized space connected to culture and art and foster it as a tourist attraction. The Namsan Renaissance Project was divided into five districts around Namsan: Jangchung, Yejang, Hoehyeon, Hannam, and N Tower, specializing in gallery parks, media art, concerts, ecology, and observation zones, and supporting artist villages, tourist accommodation villages, and musical instrument shops. The foot of Namsan Mountain, which served as Ansan in Seoul, continues to a waterfall in Myeong-dong, which has become a new city center since modern times, forming the foot of Yejang. With the establishment of the Angibu and the opening of the Namsan Tunnel, the spatial meaning as the original foot was lost, and furthermore, the Myeong-dong area, which the Yejang foot has, has also disappeared as a public center. At the natural, historical, and urban levels, Yejangjari can be understood as a space with various lights. Although traces of each era were left from the lights of nature cut off to the city center, complex and diverse lights are intertwined and disconnected from the lights of the city that can function as a central public space in Myeong-dong.Through this project, we tried to restore the edge by

사업부지는 서울특별시 중구 퇴계로 26길에 위치하며, 대지면적은 13,135m², 건축면적은 348.56m², 연면적은 9,352.28m², 규모는 지하1층, 지상1층, 용도는 공원, 버스주차장, 전시관이며, 서울시 도시재생과에서 발주하고, 사업기간은 2016년 2월~2020년 12월 준공하였다.

서울의 안산 역할을 한 남산 예장자락은 근대 이후 새로운 도심으로 자리잡은 명동 한복판까지 이어진 자락을 형성하고 있다.

그림 1. 준공 후 남산 예장자락의 전경

forming a self-sustaining space that can be expanded by three-dimensionally linking various lights of nature, history, and cities that cannot be connected.

Namsan Yejang Jarak was a martial arts training ground during the Joseon Dynasty, and it remained a historically isolated place with the official residence of the Japanese Government-General of Korea during the Japanese colonial period and the six Central Intelligence Agency after liberation. In particular, the underground road tunnel passing from Toegye-ro to the entrance of Namsan Tunnel_1 cut off pedestrian access from the city to Namsan Mountain.

The site of Yejang Jarak, promoted as part of the Seoul Urban Regeneration Project, will include an underground bus transfer parking lot, cultural facilities, and a park at the top.

Yejang Jarak was selected as the winner of the Seoul Urban Regeneration Project Competition, started with a way of interpreting the meaning of procedural regeneration. Respect for what already exists and replacing the existing urban infrastructure with a space where nature and people coexist were very meaningful tasks.

It is expected to create another meaning by colliding with the new inserted content and the memories of the original history. The direction of the design is to restore the natural landscape by recovering the old ridge at the Yejang Jarak through the interpretation of the procedural space of Namsan, which has numerous relationships, and to create a space of communication that encompasses the coexistence of the city and nature and the strata of various histories. The first button of such a system was planned as a flow in which the strata of various stations coexist by replacing the 100m section of the vehicle's underground road from Myeong-dong to the existing traffic

하지만 장소성과 역사성을 갖고 있는 이 자락의 터는 무분별한 개발과, 시대적이고 폐쇄적인 공간으로 도시와 단절되면서 중앙정보부(안전기획부), 소방재난본부, 교통방속국등이 무분별하게 자리잡아 훼손되고, 남산제1교통터널이 개통하면서 본래의 자락으로서의 공간적의미를 잃어버렸다. 나아가 예장자락이 가지고 있는 명동 일대는 공공의 구심장소로서 정체성 또한 사라졌다. 자연, 역사, 도시적 차원에서 남산 예장자락은 다양한 켜를 가진 터의 공간으로 이해할 수 있다. 도심으로 단절된 자연의 켜로부터 각 시대별로 흔적을 남겼지만 흐름이 끊어진 역사의 켜, 명동 일대의 중심 공공 공간으로 연계되지 못하는 도시의 켜까지 복잡하고 다양한 켜가 얽히고 단절되어 있다. 우리는 이 사업을 통해 훼손된 자락을 복원하고 치환하여 자연, 역사, 도시의 다양한 켜를 입체적으로 연계하면서, 자생적 공간이 될 수 있도록 문화의 공간을 형성하여 기존의 남산 예장자락을 시민들의 품으로 되돌리고자 하였다.

남산 예장자락은 조선시대 무예 훈련장(예장)으로, 일제강점기에는 조선총독부 관사, 광복 후 중앙정보부(안전기획부) 6국이 있던 역사적으로도 고립된 장소로 남아 있었고, 퇴계로와 삼일로의 큰 교차로로 인하여 남산과 도시의 흐름이 단절되었던 자락이다. 특히 퇴계로에서

그림 2. 기존의 예장자락 모습

그림 3. 도시재생 계획안

broadcasting with pedestrian space.

It is built as a tunnel of light connected to the exhibition hall installed in the underground space and plays a role in connecting Namsan Mountain, cities, and cultures in the existing disconnected urban space structure.

In addition, ecological bridges and pedestrian bridges connected to the existing topography create a landscape continuity into urban and natural spaces, and become new topography that creates a cycle that continues the visual and environmental flow into natural and urban spaces.

It is expanded and linked to the surroundings without breaking off each other beyond the limits of the land boundary. Also, the urban context and flow that had been cut off so far were connected from Namsan Mountain to Myeong-dong.

Through these actions, it is ultimately possible to construct a three-dimensional space with a multi-layered structure, and has the potential to extend into the city. In the lower part of the green park, a bus parking lot will be created to relieve the inconvenience and parking difficulties of tourists visiting Myeong-dong, and will function as an eco-friendly bus transfer center.

남산1터널 입구를 지나가는 지하차도 터널은 도시에서 남산으로의 보행자 접근을 단절시키고 있었다.

서울시 도시재생사업으로 추진된 예장자락 부지는 도시의 기반시설인 지하 버스환승주차장과 문화시설(전시장)을 상부에는 공원을 조성하는 것이다.

설계의 방향은 수많은 관계를 가진 남산의 과정적 공간의 해석을 통해 예장자락의 옛 능선을 따라 자연경관을 회복하고 도시와 자연의 공존, 다양한 역사의 지층을 아우르는 소통의 공간을 만드는 것이다.

그러한 장치의 첫 단추는 명동에서 기존의 교통방송으로 이어지는 차량의 지하차도 약 100m 구간을 보행자공간으로 치환하여 다양한 역사의 지층이 공존하는 흐름으로 계획한 것이다. 지하공간에 설치되는 전시관과 연계된 빛의 터널을 구축하여 단절된 기존 도시공간구조에서 남산과 도시, 문화를 잇는 역할을 하는 것이다.

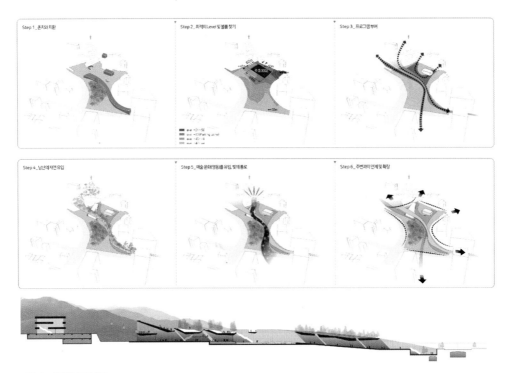

그림 4. 디자인 프로세스

그림 5. 기존의 차량 터널에서 보행자 터널로 치환된 빛의 터널

또한 기존 지형에 연결된 생태교량과 보행교는 도시공간과 자연공간으로의 경관적 연속, 반대로 자연공간과 도시공간으로의 시각적·환경적 흐름을 이어가는 순환고리를 만들어내는 새로운 지형이 된다.

대지경계의 한계를 벗어나 서로를 단절시키지 않고 주변으로 확장 연계시켜 그동안 단절되었던 도시적 맥락과 흐름을 연결, 남산에서 명동까지 이어지도록 했다.

이런 행위를 통해 궁극적으로 다층의 구조를 지닌 입체 자락의 구성이 가능해지며 도시로 뻗어 나갈 가능성을 지니게 된다. 녹지공원 하부엔 그동안 명동을 방문하는 관광객들의 불편과 주차난을 해소하기 위한 버스 주차 40면이 운영되어 친환경버스 환승센터로서의 기능을 담당하게 된다. 지금은 코로나로 인해 관광버스의 행렬을 볼 수 없으나 코로나 이전 국내외 관광객으로 인해 남산의 주변 도로는 버스주차장이 되어버렸고 차량의 매연으로 인해 보행자들이 남산으로 접근하기 힘든 상황이었다. 앞으로 친환경 녹색버스 충전소와 주차장 내 공기 정화시스템(배기팬 및 탈취유닛)을 통해 배기가스 탈취 후 공기를 배출하여 보행자들의 남산

그림 6. 남산에서 도심으로 바라 본 예장자락공원

Namsan Yejang Jarak is a place where traces of past history remain and returned to the side of citizens after 115 years. During the construction, a 'yugu-teo' was created to preserve the basic part of the government office site of the Japanese Government-General of Korea excavated as it was, With the relocation of the Ministry of Health and Welfare, the underground space of the six Central Intelligence Agency buildings purchased by the Seoul Metropolitan Government in 1995 was preserved and created as an exhibition space in the shape of a red mailbox to communicate with past history. This exhibition space is built as a Memorial Hall to look back on the painful history of modern history, and the old Central Intelligence Agency's underground torture room is reproduced as it was.

The entrance hall of the 'Udang Memorial Hall' at the bottom of the green park, which will be installed underground, was designed as a symbolic sculpture in honor of

그림 7. 도심에서 바라 본 예장자락공원

그림 8. 우당기념관

접근성은 더욱 활성화될 것이다.

115년 만에 시민의 품으로 돌아온 남산 예장자락은 과거 역사의 흔적이 남아 있는 곳이다. 공사를 진행하면서 발굴된 조선총독부 관사터의 기초 일부분을 그대로 보존한 '유구터'를 조성하였고, 안기부가 이전하면서 서울시가 1995년 매입했던 중앙정보부 6국 건물은 철거공사 전 지하공간을 보존하여 과거 역사와 소통하는 의미를 담아 빨간 우체통 모양으로 전시공간화 하여 현대사의 아픈 역사를 돌아보는 메모리얼 홀로 조성되었다. 옛 중앙정보부의 지하 고문실이 그대로 재현된 것이다.

지하에 설치되는 녹지공원 하부의 '우당기념관' 입구 홀에는 원형 모양의 테라코타 루버를 설치하여 봉오동전투의 주역인 신흥무관학교 학생 3,000여 명을 기리는 상징적 조형물로 설계되었다.

이처럼 남산 예장자락 재생사업은 각 분야별 전문가의 의견수렴과 현황 검토가 수반되었

그림 9. 신흥무관사관학교 학생 3000명을 상징화한 우당기념관 홀

3,000 students from Sinheung Military Academy, the hero of the Battle of Bongoh town Battle.

The Namsan Yejang Jarak regeneration project involved collecting opinions from experts in each field and reviewing the current status. Even during the construction period, it was a difficult task to carefully consider the traces of the land's history and

그림 10. 도심에서 바라본 남산예장자락 야경

고, 공사기간 중에도 땅이 갖고 있는 역사의 흔적과 주변과 연계된 물리적 환경을 세심하게 고려 해야 하는 어려운 과제였으나, 장소성과 역사성 회복, 주변 지역과 연계, 역사의 현장을 느끼고 기억할 수 있는 공간으로 치환되어 시민들의 품으로 돌아온 남산을 기대해 본다.

지금의 남산 예장자락이 시간이 더해지면 시민들의 장소로 하나씩 채워져질 것이다. 도시재생은 한 번에 완성되는 것이 아닌 도시가 갖고 있는 자생적 특성을 통해 유기적으로 변모하고 만들어질 것이기에 많은 고증과 협의를 통하여 장기적인 기획과 지속적인 관리가 이뤄지도록 시민. 전문가, 기관의 노력이 필요한 시점이다.

the physical environment linked to the surrounding area, but it is replaced with a space to feel and remember the site of history, and we look forward to returning Namsan Mountain to citizens.

서울 경의선숲길, 우연한 도시교통환승체계

한광야 (동국대학교 건축공학부 교수)

들어가며

서울 경의선숲길공원(Gyeongui Line Forest Park)은 버려진 폐철도선이 도시공원으로 변화하는 과정에서, 그 공원의 지하공간이 대중교통 환승거점을 수용하며 공원 전체 6km 구간이 서울 시민의 거대한 선형 환승체계로 기능하고 있는 흥미로운 사례이다. 이 사례에서 공원과 환승체계의 조성과정과 계획과정을 고찰해 본다.

현대 도시 서울의 지역거점이며 대표 방문명소로 자리잡은 경의선숲길공원은 서쪽의 홍제천 주변(모래내시장)의 가좌역(경의중앙선)부터 동쪽의 용산문화체육센터(옛 용산구청)까지 연남동, 홍대입구역, 공덕역, 효창역, 남영동을 관통하는 지상부의 선형공원(총길이 6.3km, 폭 10~60m, 총면적 102,000m²)이다.

경의선숲길은 서울 북서부와 도심을 연결하는 통근열차인 경의중앙선의 지하부에 조성된 가좌역-(연남동 구간)-홍대입구역-(와우교 구간)-서강대역-(신수동 구간)-대흥역-(대흥동 구간)-(염리동 구간)-공덕역-(새창고개 구간)-새창역-(원효로 구간)을 중심으로 연결되어 기능한다. 지난 2014년 지상부의 경의선 철로(폭 9.8~11.2m)가 지하 30~45m에 건설되어 기능하는 공항철도(폭 10.7~11.9m) 위에 이선되었다.

경의선은 대한제국 말기인 1905년 경의선-본선(1905, 서울역-개성-사리원-평양-신의주,

Unplanned Metro Transfer Hub under Gyeongui Line Forest Park, Seoul

GwangYa Han (Professor, Dongguk University, Department of Architecture)

Introduction

Gyeongui Line Forest Park—a regional hub in contemporary Seoul, as well as a visitors' attraction—is a linear park that runs through Yeonnam-dong, Hongik University Station, Gongdeok Station, Hyochang Station, and Namyeong-dong from Gajwa Station (Gyeongui–Jungang railway line) in the west to Yongsan Culture and Sports Center (formerly Yongsan-gu Office) in the east (total length of 6.3 km, width of (10–60) m, and total area of 102,000 m²).

Gyeongui Line Forest Park is located on the site of the abandoned Gyeongui–Jungang railway line, a commuter train that connected the northwest of Seoul to the city center. This was realized when in 2014, the Gyeongui Line railroad on the ground was relocated to the airport railroad that functions at (30–45) m underground.

The Gyeongui–Main Line (1905, Seoul Station–Gaeseong–Sariwon–Pyongyang–Sinuiju, 499 km) was first opened in 1905, at the end of the Korean Empire, followed by the Gyeongui–Yongsan Line (1905, Yongsan Station–Gajwa Station converted for cargo in 1975). Afterwards, the Gyeongui–Main Line and Gyeongui–Yongsan Line

499 km)이 처음 개통했고, 뒤이어 경의선-용산선(1905, 용산역-가좌역 구간은 1975년 화물 용으로 전환)이 개통했다. 이후 경의선-본선과 경의선-용산선은 일본지배기(1910~1945)에 군사작전과 산업철도로 이용되었다. 경의선-본선은 1950년 한국전쟁(1950~1953)이 발발 하며 중단되었고, 이를 대신한 경의선 단선 통근열차(서울-파주)가 1951~2009년 운행되었 고 2009년 복선전철로 확장되었다. 한편 경의선-용산선은 1975년부터 화물선으로 이용되어 왔다. 경의선-본선과 경의선-용산선 주변블록은 2000년대 중반까지 굴다리, 오막살이 등의 일본지배기의 흔적을 갖고 노후화되었다.

그림 1. 서울 경의선숲길 철로복원구간(출처: 한광야, 2021)
Railroad Section of Gyeongui Line Forest Park Restored, Seoul

한국전쟁 이후 남한과 북한 구간으로 오랫동안 단절되었던 경의선은 2007년 노무현 정부 (2003~2008)의 주도로 연결되었다. 이 과정에서 경의선은 2014년 중앙선과 연결되어 경의 중앙선(2009, 2012, 2014)으로 모두 지하로 이선되었고, 그 아래 공간은 인천공항과 서울역

were used during the Japanese colonial period (1910–1945) for military operations, and as industrial railroads. When the Korean War (1950–1953) broke out in 1950, the Gyeongui–Main Line was suspended, and the Gyeongui Line single-track commuter train (Seoul–Paju) was instead operated in 1951–2009. Meanwhile, the Gyeongui–Yongsan Line has been used as a cargo train since 1975. The blocks around the Gyeongui–Main Line and the Gyeongui–Yongsan Line had degraded until the mid-2000s with traces of Japanese rule, such as underpasses and huts.

그림 2. 서울 경의선숲길 공원, 새창고개-효창공원역 구간(출처: 김대석, 2015)
Gyeongui Line Forest Park SaeChang-Gogae-HyoChang Pakr Station seegment

The Gyeongui Line, which after the Korean War, had long cut off South Korea from North Korea, was reconnected under the lead of the Roh Moo Hyun government (2003–2008). In the process, the Gyeongui Line was connected to the Jungang Line in 2014, and relocated underground as the Gyeongui–Jungang Railway line (2009, 2012, 2014),

을 연결하는 인천공항철도선(2010, 인천공항-서울역)이 개통되었다. 이에 따라 그 지상의 경의선 철도선부지의 대체용도가 사회적인 이슈가 되었다.

서울시는 경의선의 역사를 담은 공원조성사업(총 예산 470억원)을 철도선 소유주인 철도청과 협상하며 추진했다. 경의선숲길 공원사업은 1단계(2011~2012), 2단계(2013~2015), 3단계(2015~2016)로 진행되어 완료되었고, 현재에도 경의선숲길 공원 주변의 역세권개발 프로젝트들이 진행 중이다.

흥미롭게도 경의선숲길공원 지하에서 기능하는 공항철도, 경의중앙선, 그리고 지하철 2호선, 5호선, 6호선은 총 7개의 역을 두고 다양한 환승활동으로 상호 연결되고 있다. 다시 말해 지상 6km 길이의 선형공원 지하에는 서울 시민의 거대한 선형환승거점이 기능하는 것이다. 특히 이들 중 1일평균이용자수의 관점에서 홍대입구역(2호선·공항철도·경의중앙선)는 205,323명(2019년 기준)과 공덕역(6호선·5호선·공항철도·경의중앙선)은 79,592명(2019년 기준)으로 환승기능을 통해서 서울의 새로운 도시거점으로 자리잡았다.

현대 도시의 핵심 이동수단이 도시철도이며, 탄소저감을 위한 보행활동의 활성화가 서울의 미래목표라며, 이러한 환승체계는 매우 유용한 사례로 이해된다. 그리고 도시재생의 관점에서도 이러한 결과가 쇠퇴한 도시 주거지역의 활성화에 크게 기여하고 있기 때문이다. 흥미롭게도 이러한 결과는 서울시나 도시계획가의 의도적인 계획과 개발의 결과가 아닌 우연한 결과였다.

경의선숲길공원은 서울의 오랫동안 쇠퇴해온 철도폐선 주변의 열악한 주거지에 주거환경을 개선하고 녹지공원을 제공하며 지역주민의 생활을 바꾸고 삶의 질 향상에 큰 기여를 해오고 있다. 하지만 개원 이후 공원의 인지도가 상승하고 방문객이 많아지면서, 주변 주거지의 지가 상승, 주거지의 상업화, 젠트리피케이션, 주차, 쓰레기 무단투기 등의 문제가 발생했다. 이를 해결하기 위해 유관 구청과 커뮤니티센터, 주민들이 캠페인을 진행하며 노력하고 있다.

and the Incheon International Airport Railway Line (2010, Incheon Int'l Airport-Seoul Station) was opened in the space below it. Accordingly, the alternative use of the sites of the Gyeongui Railway Line on the ground has become a social issue.

The Seoul Metropolitan Government negotiated and promoted a park construction project (total budget of 47 billion KRW) with the Korean National Railroad, the owner of the railway line. The Gyeongui Line Forest Park project was completed in stages 1 (2011–2012), 2 (2013–2015), and 3 (2015–2016), and the development projects for the areas around Gyeongui Line Forest Park are still underway.

Interestingly, the airport railroad, Gyeongui–Jungang Line, and Subway Lines 2, 5, and 6 functioning underground of the Gyeongui Line Forest Park, have a total of seven stations, and are interconnected through various interconnected transfer activities. In other words, a huge linear transfer hub is located under the 6 km long linear park. In particular, from the perspective of the average number of users per day, in which there were 205,323 users (as of 2019) for Hongik University Station (Subway Line 2, airport railroad, and Gyeongui–Jungang Line), and 79,592 people (as of 2019) for Gongdeok Station (Subway Lines 6 and 5, the airport railroad, and the Gyeongui–Jungang Line), the park has become a new urban hub in Seoul with the transfer function underneath.

Since the main means of mass transportation in contemporary cities are urban railroads, and the future goal of Seoul government is to increase walking activity to reduce carbon emissions, this transfer system is believed to be a very informative case. Also, from the perspective of urban regeneration, these results greatly contribute to the revitalization of urban residential areas that have previously long declined. Interestingly, these are unintended results, and not intentionally planned or promoted by the Seoul Metropolitan Government or urban planners.

Gyeongui Line Forest Park has contributed significantly to improving the quality of

현대 도시에서 폐선철도와 그 변신

한국 정부가 고속철도 건설을 추진하기 시작한 시점은 1980년대이다. 이미 1970년대부터 경제활동의 확장에 따른 철도 용량의 포화가 예측되어 왔으며, 이를 해결하기 위한 방안으로 고속철도 건설이 가장 효율적이라 판단되었다. 철도 운행속도의 고속화를 위해 기존의 구불한 철도선은 직선화하고 휘어진 철도선 구간은 더 크게 곡선화해 왔다. 이 과정에서 철도선의 이선에 따른 기존 철도의 폐선 구간이 다수의 한국 도시의 중심부에 남게 되었다.

그림 3. 서울 경의선숲길 공원조성 전의 쇠퇴한 경의선 주변부, 2004(출처: 서울시)
Gyeongui Line Neighborhood Condition before the formation of Gyeongui Line Forest Park, 2004

이에 따라 도시중심부를 관통하는 폐기된 철도선 구간에 대한 대안적 기능을 찾는 도시정책이 사회적 이슈와 지방자치단체의 선거공약의 일부가 되었다. 이러한 대안은 시민의 휴식을 위한 선형공원으로 조성되거나 관광열차, 레일바이크, 테마파크 등 관광명소로 변화해 왔다. 여기서 철도선과 철도선 구간에 조성된 공원의 토지를 소유하고 관리하는 주체는 한국철도시설공단이다. 그리고 지방자치단체는 이러한 폐선구간을 임대하여 공원으로 조성하여 민간기업 또는 시민조직과 함께 운영하고 있다.

the living environment, creating green open spaces, and changing the lives of local residents in poor residential neighborhoods around abandoned railroads that have long been in decline in Seoul. However, since its opening, the park has become more popular, the number of visitors has increased, and problems, such as rising land prices, rapid commercialization of the residential areas, gentrification, parking, and unauthorized dumping of garbage, have arisen. Related district offices, community centers, and residents have been striving to solve these problems by conducting campaigns.

Abandoned Railway and Its Transformation in Modern Cities

The South Korean government began to push for the construction of high-speed railways in the 1980s. Since the 1970s, saturation of railway capacity had already been predicted due to the expansion of economic activities, and the construction of high-speed railways was evaluated as being the most efficient way to solve the problem. To speed up railroad operation, the existing winding railroads have been straightened, while railroad areas that have bends have been relocated for more efficient and safe train movement. In this process, the abandoned sections of the existing railways following the relocation of railroad lines ended up remaining in the center of many cities in South Korea.

Accordingly, urban policies to find alternative functions for the abandoned railway areas passing through the center of cities have become part of the social issues and election pledges of local governments. These alternatives include creating linear

한국 내에서 최근 추진되어온 폐선철도의 공원 및 대안기능 조성 사례들은 대구 대구선 구간(7.5km)의 아양-동촌-반야월공원, 광주 경전선 구간(7.9km)의 푸른길 공원(2003~2013), 나주 호남선 나주역-영산포역-구진포 구간(3.6km)의 자전거 테마파크(2006~), 전남 곡성 전라선-섬진강 구간(13.2km)의 관광열차선, 마산 임항선(화물수송) 구간(1km)의 공원화 등이며, 서울의 경의선숲길공원도 그 중 하나이다. 그럼에도 서울 경의선숲길공원은 지하에 조성된 대중교통체계와 일체화되어 기능한다는 점에서 기타 사례들과 차별성을 갖고 있다.

경의선숲길공원 조성의 추진배경

경의선숲길공원 조성 계획은 흥미롭게도 1970년대부터 현재까지 진행된 서울의 서부 및 서북부 확장과 이를 위한 대중교통체계의 조성, 그리고 2000년대 초에 새롭게 건설된 인천공

표 1. 서울 경의선숲길공원 조성 프로젝트 진행과정
Planning and Development Process of Gyeongui Line Forest Park, Seoul

	2005	2010	2015	2020
경의선(용산-문산) 복선전철공사 실시계획 승인(2006, 국토부) 한국철도시설공단+마포구 협약(2007) 경의선 2단계 DMC-공덕 지하화(2012, 6.1 km)	▬	▬▬▬		
공항철도 인천국제공항역-김포공항역 구간(2007) 공항철도 김포공항역-서울역 구간(2010)	▬	▬		
경의선공원 조성 추진 및 기본실시설계(2009-2011) 한국철도시설공단+서울시 경의선공원 조성 협약(2010) 서울시가 철도부지 사용권의 무상 획득 서울시의회 도시관리위원회 의견 청취(2011) 서울시 도시계획위원회 심의 및 도시계획시설(공원) 고시 (2011)		▬		
경의선공원 1단계 완료(대흥동 구간) (2012) 경의선 포럼(3회), 분야별 관련전문가 17명 (2012) 1차 주민설명회(3회) 및 2차 주민설명회(4회) (2012-2013) 기본 구상 변경(2012-2013) 경의선공원 2단계(새창고개-연남동) 완공 (2013-2015) 경의선공원 3단계(신수동-동교동) 완료 (2015-2016)			▬▬	▬

(출처: 한광야 김민지, 2020)

parks for citizens to rest and recreate, or transforming them into tourist attractions, such as tourist trains, rail bikes, and theme parks. In this, the Korea Rail Network Authority owns and manages the land of the parks created from disused railroad land. In addition, local governments rent these abandoned areas to create parks, and operate them together with private enterprises or civic organizations.

Examples of parks and the alternative functions of abandoned railways in South Korea are the Ayang–Dongchon–Banyawol Park in the Daegu Line area (7.5 km), Pureungil Park in the Gwangju–Gyeongjeon Line area (7.9 km), Biking theme park in the Naju–Honam Line Naju Station–Yeongsanpo Station–Gujinpo area (3.6 km), the tourist train railway in the Gokseong Jeolla Line Seomjingang River area (13.2 km), and the parkification of the Masan Imhang Line (cargo transport) area (1 km). Gyeongui Line Forest Park in Seoul is one of these parks. Nevertheless, the Seoul Gyeongui Line Forest Park is different from other cases, in that it functions in integration with the public transportation system created underground.

Background of the Construction of Gyeongui Line Forest Park

Interestingly, the planning process to create Gyeongui Line Forest Park built up momentum while the west and northwest of Seoul were expanding from the 1970s to the present with ever-increasing demand for a public transportation system for this, and the Incheon Int'l Airport Railway, which connects the newly built Incheon Int'l Airport and the city center, was opened in the early 2000s.

With this background, the construction process of Gyeongui Line Forest Park was

항과 도시중심부를 연결하는 인천공항철도의 건설이 그 계기가 되었다.

이러한 배경에서 경의선숲길공원의 조성과정은, 첫째 서울-신의주 화물철도 폐선구간의 지상부의 대안적 활용, 둘째, 서울 서부-서북부 확장과 통근열차의 현대화와 확장, 셋째, 인천 국제공항(1992~2001)과 서울 중심부를 연결하는 공항철도의 건설이라는 세 가지 요소들과 함께 진행되었다. 이러한 이유로 경의선숲길공원의 지하에는 이러한 세 요소들이 구체화된 3개의 열차들이 운행되고 있다.

서울-신의주 경의선 철도

경의선(京義線)은 대한제국시기에 서울역과 신의주역을 연결하는 철도로 처음 조성되었다. 당시 일본 세력은 프랑스 세력에 뒤이어 1904년 대한제국으로부터 50년간의 경의선 임대조약 및 경의선철도선 부설권을 획득했다. 이후 경의선철도는 1905년 서울과 평안북도 신의주를 연결하는 경의선-본선(서울-개성-사리원-평양-신의주, 499km)으로 개통되었고, 현재 남가좌역에서 분기하여 한강을 따라 용산을 연결하는 경의선-용산선(계획, 1896; 개통, 1906, 용산-개성-사리원-평양-신의주, 499km)이 신설되었다. 경의선은 이후 한일병합조약(1910)에 이어 압록강철교(1911)를 통해 만주로 연결되었다.

경의선은 한국전쟁(1950~1953) 이후 한반도 분단에 따라 문산-개성 구간이 폐지되면서, 그 본선 구간이 서울-문산으로 축소되었으며, 이후 가좌역에서 분기하는 경의선-용산선(1905, 가좌-용산)은 1975년 화물용으로 전환되어 사실상 기능을 잃게 되었다. 이때부터 경의선-용산선의 가좌역-동교동삼거리-대흥동-공덕동-원효로 구간은 주변 지역사회를 나누고 주거환경을 쇠퇴시키는 폐선으로 남겨져 왔다.

서울 서부-서북부 확장과 통근열차

서울이 서쪽으로 확장되기 시작한 시점은 1960년대 초이다. 이 즈음 서울은 동교동삼거리에서 뻗어 나온 양화로(1972)를 지나 양화교(구교, 1962; 신교, 1979)를 넘어 영등포로 확장했다. 이 시기에 용산역과 가좌역을 오가며 석탄열차로 불리던 경의선-용산선 화물열차는 동

carried out with three elements: first, alternative utilization of the abandoned areas of the Seoul-Shinuiju cargo railway; second, the expansion of the west--northwest areas and modernization and expansion of commuter trains; third, the construction of an airport railway that connects Incheon Int' Airport (1992-2001) and central Seoul. For this reason, three trains embodying these three elements are in operation underground of the Gyeongui Line Forest Park.

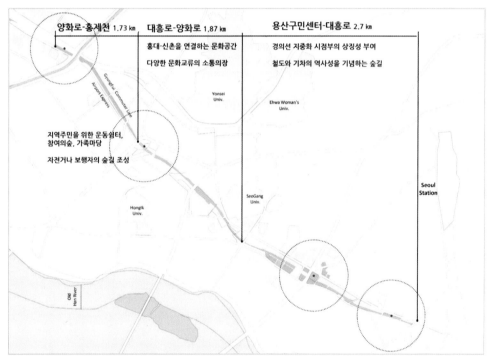

그림 4. 서울 경의선숲길 공원의 1,2,3 단계 조성구간(출처: 한광야 김민지, 2020)
Planning and Development Process of Gyeongui Line Forest Park, Seoul

Seoul-Shinuiju Gyeongui Line Railroad

The Gyeongui Line was first constructed as a railroad connecting Seoul Station to Sinuiju Station during the Korean Empire. In 1904, Japanese forces, replacing French

교동삼거리를 지나 당인리발전소(현재 서울복합화력발전소)까지 석탄을 운송했다. 이곳에는 이러한 이유로 이 지역의 오래된 랜드마크였던 청기와주유소(1969, 유공/SK에너지의 최초 현대식 주유소; 현재 롯데호텔, 2010)가 입지했고, 뒤이어 강북과 강남을 연결하는 지하철2호선 동교동역(1980, 현재 홍대입구역)이 조성되었다. 이후 홍대입구역은 5호선(1995)과 6호선(2000)이 더해지며 환승거점으로 기능하기 시작했다.

서울의 1980년대 인구증가로 대규모 주택시장이 한계에 이르며 부동산 투기와 급등이 심각한 사회문제로 대두되었다. 이에 노태우정부(1988~1993)는 1988년 '주택 200만호 건설 계획'을 발표했으며, 이에 따라 수도권 1기 신도시(군포시 산본신도시, 부천시 중동신도시, 안양시 평촌신도시, 성남시 분당신도시)의 하나로 고양시 일산신도시(1989~1992)가 건설되었다. 일산신도시는 서울 도심에서 약 20km에 개발된 베드타운(면적 1,573ha, 계획인구 275,000명)으로 건설되었다. 이후 노무현정부(2003~2008)도 서울의 집값 급등을 막기 위해 수도권 10곳과 충청권 2곳에 2기 신도시 건설을 추진하였다. 이때 일산신도시의 배후지역에 파주시 운정신도시(면적 1,660ha, 계획인구 217,000명, 2003~2020)가 계획되어 최근까지 건설 중이다.

이러한 배경에서 서울의 북서쪽에 조성된 일산신도시와 운정신도시의 대중교통 접근성을 확보하기 위해 대중교통체계가 건설되었다. 먼저 일산신도시의 통근여객운송을 위해 고양 지축역에서 일산 대화역을 연결하는 일산선(고양 지축-일산 대화, 1996, 3호선으로 흡수)이 완공되었다.

또한 고양과 파주의 교통체계와 도시접근성을 개선하기 위해 경의선 복선전철 건설이 2000년부터 계획되었다. 이에 따라 1951년 이후 약 60년 동안 경의선 단선 통근열차로 기능했던 경의선 단선구간(문산-DMC역, 2000-2009, 40.6km)이 복선으로 확장되어, 2개의 환승역인 DMC역(지하철6호선)과 대곡역(지하철3호선)을 갖게 되었다. 이후 경의선 DMC역-공덕역(2012, 6.1km) 구간도 복선으로 지하에 개통되었다. 그리고 경의선은 수도권의 동쪽 통근열차인 중앙선이 용산선 공덕-용산 구간(2014, 1.9km)의 개통으로 하나의 경의-중앙선(2014, 경의선+중앙선)으로 통합되었다.

forces, acquired a 50-year Gyeongui Line lease treaty from the Korean Empire and the right to install the Gyeongui railroad line. The Gyeongui Line railroad was opened in 1905 as the Gyeongui Line–Main Line connecting Seoul to Sinuiju in Pyeonganbuk-do (499 km), and at that time, the Gyeongui Line–Yongsan Line (planned in 1896; opened in 1906, 499 km), which diverges from Namgajwa Station and connects to Yongsan along the Han River, was also established. Following the Korea–Japan Treaty of 1910, the Gyeongui Line was connected to Manchuria through the Yalu railroad bridge (1911).

After the Korean War (1950–1953), the Munsan–Gaeseong area was abolished, and the Main Line section was reduced to Seoul–Munsan; afterwards, the Gyeongui Line–Yongsan Line (1905, Gajwa–Yongsan) branching from Gajwa Station was converted for cargo in 1975, and thus actually lost its function. Since then, the Gajwa Station–Donggyo-dong Samgeori–Daeheung-dong–Gongdeok-dong–Wonhyoro section of the Gyeongui Line–Yongsan Line has remained as an abandoned railroad that divides the surrounding local communities, and degrades the residential environment.

Expansion of the West-Northwest of Seoul and Commuter Trains

Seoul began to expand westward in the early 1960's. Around this time, Seoul opened Yanghwa-ro (1972) extending from Donggyo-dong Samgeori, and expanded to Yeongdeungpo beyond Yanghwa Bridge (Gugyo, 1962; Shingyo, 1979). During this period, the Gyeongui Line–Yongsan Line cargo train, which was a coal train running between Yongsan Station and Gajwa Station, transported coal to the Dangin-ri Power Plant (current Seoul Thermal Power Plant) through Donggyo-dong Samgeori. For this reason, the region's old landmark, Cheonggiwa Gas Station (1969, Korea Oil Company/ SK Energy's first modern gas station; current Lotte Hotel, 2010), was located here, followed by Donggyo-dong Station (1980, current Hongik University Station), which is

인천국제공항 공항철도

한국의 관문으로 김포공항을 대신하는 인천공항(2001)이 개항하면서, 신공항과 서울의 도시 중심부를 연결하는 인천공항철도선(2001~2007, 2010, 63.8km, 14역, 운행속도 110km/h)이 수도권 도시철도의 일부로 지하에 건설되었다. 인천공항철도는 2001년 공사를 착공하여 인천국제공항역-김포공항역 구간(2007)이 먼저 개통되었고, 이후 김포공항역-서울역 구간(2010) 그리고 공항1터미널역-공항2터미널역(2018)이 개통되었다. 이러한 공항철도의 개발과 함께 공항철도선의 종착역인 서울역 전의 정거장인 마곡역, 상암역, 공덕역, 그리고 홍대역을 중심으로 서울의 신도심 거점개발이 진행되었다.

경의선숲길공원의 구상과 추진

경의선숲길공원(2009~2016)은 서울시 마포구 연남동에서 용산구 원효로까지 지하화된 철도선 상부의 지상에 조성된 공원이다. 경의선-용산선의 DMC역-공덕역 구간이 특히 2012년 지하로 이설되면서, 기존의 폐선 지상공간이 긴 선형의 산책공원으로 조성되어, 효창동의 아파트 단지, 마포의 고층빌딩 블록, 서강대와 홍대 앞 상업구역, 연남동 주택지 사이를 지나 홍제천으로 연결되고 있다. 이로 인해 철도선 주변의 저층주거지는 급속한 대규모 아파트 재개발의 대안을 찾을 수 있게 되었고, 오히려 인접 공원으로 주거환경이 획기적으로 개선되었다.

경의선 폐선지의 대안적 구상이 서울시 사업으로 본격적으로 논의되기 시작한 시점은 박원순 시장(2011~2020) 집무 초기였다. 물론 마포구간의 지상부 공원화사업은 오세훈 시장(2006~2011) 집무기에 마포구의 주도로 대흥로-용산구민센터구간의 공원조성사업이 시작되었다. 당시 서울시는 경의선 폐선지의 난개발을 방지하기 위해 대안적 계획안을 준비했다. 경의선숲길공원의 계획수립 당시 서울시 계획예산은 2014년 기준 457억원이었다.

당시 서울시가 추진한 경의선 폐선지사업의 계획목적은, 첫째 철도운행으로 인한 피해와 불편을 겪은 지역주민의 주거환경 개선, 둘째 경의선 지상구간을 공원으로 결정(도시관리계

on subway line 2 connecting Gangbuk and Gangnam. Since then, Hongik University Station has begun to function as a transfer hub with the addition of neighboring subway lines 5 (1995) and 6 (2000).

As the large-scale housing market reached its limit due to the population growth in Seoul in the 1980s, real estate speculation and price hikes emerged as serious social problems. Accordingly in 1988, the Roh Tae-woo administration (1988–1993) announced the 'plan to build 2 million houses', and therefore, Ilsan Newtown (1989–1992) in Goyang was built as one of the first new cities in the Seoul metropolitan area (Sanbon Newtown in Gunpo, Pyeongchon Newtown in Anyang, and Bundang Newtown in Seongnam). Ilsan Newtown was built as a commuter town (area of 1,573 ha, planned population of 275,000), and developed about 20 km from downtown Seoul. After that, the Roh Moo-hyun government (2003–2008) also pushed for the second development of 10 new cities in Seoul metropolitan areas and two Chungcheong areas, to prevent a surge in housing prices in Seoul. Here, Unjeong Newtown in Paju (area of 1,660 ha, planned population of 217,000, 2003–2020) was planned in the area behind Ilsan Newtown, and has until recently been under construction.

With this background, a group of public transportation systems were built to secure public transportation accessibility for Ilsan Newtown and Unjeong Newtown, which were created in the northwest of Seoul. First, the Ilsan Line (Goyang Jichuk–Ilsan Daehwa, 1996, combined into Line 3), which connects Goyang Jichuk Station to Ilsan Daehwa Station for commuting passenger transportation in Ilsan Newtown, was completed.

In addition, the construction of the double-track subway on the Gyeongui Line was planned from 2000 to improve the transportation system and urban accessibility of Goyang and Paju. Accordingly, the Gyeongui Line single-track section (Munsan–DMC Station, 2000–2009, 40.6 km), which had functioned as a single-track commuter train

그림 5. 서울 경의선숲길공원 새창고개 구간(출처: 한광야, 2020)
Saechang Gogae Section of Gyeongui Line Forest Park Restored, Seoul

획안, 제21차 서울시 도시계획위원회)이었다. 이러한 경의선 지상구간의 공원조성의 결정의 도는 폐선 내의 지상구간 및 주변지에서 민간기업이 주도하는 토지매각이나 임대 후의 불필요한 난개발을 방지하겠다는 것이었다. 이를 위해 서울시는 철도부지의 소유권자인 철도시설공단과 무상으로 공원부지 사용권을 확보했다.

발표된 서울시 보도자료에 의하면, 이러한 서울시의 사업목표가 "경의선 사업 구간은 1906년 개통 후 국민의 애환과 추억이 깃든 경의선철도의 시발점 구간이다. 이 구간은 당시 서울 도시환경과 여건변화로 철도지하화(인천국제공항철도+경의선철도)가 추진되고 있었다. 이에 지상부 철도구간의 난개발을 방지하고 시민들의 여가 휴식공간을 조성"으로 확인된다.

on the Gyeongui Line for about 60 years since 1951, was expanded to a double track, resulting in having two transfer stations, DMC Station (subway line 6), and Daegok Station (subway line 3). Since then, the section (2012, 6.1 km) between DMC Station and Gongdeok Station on the Gyeongui Line has also opened underground as a double track. Also, with the opening of the Gongdeok–Yongsan section (2014, 1.9 km) of the Yongsan Line, the Gyeongui Line, and the Jungang Line, the eastern commuter train of the metropolitan area, were connected into the one Gyeongui–Jungang Line (2014).

Airport Railroad for Incheon International Airport

With the opening of Incheon Int'l Airport (2001), which replaces Gimpo Airport as the premier gateway to South Korea, the Incheon Int'l Airport railroad line (2001–2007, 2010, 63.8 km, 14 stations, and 110 km/h) which connects the new airport and the city center of Seoul, was built underground as part of the Seoul Metropolitan City railroads. The construction for Incheon Int'l Airport railroad began in 2001, and the section of Incheon Int'l Airport Station–Gimpo Airport Station (2007) opened first, followed by the section of Gimpo Airport Station–Seoul Station (2010), and the section of Airport Terminal 1 Station–Airport Terminal 2 station (2018). Along with the development of the airport railroad, the development of new hubs in Seoul was carried out, centering on Seoul station, which is the terminal station of the airport railroad line, and the stations before—Magok Station, Sangam Station, Gongdeok Station, and Hongik University Station.

당시 서울시는 경의선의 역사 스토리를 담은 숲길로 재탄생시키고 남산과 용산, 월드컵공원을 자전거와 보행통로로 연결하는 광역 그린웨이를 구축할 계획이며, 인위적인 시설의 최소화를 통해 시간의 흐름을 그대로 담은 경관의 형성하고자 했다.

제안된 공원의 총 면적은 102,000m²로 여의도공원의 1/2이며, 뉴욕시 센트럴파크의 1/33의 규모이며, 총길이는 길이 6.3km이고 이 중 공원 구간은 4.3km, 그리고 공원의 폭은 10~60m이다.

경의선숲길공원은 3개 구간으로 나누어 계획되었다. 첫째, 용산구민센터-대흥로 구간 (2.7km)은 경의선 지중화 시작점으로서의 상징성을 부여하고 시민들에게는 철도와 기차가 주는 과거의 기억과 역사성을 기념할 수 있는 숲길로 계획되어 조성되었다. 둘째, 대흥로-양화로 구간(1.87km)은 홍대지역과 신촌지역을 연결하는 젊은이들의 열린문화공간과 다양한 문화교류가 이루어지는 소통의 장으로 조성되었다. 셋째, 양화로-홍제천 구간(1.73km)은 지역주민을 위한 운동 쉼터, 참여의 숲과 가족 마당이 조성되었다.

이후 경의선숲길공원은 다음 세 단계로 나뉘어 단계적으로 조성되었다. 1단계 구간은 대흥동-염리동 구간(2012, 길이 920m)으로 공덕역을 중심으로 업무구역 내 메타세콰이어길, 느티나무숲의 산책구간, 염리동 늘장이 조성되었다. 2단계 구간은 도화동 새창고개 구간과 연남동 구간(2013~2015, 1300m+990m)으로 공덕시장, 새창고개, 소나무숲이 조성되었다. 3단계 구간은 와우교 동교동 구간과 신수동 구간(2015~2016, 360m+420m)으로 홍대입구역을 중심으로 미술관, 책거리 등이 조성되었다.

경의선숲길공원 조성사업은 주변 커뮤니티 관점에서, 주민, 지역단체, 전문가 등의 참여를 통한 공원만들기를 위한 거버넌스를 도입했다. 경의선을 사랑하는 친구들, 경의선 포럼 등 자발적 운영과 관리주체가 발굴되어 구축했고, 주민마당, 커뮤니티 광장, 참여정원 등의 주민 참여공간을 조성했다. 이를 위해 경의선숲길을 시민 주도로 운영하는 공원으로서 기능하도록 운영과 관리를 전담할 비영리단체 '경의선숲길지기'가 2015년 시민, 전문가, 기업들로 구성되어 발족되었다. 이후 연남동지기, 창전동지기, 대흥동지기, 염기동지기, 도화원효지기 등 지역별로 담당단체가 세분화되었다.

Concept and Implementation of Gyeongui Line Forest Park

Gyeongui Line Forest Park (2009–2016) is built on the ground above the underground railroad line from Yeonnam-dong in Mapo-gu, Seoul to Wonhyo-ro of Yongsan-gu. As the DMC Station–Gongdeok Station section of the Gyeongui Line-Yongsan Line was relocated to the underground in 2012, the existing abandoned above-ground space was built as a long linear walking park, connecting Hongje Stream through apartment complexes in Hyochang-dong, high-rise building blocks in Mapo, and commercial districts in front of Sogang University and Hongik University. As a result, it was possible to find an alternative to the rapid large-scale apartment redevelopment in low-rise residential areas in and around the railroad line, which is accompanied by drastic changes in the residential environment.

In the early days of Mayor Park Won-soon's office (2011–2020) of Seoul Metropolitan Government, alternative plans for the abandoned railroads of the Gyeongui Line began to be discussed as one of his projects for Seoul. However, the above-ground park project in the Mapo section was originally initiated under the leadership of Mapo-gu during Mayor Oh Se-hoon's tenure (2006–2011), to create a park in the section between Daeheung-ro and Yongsan-gu Community Center. Along with this local government effort, the Seoul Metropolitan Government prepared an alternative plan to prevent reckless development of the abandoned land of the Gyeongui Line. At the time of planning of Gyeongui Line Forest Park, the Seoul Metropolitan Government's planned budget was 45.7 billion KRW as of 2014.

The purpose of the Gyeongui Line abandoned land project promoted by the Seoul Metropolitan Government at the time was: first, to improve the residential environment of local residents, who suffered due to the damage and inconvenience caused by

그림 6. 서울 경의선숲길공원, 공덕동 구간(출처: 김대석, 2015)
Gyeongui Line Forest Park, GongDuk-Dong segment

그림 7. 서울 경의선숲길공원, 동교동삼거리-연남동 구간(출처: 김대석, 2015)
Gyeongui Line Forest Park DongGyu-Dong Intersection-YonNam-Dong seegment

railroad operation; and second, to decide the ground sections of the Gyeongui Line as a park (Urban Management Plan, the 21st Seoul Urban Planning Committee). The intention to create a park in the ground section of the Gyeongui Line was to prevent unnecessary reckless development after the sale or lease of land led by private companies in the ground sections and surrounding areas within the abandoned line. To this end, the Seoul Metropolitan Government secured the right to use the park site free of charge from the Korea Rail Network Authority, the owner of the railway site.

According to the press release by the Seoul Metropolitan Government, the Seoul Metropolitan Government's goal is: "The section under the Gyeongui Line project is the section of the starting point of the Gyeongui Line Railway, which has the joys and sorrows and memories of the people since its opening in 1906. In this section, railroad undergroundization (Incheon Int'l Airport Railroad + Gyeongui Line Railroad) was underway due to changes in the urban environment and conditions of Seoul at the time. Accordingly, it prevents reckless development of railroad sections on the ground, and creates leisure resting areas for citizens."

At that time, the Seoul Metropolitan Government planned to recreate the area as a forest path illustrating the history of the Gyeongui Line, build a large-scaled greenway connecting Namsan Mountain, Yongsan, and the World Cup Park with bicycle and walking paths, and form a landscape that contains the passage of time by minimizing artificial facilities. The total area of the proposed park is 102,000 m^2—one half of the Yeouido Park area, and 1/33rd the area of Central Park in New York City—with the total length of 6.3 km, of which 4.3 km length of the park was of (10–60) m width.

Gyeongui Line Forest Park was planned to be divided into three sections. First, the Yongsan-gu Community Center–Daeheung-ro section (2.7 km) was planned as a forest path that symbolizes the starting point of under-groundization of the Gyeongui Line,

경의선숲길의 교통환승체계

현대 도시의 핵심기능은 서로 다른 체계의 대중교통체계 간의 효율적인 연계이다. 이에 도시의 교통체계는 도시와 공항항구, 도시와 도시, 중심부와 교외지 및 위성도시의 고속연결과 더불어 보행가로와 마을버스로 중심부와 철도역 및 메트로역, 철도가 나누었던 학교와 같은 네이버후드들의 보행연결이 주요 과제이다.

이러한 도시의 교통체계 간의 연결은 상이한 교통수단들 간의 일체화된 요금지불방식을 전제로 물리적인 보행연결로 완성될 수 있다. 이에 현대 도시의 교통환승체계는 서로 다른 2개 이상의 교통체계가 하나의 거점, 즉 역 또는 부두의 내부 또는 외부공간에서 연결되는 체계로 정의될 수 있다. 이는 서로 다른 보행 기준층을 가진 교통수단들이 환승거점을 중심으로 보행권 내에서 수평 및 수직이동으로 연결된 보행체계를 가지고 있어야 함을 말해준다.

경의선숲길공원은 지상부의 공원기능을 넘어 흥미롭게도 이러한 현대 도시의 서로 다른 대중교통체계를 연결해주는 교통환승체계의 대안적 모델을 제시해주고 있다. 그리고 이러한 지상부 공원과 지하부 대중교통환승체계는 초기부터 계획에 의한 결과가 아니라는 점에서 더욱 흥미롭다. 이렇게 조성된 지상부 선형공원 아래 대중교통 환승체계는 과거 서울 도심에서 을지로를 따라 지하에 조성된 국내 최장의 을지로 지하보도와 시민의 이용도와 만족도 그리고 주변변화라는 관점에서 매우 비교된다. 을지로 지하보도는 지하철 2호선 을지로입구역에서부터 을지로 3, 4가역을 거쳐 동대문역사문화공원역을 연결하는 총길이 2.8km의 지하거리로서 여전히 그 활성화가 이슈화되어 왔다.

현재 경의선숲길공원의 지하 30+ m에 조성된 인천공항철도, 지하 20+ m에 조성된 지하철 2, 5, 6호선, 그리고 공원 바로 밑 지하 10+ m에 조성된 경의중앙선의 정차역들은 상호 상이한 조합의 교통환승이 이루지고 있으나, 하나의 선형공원 지상부를 통해 보행체계로 연결되어 서울 서부권의 거대한 대중교통환승체계를 완성하고 있다. 현재 경의선숲길 공원 지하에서 기능하는 환승역은 서쪽부터 가좌역(경의중앙선), 홍대입구역(2호선·공항철도·경의중앙선), 서강대역(경의중앙선), 대흥역(6호선), 공덕역(6호선·5호선·공항철도·경의중앙선), 효창

and commemorates the memories and history of the past given by railroads and trains to citizens. Second, the section between Daeheung-ro and Yanghwa-ro (1.87 km) was created as a place for communication where various cultural exchanges would take place with an open cultural space for young people, connecting the Hongik University and Sinchon areas. Third, in the Yanghwa-ro–Hongje river section (1.73 km), an outdoor exercise zone for local residents, a participation forest, and a family yard were created.

Since then, Gyeongui Line Forest Park has been divided into the following three stages. The section in the first stage is the Daeheung-dong–Yeomni-dong section (2012, 920 m in length), and around Gongdeok Station, Metasequoia-gil in the business area, Zelkova Forest Path, and Yeomri-dong Neuljang were created. The sections in the second stage include the Saechanggogae section in the Dohwa-dong section and Yeonnam-dong section (2013–2015, 1,300 m + 990 m), and Gongdeok Market, Saechanggogae, and pine forests were created. The sections in the third stage are the Waugyo Donggyo-dong section and the Sinsu-dong section (2015–2016, 360 m + 420 m), and art galleries and book streets were created around Hongik University Station.

The Gyeongui Line Forest Park Project called for governance to create a park from the perspective of the surrounding community through the participation of residents, local organizations, and experts. Voluntary operation and management subjects, such as 'folks who love Gyeongui Line' and Gyeongui Line Forum, were devised and created, and residents' participation spaces, such as residents' agorae, community squares, and participation gardens, were created. To this end, the Friends of Gyeongui Line Forest Road, a non-profit organization dedicated to the operation and management of the citizen-led park, was launched in 2015 with citizens, experts, and companies. Since then, the subdivided groups in charge of each region, such as Yeonnamdong-jigi,

표 2. 경인선숲길의 환승체계 / Metro Station for Metro Transfer underneath of Gyeongui Line Forest Park, Seoul

	경의중앙선	2호선	5호선	6호선	공항철도선
가좌역	●				
홍대입구역	●	●			●
서강대역	●				
대흥역				●	
공덕역	●		●	●	●
효창공원역	●			●	

(출처: 한광야 김민지, 2020)

공원역(6호선·경의중앙선), 삼각지역(4호선·6호선)이다.

특히 연남동의 홍대입구역과 공덕동의 공덕역은 이러한 환승기능을 통해서 서울의 새로운 도시거점으로 자리잡았다. 이 2개의 환승역에서는 지하철 2호선, 5호선, 6호선이 인천국제공항과 김포공항으로 연결되며, 또한 지상철도거점인 서울역과 용산역으로 연결된다. 현재 홍대역의 1일 사용인구는 205,323명(2019년 기준)이며, 공덕역의 경우 79,592명(2019년 기준)이다. 이러한 교통환승거점과 주변 동네는 일군의 마을버스(마포 05, 06, 07)로 연결되고 있다.

Changjeondong-jigi, Daeheungdong-jigi, Yeomgidong-jigi, and Dohwawonhyo-jigi, have been created.

Mass Transit Hub

A core function of modern cities is efficient connection between the various modes of public transportation systems. In this regard, the main goals of the transportation systems in cities include effective connections between cities and airports, cities and cities, centers and suburbs or satellite cities, as well as pedestrian connection between center, metro stations or railway stations, schools, and neighborhoods through pedestrian and local streets.

The connection between the urban transportation systems may be completed with a physical walking connection on the premise of an integrated fare payment method between different means of transportation. Accordingly, the transit transfer system of a modern city can be defined as a system in which two or more different transportation systems are connected at one hub, that is, at the inner or outer space of a station or a pier. This suggests that public transportation with different platforms should be connected horizontally or vertically within the walking area around the transfer hub.

Gyeongui Line Forest Park is interestingly presenting an alternative model of the transportation transfer system that connects the different public transportation systems of these modern cities beyond the ground park function. In addition, it is intriguing that this underground public transit transfer system below ground park is not the result of planning from the beginning. The public transportation transfer hub under the ground linear park is very comparable to the nation's longest Euljiro Pedestrian Underpass built along Euljiro from the center of Seoul in the past, in terms also of the level of

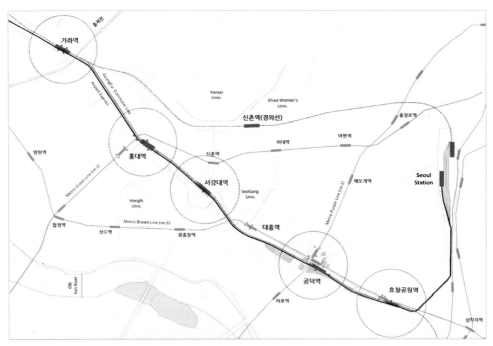

그림 8. 서울 경의선숲길 지하의 경의중앙선+공항철도+지하철 2, 5, 6호선 노선과 정차역(출처: 한광야 김민지, 2020)
Gyeongui-Jungang Line, Airport Line and Subway 2, 5 and 6 Lines and the stations underneath of Gyeongui Line Forest Park, Seoul

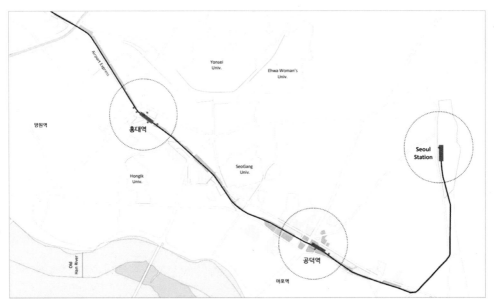

그림 9. 서울 경의선숲길공원 지하의 공항철도 노선과 정차역(출처: 한광야 김민지, 2020)
Incheon Int'l Airport Line and the stations underneath of Gyeongui Line Forest Park

usage by citizens, their satisfaction, and the changes of the surroundings. Euljiro Pedestrian Underpass is a 2.8 km long underground passage connecting Euljiro Station on subway line 2 through Euljiro 3 Station and 4-ga Station to Dongdaemun History and Culture Park Station, and its activation is still an issue.

Currently, although there has been a mutually discussed combination of traffic transfers at Incheon International Airport Railroad, which was built more than 30 m underground of Gyeongui Line Forest Park, subway lines 2, 5, and 6 built more than 20 m underground, and the stops on the Gyeongui-Jungang Line built more than 10 m underground of the park, they are also connected by a walking system below the ground linear park, thus completing a huge public transportation transfer system in western Seoul. Currently, the transportation and transfer stations functioning underground of Gyeongui Line Forest Park are, from the west, Gajwa Station (Gyeongui-Jungang Line), Hongik University Station (Line 2, Airport Railroad Line, and Gyeongui-Jungang Line), Daeheung Station (Line 6), Gongdeok Station (Lines 5 and 6, Airport Railroad Line, and Gyeongui-Jungang Line), Hyochang Park Station (Line 6 and Gyeongui-Jungang Line), and Samgakji Station (Lines 4 and 6).

In particular, Hongik University Station in Yeonnam-dong and Gongdeok Station in Gongdeok-dong have become new urban hubs in Seoul through these transfer functions. At these two transfer stations, subway lines 2, 5, and 6 connect to Incheon Int'l Airport and Gimpo Airport, and also connect to Seoul Station and Yongsan Station, the hub of the ground railways. Currently, the population using Hongik University Station per day is 205,323 (as of 2019), and in the case of Gongdeok Station, 79,592 (as of 2019). These transportation transfer hubs and surrounding neighborhoods are connected by a group of local buses (Mapo 05, 06, & 07 lines).

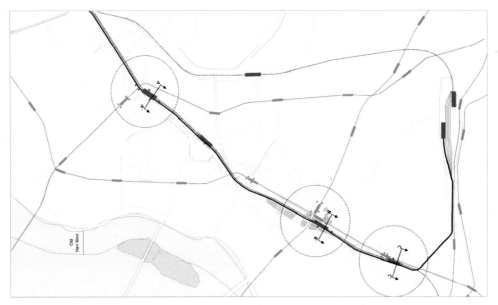

그림 10. 서울 경의선숲길공원 지하의 교통환승거점 수직연결체계(출처: 한광야 김민지, 2020)
Metro Transfer System under Gyeongui Line Forest Park

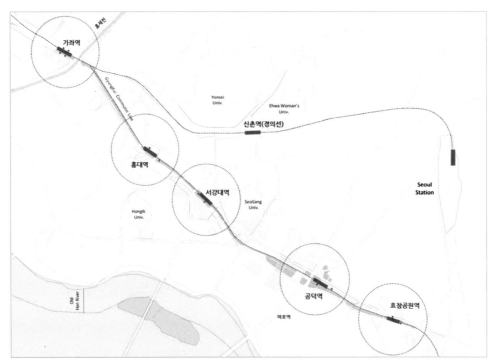

그림 11. 서울 경의선숲길공원 지하의 경의중앙선 노선과 정차역(출처: 한광야 김민지, 2020)
Gyeongui-Jungang Line and the stations underneath of Gyeongui Line Forest Park, Seoul

A-A' 홍대역입구역 단면 A-A' Hongdae Station Area Section

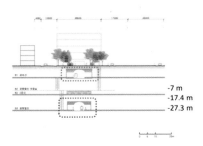

B-B' 공덕역 단면 B-B' GongDuck Station Area Section

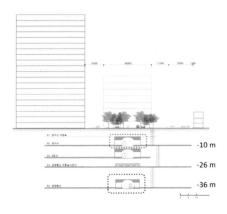

C-C' 효창공원앞역 단면 C-C' Hyochang Park Station Area Section

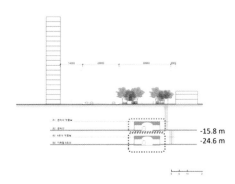

(출처: 한광야 김민지 유지인, 2020)

맺으며

경의선숲길공원은 정치적 성향이 다른 두 서울 시장의 임기기간에 연이어 진행된 사업으로서 의미를 갖고 있다. 먼저 마포구 주도로 2006년 염리동·대흥동 구간(1단계 760m, 2006~2012)이 공원계획을 수립하여 추진되었고, 뒤이어 공원조성은 전 구간으로 확장되어 2016년에 최종 완공되었다.

또한 이렇게 조성된 경의선숲길공원은 지하의 3개 레벨(30+, 20+, 10+ m)에 조성된 공항철도, 지하철, 경의중앙선 역들과 일군의 마을버스, 그리고 일군의 교통환승거점들이 어우러져 현대 도시의 새로운 교통환승거점으로서 기능하고 있다. 흥미롭게도 이러한 선형공원과 기능하는 교통환승체계는 계획에 의해 조성된 결과가 아닌 계획되지 않은 결과물이다.

한편 지상부의 공원이 서울 서남부의 새로운 명소로 떠오르면서, 주변의 골목상권이 활성화되었다. 특히 선형의 긴 공원 주변과 배후 골목에는 레스토랑, 카페, 술집들이 손님들로 북적이고 있다. 이에 따라 자연히 토지와 건물의 가치도 상승되었다. 단점으로는 몰리는 사람들로 인해 야간 고성방가나 음주폭행 등 사건사고가 끊이지 않는다는 점이다. 이러한 변화와 함께 거대한 선형공원의 향후 과제로 무엇보다 중요한 것은 공원의 지속적인 관리주체와 운영이다.

또한 부동산의 가치 상승과 지속적인 주거지의 상업화에 따라, 동네의 중심부 기능을 갖고 있던 주변의 초·중·고교의 이전 이슈와 동네 커뮤니티의 와해가 우려되고 있다. 특히 이 지역의 임대료가 상승하고 외지인의 부동산 투자가 진행되면서 재건축이나 부동산 매매를 통해 기존 원주민이나 세입자가 내몰리는 현상이 발생하고 있다.

Closing Remarks

Gyeongui Line Forest Park is meaningful as a project that has been carried out in stages during the term of offices of two Seoul mayors with different political tendencies. First, under the leadership of Mapo-gu, the Yeomri-dong–Daeheung-dong section (1st stage, 760 m, 2006–2012) was established and promoted in 2006, and then the park construction expanded to the entire section, finally completed in 2016.

In addition, this Gyeongui Line Forest Park functions as a new transportation hub of a modern city that encompasses the Airport Railroad, subways, and the Gyeongui–Jungang Line stations built at three levels (30+, 20+, 10+ m), as well as village buses, and transportation transfer hubs. Interestingly, this linear park and functional transportation transfer systems are not the results of a plan, but unplanned results.

Meanwhile, the emergence of the ground park as a new attraction in southwestern Seoul revitalized the surrounding commercial districts. In particular, restaurants, cafés, and bars crowd the long linear park and the alleys behind it. Accordingly, the value of land and buildings has naturally increased. However, the disadvantage is that due to crowds, incidents, such as loud noises at night or drunk assaults, continue. Along with these changes, the most important thing as a future task of the huge linear park is the continuous management and operation of that park.

In addition, due to the rise in value of real estate and the continued commercialization of residential areas, there are concerns about the relocation of the surrounding elementary, middle, and high schools that had the central functions of the neighborhood, and the collapse of the local community. In particular, as rents in the region rise and outsiders invest in real estate, reconstruction or real estate sales have driving out local residents and tenants.

참고문헌

1. 경의선 숲길지기 홈페이지

2. 박순욱. 조선닷컴. 걷다, 서울: 서울에서 가장 긴 공원, 경의선숲길, 2018

3. 변진석. KBS. '경의선숲길공원' 전구간 완공…오늘 개원행사, 2016

4. 손효숙. 한국일보. 버려진 철길, 연트럴파크로 신분 상승, 2015

5. 내 손안의 서울. 버려진 '철길'이 시민의 '숲길'이 되어, 2015

6. 이영근. 매일경제. 새로 생긴 서울 공원길…경의선숲길 연남동에서 공덕동까지, 2015

References

1. Gyeongui Line Forest Park website.

2. https://www.facebook.com/gyeonguiline/

3. Park, Soon-wook. 2018. "Walking, Seoul: The Longest Park in Seoul, Gyeongui Line Forest Park", Chosun Ilbo, 2018.9.21.

4. Son, Hyo-suk. 2015. "Abandoned Railroad, Rising status as Yeontral Park", Hankook Ilbo, 2015.10.11.

5. Seoul in my Hands. 2015. "Abandoned Railroad has become a Citizen's Forest Park".

6. Lee, Young-geun. 2015. "New Seoul Park Road, Gyeongui Line Forest Road from Yeonnam-dong to Gongdeok-dong", Maeil Business News, 2015.7.8

대구 동대구역 복합환승센터

임동원 (도화엔지니어링 상무)

그림 1. 남쪽에서 바라본 동대구역 복합환승센터 / South birdview of Dongdaegu Multi Transit Complex

들어가며

동대구 복합환승센터는 대구시 동구에 위치한 연면적 296,841m², 지하 7층 지상 9층의 대규모 복합환승센터이다. 이 프로젝트의 의미는 첫째는 거대한 인공지반을 만들어 광역고속

Dongdaegu Multi Transit Center, Daegu

Dongwon Lim (Senior Director, Dohwa Engineering)

Introduction

Dongdaegu Multi Transit Center (herein after DMTC) is a large-scale complex building located in Dong-gu, Daegu, with a total floor area of 296,841m², 7 stories below the ground and 9 above. The meaning of this project is that, firstly, it shows a huge artificial ground and connects the areas of express bus terminal and the railway station in order to realize the true 'multi transit center'. Through this, it contributed to the improvement of urban traffic condition by connecting the inter-city main road that had been cut off. It is a mega-scale urban architecture project.

The Site and Daegu City

Daegu Station and Dongdaegu (East Daegu) Station

Why Dongdaegu? Daegu Station and Dongdaegu Station are not very far apart. However, the reason why there are two stations is that it became difficult to increase

버스터미널과 철도역사를 연결시켜 명실상부한 '복합환승센터'를 구현했고, 이를 통해 그동안 단절되었던 도시간선도로를 연결시켜 도시교통개선에 기여한 메가 스케일의 도시건축 프로젝트이다.

대상과 도시와의 관계

대구역과 동대구역

왜 동대구역일까? 대구역과 동대구역은 그리 멀리 떨어져 있지 않다. 그런데 역이 2개인 이유는 대구역이 생기고 나서 시가지가 들어서는 통에 역 확장(노선 확장)이 어려워져 1969년에 역을 하나 더 만든 철도역이 동대구역이라고 한다(역의 규모 또한 동대구역이 더 크다. 대구역은 4선인데 반해 동대구역은 8선이다). 이후 향후 철도역의 위상으로 보면 동대구역이 압도하는데 KTX의 정차역은 대구가 아닌 동대구역을 지난다.

철도역+고속터미널 : 초광역 교통거점

이 건물의 가장 큰 특이점은 대도시의 메인 철도역과 메인 고속터미널이 함께 입지한다는 것이다. 서울, 부산 등 어느 대도시를 가도 이렇게 두 메인 교통거점들이 붙어있는 경우는 찾아보기 힘들다. 그렇기에 여기는 몇 개의 지하철 노선이 만나는 역세권의 개념을 넘어서 공공교통의 양대 축인 철도와 버스가 한 곳에 붙어있는 초광역 스케일의 교통거점이라 할 만하다.

대상의 계획 및 조성과정

동대구역 복합환승센터는 사실 대구 고속터미널 부지였다. 먼저 이 계획의 효시가 될 만한 고속터미널 복합개발의 개략사부터 살펴본다.

그림 2. 대구역과 동대구역 / Deagu station and Dongdeagu (East Deagu) station

the railway track number as the city grows after Daegu Station was built. (The size of the station is also larger at Dongdaegu Station. Daegu Station has 4 lines, whereas Dongdaegu Station has 8 lines.) From the perspective of the future railway station hierarchy, Dongdaegu Station dominates Daegu station regarding that KTX (Korea express train) stops through Dongdaegu Station, not Daegu.

Regional Transportation Hub: Railway Station + Express Terminal

Main character of the building is that the integration of main railway station with main express terminal together. No matter where you go to any big city such as Seoul or Busan, there is few case where the two main transportation hubs are tie together. Therefore, it can be said that this is an regional scale transportation hub much bigger than the conventional notion of hub as one or two subway station meets.

고속터미널 부지 : 국토 광역교통의 중요 거점

1975년 강남 고속터미널이 강남이 맨 흙바닥인 시절 건립된 것은 어찌보면 한국 도시사에 굉장히 중요한 터닝 포인트였다(1975년에 구자춘 당시 서울시장은 도심 집중을 완화하고 강남을 개발하기 위해 반포동 종합버스터미널 계획을 세우고, 1976년 4월 8일에 기공식을 가졌다). 이는 당시 강남개발의 성공요인으로 부처 이전, 명문고 이전과 함께 고속터미널 건설이라는 교통거점을 일종의 도시계획적 앵커(anchor)시설로 인식하였던 것이다. 그렇기에 이후 분당 야탑 고속터미널 (2001), 광주 고속터미널 (유스퀘어, 2005), 천안 고속터미널(신세계, 2010) 등 지방도시도 새로운 택지개발 및 신도시의 거점으로 고속터미널 건립을 활용하는 계기를 마련하게 된다.

고속버스 및 시외버스 터미널 등의 광역버스터미널 등은 기차 선로에 절대적인 영향을 받는 철도역과는 다르게 비교적 사방으로 연결되는 주요 교통 광로에 연계되는 '길목'에 자생적으로 많이 건립되었다. 그러나 이러한 소위 말하는 '종합터미널' 개념의 중추 터미널역할이 강화되면서 한번에 자금력을 모은 운수회사들은 그들의 사업모델로 복합개발의 중요 투자자로 참여하면서 새로운 복합종합터미널에 만들고, 이를 다시 복합환승센터라는 이름으로 지하철역 혹은 철도역으로 연계될 수 있는 법적인 단초까지 마련하였다. 국가통합교통체계효율화법, 복합환승센터 개발실시계획 수립지침(2016)은 그 예이다.

개발경위를 살펴보면, 기존 대구 고속터미널이 기존 동대구역과 함께 공존하고 있었다는 것은 어찌 보면 매우 큰 개발적인 호재였던 것으로 보인다. 사실 공공(대구시청, 철도청)이 먼저 이 개발기획을 한 것인지, 민간자본이 먼저 기획을 한 것인지는 좀더 문헌조사가 필요한 사항이지만, 여기에 대구 도시철도 1호선이 들어오고(2003) 이후 롯데와 함께 치열한 역세권 싸움을 전개하고 있던 신세계는 기존의 터미널 개발의 경험 등을 토대로 과감한 투자를 할 수 있었던 것으로 보인다. 문헌조사에 따르면, 동대구 사업비 총 8,800억 중 신세계측이 55%(4,500억) 부담, 나머지는 외부 차입(3,700억)으로 조달하였음을 알 수 있다.

Planning Process

DMTC site was actually the area of the Daegu Express Bus Terminal. Thus, it is valuable to note that the historical outlook of micro history of multi-functional bus terminal development in Korea.

Historical Outlook of Multi Functioned Express Bus Terminal in Korea

The fact that Gangnam Express Bus Terminal was built in 1975 when Gangnam was on the bare ground was a significant point in Korean urban development history. This kind of new terminal development as a anchor facility deemed a important success factor for Gangnam as green field urban development. Thereafter, another terminal building in green field urban development was followed, such as Bundang Yatap Express Terminal (2001), Gwangju Express Bus Terminal (U-Square, 2005), and Cheonan Express Terminal (Shinsegae, 2010). Those are also the same anchor tenants for each new cities.

Unlike railway stations, strongly influenced by railway tracks, bus terminals such as express/ intercity bus was built autonomously on the main transportation point along the main arterial road. However, as the role of the main bus terminal of the so-called 'multi-functioned terminal' concept has been strengthened, major bus running companies that have accumulated financial power at once participated as important investors in the terminal complex building development with their business model and made it into a new complex terminal, which was renamed the 'Multi Transit Center.' It even was provided a legal foundation recently.

또한 이 프로젝트는 고속버스터미널을 새로 만들면서 각 회사별로 분산되어 있던 고속버스터미널과 시외버스터미널인 대구 동부, 남부정류장을 모두 합쳐 본 계획으로 역량을 집결하였다. 여기에 신세계의 자본과 대구시와 중앙정부의 공공자본이 합쳐져 다른 계획과는 좀더 통이 큰 공공시설- 인공데크를 설치하였다. 이러한 거대한 인공대지를 통해 이 복합 환승센터는 철도역과의 연계, 나아가 대구시의 중요 남북교통로까지 제공하고 있다.

거대한 인공대지 : 연계되는 철도역, 그리고 중요 missing link의 연결

그림 3. 개발 전과 후 Note: 2008년(왼쪽), 2019년 현재(오른쪽) / Before and after the project(2006, left/ 2019, right)

철도역사가 들어올려지고, 그 역과 함께 복합쇼핑몰이 들어오는 개발방식은 역사가 매우 오래되었다. 서울역(1988, 한화, 현 롯데), 영등포(1990, 롯데), 용산역(2004, 현대산업개발), 청량리(2010, 롯데) 등이 소위 1세대들이다. 이들 역 중 특히 서울역은 이제 30년의 1차 임대기간도 지나 새로운 계약갱신의 시기로 접어들었으니, 이제 한 generation이 법적으로는 끝나는 셈이 되었다. 그렇게 역세권 하면 '들어올려진 철도+상업시설'의 조합이 떠오른다. 상기의 사진을 살펴보면 여기에 동대구역 환승센터는 거대한 인공대지를 붙여 기존 철도역과 복합터미널을 연결하고 더 나아가 단절된 남북로를 신안남로라는 도로를 통해 연결하고 있다.

Investigating the DMTC development process, the fact that the existing Daegu Express Bus Terminal coexisted with the existing Dongdaegu Station seems to be a great developmental advantage. Whether the public side (Daegu City Hall, Korean Railways) planned this development first or the private capital carried out the plan first is a matter that requires more literature research, but it is here that Daegu Urban Railway Line 1 entered (2003) and later, together with Lotte, Shinsegae, which was fighting a fierce battle in the station area, seems to have been able to make bold investments based on their experience of existing terminal development.

All inter and intra-city bus terminals as Daegu eastern and southern bus terminals intended to join in the project site. Then Shinsegae's capital investment comes in with the idea of public contribution plan. It provides a linkage plan with the railway station through a huge artificial land.

Massive Artificial Land : Connecting Railway Station and Important Missing Links

The approach of this kind of development that the lifted railway station with shopping mall complex has quite long history. Seoul Station (1988, Hanwha, now Lotte), Yeongdeungpo (1990, Lotte), Yongsan Station (2004, Hyundai), and Cheongnyangni (2010, Lotte) are the so-called first-generation. Among these stations, Seoul Station, in particular, has now passed the first lease period of 30 years and has entered a new contract renewal period. When we think of the station area like that, the combination of 'raised station + commercial facilities' comes to mind. If you look at the photo above, the Dongdaegu Station Transit Center attaches a huge artificial land to connect the existing railway station and the complex terminal, and further connects the North-South road through a new road called Sin-Annam-ro.

의외로 심플한 내부 평면

밖에서 보여지는 외양은 다양한 직선의 조합이 만들어낸 매우 기하학적인 힘과 복잡함이 느껴진다. 그렇기에 매우 복잡한 평면구성을 예상하였으나, 이 건물이 가진 다이어그램은 의외로 매우 단순하다.

그림 4. 단면 다이어그램 / Section diagram

일반적으로 이런 PPP 방식의 복합개발은 소위 말해 salable한 면적을 얼마나 많이 확보하느냐가 사업성하고 직결되기에 최대한 매장면적을 많이 확보하는 전략을 추구한다. 여기도 그런 느낌은 없지 않다. 불편함 없이 터미널 이용을 할 수 있는 충분한 면적은 확보하고 있으나, 통합된 상업공간의 대규모 개발은 터미널 면적을 압도한다.

1층의 평면을 보면 백화점과 터미널의 평면은 거의 균등하게 분할되고 있는데, 특히 터미

Simple form of Interior Space

The first appearance of DMTC shows the very complicated form with diverse geometric shape created by the combination of straight lines. However, the interior diagram of this building looks very simple.

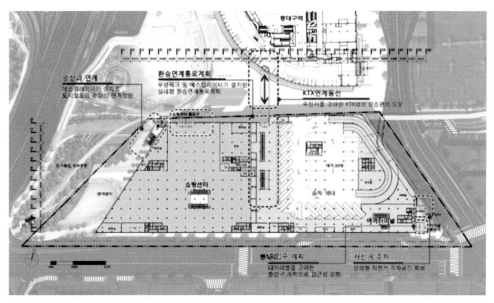

그림 5. 지상1층 평면도 / Ground floor plan

In general, this PPP-type multi-functional building development pursues a strategy to secure maximum salable area for business viability. Although sufficient area is secured to use the terminal without inconvenience, the large-scale development of integrated commercial space overwhelms the terminal size.

s per the first floor (ground floor), the department store and the terminal are divided almost equally. In particular, the terminal space is located on the east side and cannot have the front square. It seems to have given the priority to the commercial side. Also, in terms of access, facing the east side of the road seems to be advantageous because

널 공간이 동쪽에 위치하여 전면광장에 면하지 못하는 건 아무래도 거대투자자인 신세계의 민자공간에 광장을 배려하여 우선순위를 준 것으로 보인다. 또한 진입의 측면에서도 동쪽 도로에 면하는 것이 철도역 환승체계와 고속버스의 진출입을 분리시키고자 하는 교통연계상 유리한 점도 감안한 것으로 보인다.

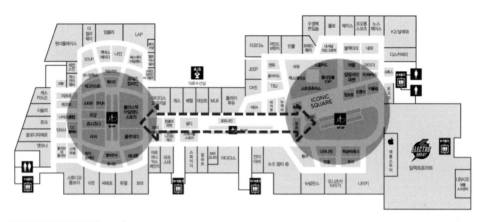

그림 6. 8층 평면도 / 8th floor plan

그림 7. 동대구역 복합환승센터 내부 아트리움
Atrium space of Dongdaegu Multi Transit Complex

5층부터는 본격적인 상업공간의 평면으로 대표적 상업공간 유형인 덤벨(Dumbel)형이 나타난다. 즉 2개의 큰 시각적, 기능적인 초점공간을 만들고 이 둘이 일종의 상업공간의 군을 이뤄 에너지를 주고받는 것이다. 특히 오른편의 중정의 크기는 매우 큰 규모여서 관람객들에겐 단번에 메인 실내이미지로 각인되고, 상업공간의 밀도적 측면이나 상품집중도 측면 또한 매우 높아 5층에 진입과 동시에 터미널의 기억은 전혀 남지 않고, 일반적 교외형 몰의 느낌을 구현하

it can separate the noise from the main route to the gate of the railway station and express buses.

From the 5th floor, a dumbbell-form, a typical commercial space appears as the floor plan of a full-fledged commercial space. Two main focal spaces are put in, then both spaces are connected by commercial corridor. In particular, the size of atrium is huge, so it is imprinted on the main interior image by visitors at once, and the density of the commercial space and the concentration of products are also very high. It succeeds in embodying the feeling of a suburban mall. This is a very surprising sense of spatial disconnection.

The sense of exterior architectural mass is very different from the surroundings. KPF's unique straight lines and hexahedral cube design are realized on a huge scale, and are in complete opposition to the dense urban masses around them. This gives a very estrange feeling. If the Dongdaemun DDP is cut off from its surroundings due to its streamlined grandeur, here it competes with the gigantic straight-line and right-

그림 8. 서쪽에서 바라본 동대구역 복합환승센터 / West birdview of Dongdaegu Multi Transit Complex

는 데 성공한다. 이는 매우 놀라운 공간적 단절감이다.

건물 외부의 매스감은 매우 주변과 이질적이다. KPF(설계사) 특유의 직선과 육면체 큐브의 느낌이 거대한 스케일은 주변의 오밀조밀한 도시형 군락 매스와 완전히 대립하고 있으며 이는 매우 생경한 느낌을 준다. 마치 동대문 DDP가 유선형의 거대함으로 주변과 단절감을 준다면, 여기는 직선과 직각의 육면체의 거대함으로 주변을 압도한다. 그렇기에 기존의 동대구역이 가졌던 반원형의 매스의 느낌과는 더욱 극명하게 부조화를 이룬다.

맺으며

이 프로젝트는 터미널이라는 도시의 교통시설이 대형 쇼핑센터와 연계되어 개발된 민간 합동 프로젝트로서, 고속터미널을 도시활성화의 앵커기능으로 활용하고자 했던 한국도시건축의 유전자를 가진 건축이라 할수 있다. 또한 이 대형 개발사업은 인근지역의 개발사업에도 영향을 미쳤음을 보여준다. 대구 도심의 낙후지역으로 꼽혔던 대구 동구는 동대구역 이용객과 신세계백화점 방문객이 증가하여 상권변화가 나타나고 있고, 인근의 재개발, 재건축 등의 정비사업의 추진이 목격되고 있다. 그러나 인근의 주변환경과의 경관상의 부조화 등은 매우 독특한 개별 건물의 아이덴티티는 확보하였음에도 불구하고 여전히 아쉬운 점으로 남는다.

참고문헌

1. Alex wall, Victor gruen: from urban shop to new city, 2007

2. 대구시. 분야별 정보〉교통〉동대구역 복합환승센터, 2020

3. https://www.daegu.go.kr/mobile/index.do?menu_id=00932977

4. 위키백과. 동대구 복합환승센터, 2021

5. https://ko.wikipedia.org/wiki/%EB%8F%99%EB%8C%80%EA%B5%AC%EB%B3%B5%ED%95%A9%ED%99%98
   %EC%8A%B9%EC%84%BC%ED%84%B0

6. 신세계 홈페이지. 신세계 대구점, 2022

7. https://www.shinsegae.com/store/main.do?storeCd=SC00013

angled hexahedron. It moreover is clearly inconsistent with the semicircular mass of the existing Dongdaegu station.

Closing Remarks

This project is a type of Korean mixed-use urban architecture, regarding a PPP project developed in connection with a large shopping center in the urban transportation facility. This project had an impact on development projects in neighboring areas. Daegu Dong-gu, considered as an underdeveloped area of downtown Daegu, is seeing changes in its business district as the number of DMTC visitors increases, and the promotion of redevelopment and reconstruction projects in the vicinity is witnessing. However, the incongruity of the landscape with the surrounding environment remains challenging, despite securing an unique individual building identity.

References

1. Alex wall, "Victor gruen: from urban shop to new city", 2007

2. Internet access, Daegu municipality, 2020

3. https://www.daegu.go.kr/mobile/index.do?menu_id=00932977

4. Internet access, Wikipedia, 2021

5. https://ko.wikipedia.org/wiki/%EB%8F%99%EB%8C%80%EA%B5%AC%EB%B3%B5%ED%95%A9%ED%99%98
%EC%8A%B9%EC%84%BC%ED%84%B0

6. Internet assess, Sinsaegae homepage, 2022

7. https://www.shinsegae.com/store/main.do?storeCd=SC00013

경주의 철도기반 교외지 개발과 역사보존의 충돌

한광야 (동국대 건축공학부 교수)

들어가며

이 원고는 '도시는 철도와 어떻게 만나는가'라는 이슈를 갖고, 한국의 대표적인 역사도시인 경주의 사례를 통해, '유네스코 세계문화유산의 역사도시에서 과연 철도는 어떻게 처음 조성되었고, 이후 이를 대신하는 고속철도는 또 어떻게 건설되어 역사도시에 어떠한 영향을 주고 있는가'의 질문을 중심으로 신도심 개발이 주도하는 도시확장과 역사구역의 보전과의 상충관계를 알아보고자 한다.

이러한 배경에서, 이 원고는 일제강점기에 철도가 처음 조성되어 한국전쟁 이후 현재까지 강력한 역사보존정책을 실행하며 유네스코 세계역사문화도시의 위상을 지켜오고 있는 경주의 사례를 통해, 단계별 신도심 개발을 유도해온 철도선의 건설과 최근 고속철도선 건설을 위한 경로 선정 및 신경주역(2010) 부지 선정과정을 고찰하며, 원·구도심의 유적보존과 도시확장의 관리정책이 주는 교훈을 확인하려고 한다.

Conflict of Railroad-driven Development and Historic Preservation, Gyeongju

GwangYa Han (Professor, Dongguk University, Department of Architecture)

Introduction

With the topic 'how a city intersects with the railway', this paper looks into the conflicting relationship between urban expansion led by new city center development, and the preservation of a heritage site, with the case of Gyeongju, a historic city in Korea, through the questions, 'how was the railway first established in a historic city designated as a United Nations Educational, Scientific and Cultural Organization (UNESCO) World Heritage Site; and how was the high-speed railway, which later replaced the railway, constructed, and how did it affect the historic city?'

With this background, this paper reviews the historical growth and relocation of railways including the latest high-speed railway and the Sin-Gyeongju Station (2010), which have led new developments through the different expansion stages. The paper identifies lessons learned from policies to preserve historic sites in an old city center and to manage urban expansion through the case of Gyeongju whose railway was first built during the Japanese colonial period and has grown after the Korean War with strong historic preservation policies to maintain its standing as a UNESCO World

역사도시와 철도선

경주가 한반도의 역사도시로서 의미를 갖는 이유는 약 1400년 이상의 왕조의 수도, 지역 행정의 중심, 관광도시로서의 일련의 변화과정과 이에 따른 도시 특성이 현재 경주 원·구도심에 밀도 있게 인접해 집중되어 있기 때문이다. 따라서 경주는 이러한 변화과정에서 각기 다른 지배그룹과 사회가치가 어떠한 도시를 실현해왔고, 또한 이 과정에서 구성원들의 일상환경은 어떻게 변화해왔는가를 밀도 있게 비교할 수 있는 대상이기 때문이다.

선사시대에 경주 분지(慶州盆地)의 주인은 마한세력(馬韓, BC 1세기~3세기)이었고, 뒤이어 낙동강의 동쪽지역을 지배한 진한(辰韓, BC 1세기~3세기)과 사로국 세력(斯盧國, ?~BC 57), 그리고 이를 계승한 신라(新羅, BC 57~935) 왕조의 수도였다. 이 시기에 경주의 상징적 중심부는 경주 김씨의 시조인 김알지(金閼智, 65~?, Primogenitor of Gyeongju Kim family and Kim Dynasty of Shilla Kingdom)의 탄생지인 계림(鷄林)이었고, 요

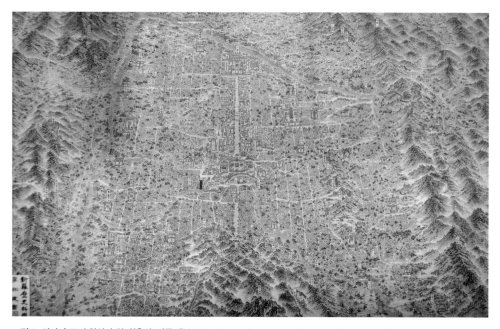

그림 1. 신라 수도의 월성과 왕경(출처: 경주시) / Shila Kindom's Wolsung Palace Citadel and Wanggyung City

Heritage Site.

Historic City and Railway Line

What makes Gyeongju a historic city in the Korean Peninsula is the series of changes and subsequent urban characteristics as a royal capital of more than 1,400 years, a hub of local administration, and a tourism city that is densely populated, and encapsulated in the current old city center of Gyeongju. Hence, the case of Gyeongju allows the author to closely compare how different ruling groups and social values shaped the city amid these changes, and how this process changed the everyday life of its members.

In prehistoric times, Gyeongju Basin was occupied by Mahan (1 c. BC – 3 AD), followed by Jinhan (1 c. BC – 3 AD) and Saro-guk (? – 57 BC), which ruled over the eastern side of the Nakdong River, and then served as the capital of the Silla dynasty (57 BC – 935 AD). Gyerim Forest, the birthplace of Kim Al-ji (65–?), whose descendants established the Kim royal clan in Silla, was a symbolic center of Gyeongju, and Wanggyeong Civilian Area, modeled after Chang'an in the Tang dynasty (618–907), with Kumseong (57 BC – 101 AD) and Wolseong (101–935), which worked as a fortress and palace, respectively, was an ideal city that sought the values of Buddhism.

Then in the Goryeo dynasty (918–1392), Gyeongju served as Donggyeong, a center of regional administration; and later in the Joseon dynasty (1392–1910), Gyeongju Eupseong and Gyeongju Hyanggyo (1492), a center of regional administration, formed the current old city center. In addition, Seoak Seowon (1563), Oksan Seowon (1573), and Gugang Seowon (1692) established in the suburbs of Gyeongju completed Gyeongju as a Neo-Confucian city.

새이며 왕궁인 금성(金城, BC 57~101)과 월성(月城, 101~935)을 두고 중국 당나라(唐朝, 618~907)의 수도인 장안(長安, Xi'an)을 모델로 건설된 왕경(王京, Wang Gyung Civilian Area)은 불교의 가치를 추구했던 이상도시였다.

이후 경주는 고려왕조(高麗王朝, 918~1392)의 지역행정거점이었던 동경도성(東京都城) 이었고, 뒤이은 조선왕조(朝鮮王朝, 1392~1910)의 지방행정거점인 경주읍성(慶州邑城)과 경주향교(鄕校, 1492)가 그 중심부로 현재 원도심을 구성했다. 그리고 경주의 교외지에 조성된 서악서원(西岳書院, 1563), 옥산서원(玉山書院, 1573), 구강서원(龜崗書院, 1692)은 경주를 주자학의 도시로 완성했다.

뒤이어 일제강점기(1909~1945)의 경주는 지역 행정거점이며 철도도시로서 경주읍성의 동헌에 조성된 경주군청(현 경주문화원 일대), 경주우편소(현 경주우체국), 경남합동은행, 경주산업조합, 경주금융조합 등이 밀집한 본정통(本町通, 현 봉황로)의 상업중심부, 그리고 대구, 부산, 포항, 울산과 연결된 경주역과 중앙시장을 중심으로 현재 경주의 구도심을 더했다.

경주에 철도가 처음 건설된 시점이 이 즈음인 1910년대에 구도심이다. 당시 민간기업인 조선중앙철도주식회사(1910년대 이후 조선철도주식회사로 흡수, 1923)는 경주와 서악을 연결하는 중앙선(1918, 현재 경동선)과 이를 연장한 대구-경주-불국사 철도선(1919) 그리고 경주-포항 철도선(1936, 현재 동해남부선)을 건설했다.

경주는 한국전쟁 이후 역사유산을 이용한 관광도시로 변화했으며, 특히 2000년 경주의 다섯 구역이 유네스코 세계문화유산으로 지정되면서 국제적 경쟁력을 갖춘 역사도시로서 변화해 왔다. 이 과정에서 1970년대 초부터 중앙정부가 역사유적이 집중되어 있는 경주 원·구도심의 문화재 유적 보존을 목적으로 주변지에 개발규제를 엄격히 적용해 왔고, 이와 반대로 대규모 레저, 주거, 산업단지들의 개발과 대학캠퍼스 조성이 경주 교외지에 집중되며 경주의 서쪽과 북쪽의 도시확장을 유도해왔다.

한편 경주를 연결하는 고속철도 건설이 논의되기 시작한 시점은 1970년대이다. 당시 서울과 부산을 연결하는 경부축은 우리나라 인구 및 지역생산과 교통량의 70%를 담당해온 경부고속도로의 근간이었으나, 기존 고속도로와 철도는 용량이 한계에 도달한 상태였고, 새로운

During Japanese colonial rule (1909–1945), Gyeongju served as again a local administration center and a railway city, adding more features to the old city center of today's Gyeongju with Gyeongju-gun Office (now the GyeongJu Culture Center and nearby), established on the east of Gyeongju Eupseong, Gyeongju Post, Gyongnam Joint Bank, Gyeongju Industrial Association, and Gyeongju Financial Association along Honmachi street (now Bonghwang-ro Road) as a commercial center, as well as Gyeongju Station, connecting the city with Daegu, Busan, Pohang, and and Ulsan.

In the 1910s when the first railway was built in Gyeongju, it formed the old city center. Joseon Central Railway Corporation (1910s), a private-sector firm at that time, built the Jungang Line (1918) connecting Gyeongju and Seoak, its extension of the Daegu–Gyeongju-Bulguksa railway line (1919) and the Gyeongju–Pohang railway line(1936).

Since the Korean War, Gyeongju transformed itself into a tourism city based on its historic heritage, and in 2000, five Gyeongju Historic Areas were designated as a UNESCO World Heritage Site, making Gyeongju a historic city with a globally competitive edge. During this process, central government since the early 1970s has imposed strict restrictions on the development of areas near the old city center of Gyeongju, which includes many historic sites, for the purposes of preserving cultural and historic heritage. By contrast, the development of large-scale leisure, housing, and industrial complexes and the establishment of university campuses have been developed on the suburbs of Gyeongju, leading to Gyeongju's urban expansion towards the west and north.

Political discussion of a high-speed railway connecting Gyeongju began in the 1970s. The discussion was sparked by the fact that although the Gyeongju axis geographically connecting Seoul and Busan was the backbone of the Gyeongbu Expressway, which accounted for approximately 70 % of the population, local production, and traffic in

그림 2. 경주 남천과 월성과 안압지(출처: 경주시)
Ancient Gyeungju's Namchun Stream, Wolsung Palace Citadel and Anapji Royal Pond

그림 3. 경주 교촌과 향교(출처: 경주시)
Gyeongju GyoChon and Hanggyo

Korea and carried the transportation system in the country, traditional highways and railways in the axis had reached their limits of capacity, and needed the construction of new transport facilities. Nonetheless, it was not until the 1990s that the construction of the Gyeongbu High-Speed Railway started.

The purpose of construction of the Gyeongbu High-Speed Railway in the 1990s was not to connect directly Seoul and Gyeongju, or Gyeongju and Busan. was not to connect Seoul and Gyeongju, or Gyeongju and Busan. At that time, Gyeongju was pursued as a middle hub connecting Seoul and Busan on the Gyeongbu High-Speed Railway (426 km, Seoul-Daejeon–Daegu–Gyeongju-Busan). As Gyeongju was a junction to connect Pohang and Ulsan, a railway with double tracks between Gyeongju and Pohang, and Gyeongju and Ulsan, was urgently needed to be built, to ensure an efficient transport system. The High-Speed Railway including Gyeongju was necessary both for Gyeongju residents, and for other people living along the east coast of the country, such as the cities of Ulsan and Pohang.

However, the construction of the High-Speed Railway and its station in the historic city Gyeongju faced many limitations. To preserve its historic heritage due to Gyeongju's symbolic meaning as the country's pre-eminent historic city, a railway passing through the center of the city was unacceptable. Furthermore, due to objections from the city's residents, a railway passing through suburbs with little access cound not obtain any public support, as it would be located independently of the old city center of Gyeongju. Consequently, unlike what was expected, it took about 20 years of planning to find a compromise that would ensure access to transport for residents, while minimizing any potential damage to historic heritage. The current Sin-Gyeongju Station (2010) of the Gyeongbu High-Speed Railway is a product of this process. Nevertheless, Sin-Gyeongju Station has been subject to criticism due to its poor access to the old city center of

교통시설의 건설이 필요한 상황이었다. 그럼에도 경부고속철도 건설이 추진된 시점은 1990년대에 들어서면서이다.

1990년대부터 추진된 경주의 고속철도 건설의 목적으로 서울과 경주 또는 경주와 부산을 연결이 아니다. 당시 경주는 서울과 부산을 연결하는 경부고속철도(총길이 426km, 서울-대전-대구-경주-부산)를 건설하는 중간 거점으로 추진되었다. 그리고 경주는 포항과 울산을 연결하는 분기점으로 효율적인 수송체계를 위해 경주-포항 그리고 경주-울산간 철도의 복선전절화가 시급했다. 이에 고속철도 경주노선은 경주 시민뿐만 아니라 울산과 포항을 포함하는 동해권 지역의 주민들을 위해서도 필수적인 것이었다.

그러나 역사도시 경주에서 고속철도선과 고속철도역을 건설하는 것에는 많은 제약을 안고 진행되었다. 한국의 대표 역사도시 경주가 갖고 있는 역사문화재 보호라는 특별한 상징적 사명 때문에 도시중심부를 지나는 철도선은 선정될 수 없었다. 또한 접근이 어려운 외곽노선은 기존 경주 원·구도심과의 독립된 입지성으로 시민들의 반발로 도입될 수 없었다. 따라서 역사문화재 훼손을 최소화하면서도, 시민들의 교통접근성을 확보할 수 있는 절충안을 찾는 작업은 초기 예상과 달리 약 20년이 걸린 계획과정으로 진행되었다. 현재 경부고속철도 신경주역(2010)은 이러한 과정을 거쳐 결정된 결과물이다. 그럼에도 현재 신경주역은 경주 원·구도심과의 낮은 근접성으로 현재까지 비판을 받아왔다.

경주 철도선의 조성과 도시외곽 개발

일제강점기의 경주는 모던 경주의 도시구조적 변화를 진행했다. 당시 진행된 경주의 대표적 물리적인 도시변화는 경주읍성의 해체(1912~1932)와 신작로와 철도선을 중심으로 한 구도심의 조성이다. 이 시기에 경주는 조선총독부가 진행한 한반도 내에서 문화재 조사와 발굴 그리고 이를 이용한 관광활성화를 위한 핵심 대상 도시이기도 했다.

당시 경주의 변화는 경주읍성의 남문이 1912년 전후로 해체되면서 시작되어, 경주와 그 주

Gyeongju.

Establishment of the Gyeongju Railway Line and Suburban Development

Under Japanese colonial rule, Gyeongju went through urban structural changes to become a modern city. One of the most noticeable physical changes in the city was the demolition of Gyeongju Eupseong (1912–1932), and the establishment of a new city center (now old city) around the new road and railway line. During this time, Gyeongju was a key city for investigating, discovering, and using cultural heritage to boost tourism in the Korean Peninsula, as sought by the Joseon Government General.

Changes in Gyeongju at that time included the deconstruction of Gyeongju Eupseong with its south gate demolished around 1912, and the construction of the new road Taejong-ro (1906), connecting Gyeongju with Daegu, a hub city of the Gyeongbu railway line, and Pohang, a seaport. At that time, Taejong-ro defined the southern border of Gyeongju, connected with Seocheongyo and Bugcheongyo, and encouraged Gyeongju's expansion into the west and north.

Taejong-ro is a section of the west–east new road (Gunsan–Daegu–Gyeongju, now Route 4) and the north–south new road (Busan–Ulsan–Gyeongju–Pohang–Donghae, Route 7), connecting the bus terminal and railway station in Gyeongju with Daegu and Pohang. For this reason, Gyeongju is a city where intercity bus business began for the first time in the country with the opening of the intercity bus lines of the Daegu–Gyeongju intercity bus (1912), Gyeongju–Pohang intercity bus (1921), and Gyeongju–Busan intercity bus (1932).

변의 경부철도선 거점도시인 대구와 바다 항구인 포항을 연결하는 신작로인 태종로(1906)의 건설로 진행되었다. 당시 태종로는 경주의 남쪽 경계를 정의하고 서천교, 북천교와 연결되어 경주의 서쪽과 북쪽 확장을 유도했다.

태종로는 경주의 버스터미널(현재 고속버스터미널)과 철도역을 대구와 포항으로 연결하는 서동방향의 신작로(군산-대구-경주, 현재 국도4)와 북남방향의 신작로(부산-울산-경주-포항-동해-함경북도, 현재 국도7)의 일부 구간이기도 했다. 이러한 이유로, 경주는 전국 최초로 자동차운수업인 시외버스운수업이 시작되어, 첫 번째 버스노선인 대구-경주 시외버스(1912), 경주-포항 시외버스(1921), 경주-부산 시외버스(1932)가 차례로 개통된 도시이기도 했다.

한편 경주읍성은 경주 철도선사업을 진행했던 조선중앙철도주식회사의 주도로 경주 철도선 개설에 필요한 석재 조달을 위해서 1912~1932년 남문부터 순차적으로 해체되었다. 당시 조선중앙철도는 1916년 조선총독부로부터 대구-경주-울산-동래와 경주-포항 구간의 경편철도 부설권을 허가받았다.

이후 경주의 철도선은 경주와 무열왕릉을 포함한 대형 고분군이 입지한 서악을 연결하는 중앙선(1918, 현재 경동선)과 이를 연장한 대구-서악-경주-불국사의 중앙선 연장 철도선(1919), 그리고 경주-포항 철도선(1936, 동해남부선)으로 건설되었다. 특히 서악-경주의 중앙선은 서천교를 중심으로 서쪽으로 무열왕릉 입구와 연결된 서악역 그리고 서천교 동쪽으로 현재 태종로를 따라 대릉원의 금관총과 천마총을 양분하며 건설되었다. 그리고 경주의 첫번째 철도역은 1918년 현재 태종로의 서라벌 문화회관 및 경주도서관 부지에 입지했다.

당시 경주 철도선의 건설목적은 경주를 대구와 부산으로 연결하여 신라 유적을 이용한 관광사업의 추진이었다. 실제로 경주는 철도역이 조성된 이후 1920년부터 일본 중·고교 학생들의 수학여행지로서 기능했다. 그럼에도 서악역과 경주역의 위치를 고려하면, 철도선을 이용해 무열왕릉, 금관총, 천마총 및 주변에서 발굴되는 다수의 문화재가 대구와 부산으로 효과적으로 운반되었을 것이라는 합리적 예측을 가능하게 한다. 첫번째 경주역은 경주 IC 서쪽에 입지했던 서악역(1918~1939, 현재 율동역)이 폐지되면서 시설이 확장되었다.

Meanwhile, Gyeongju Eupseong was demolished by the Joseon Central Railway Corporation from 1912 to 1932 to obtain the stone 1932, beginning with its south gate, to obtain the stone materials required for the construction of the Gyeongju railway line. In 1916, Joseon Central Railway Corporation gained a permit from the Joseon Government General to construct a light railway in the Daegu–Gyeongju–Ulsan–Dongrae section and the Gyeongju–Pohang section.

Then, additional railway lines in Gyeongju were constructed, including the Jungang Line (1918) connecting Gyeongju and Seoak, which houses large tombs, such as the Royal Tomb of King Taejong Muyeol; the Jungang Line extension (1919) between Daegu and Bulguksa; and the Gyeongju–Pohang railway line (1936). In particular, the Jungang Line between Seoak and Gyeongju was constructed with Seocheongyo at the center: Seoak Station connected to the entrance of the Royal Tomb of King Taejong Muyeol west of Seocheongyo, and Geumgwanchong and Cheonmachong in Daereungwon along Taejong-ro, east of Seocheongyo. In 1918, the first railway station in Gyeongju was built in the area, which has the Seorabeol Cultural Center and Gyeongju Library in today's Taejong-ro.

The construction of the Gyeongju railway line at that time was to pursue tourism business based on Silla heritage by connecting Gyeongju to Daegu and Busan. After the railway station was built, Gyeongju became a tourism destination for Japanese middle and high school students from 1920. Given the locations of Seoak Station and Gyeongju Station, it is reasonable to assume that the railway line was used to effectively transport many types of cultural heritage discovered in the Royal Tomb of King Taejong Muyeol, Geumgwanchong, Cheonmachong, and other surrounding areas to Daegu and Busan. Facilities in the first Gyeongju Station were expanded as Seoak Station (1918–1939, now Yool-dong Station) west of the Gyeongju Interchange (IC) was closed down.

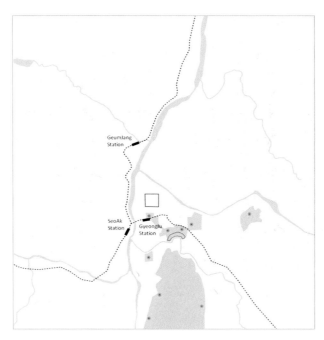

그림 4. 경주의 철도체계, 1910 (출처: 한광야, 곽혜빈, 2018) / Railroad System in Gyeungju, 1910

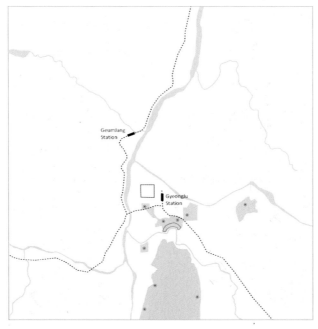

그림 5. 경주의 철도체계(1936) / Railroad System in Gyeungju, 1936

In 1936, Gyeongju Station in Taejong-ro was relocated and newly built in the area in Hwango-dong, through which the north–south Donghae Nambu Line connecting the city with Pohang, Gyeongju, and Busan. In 1943, the Taejong-ro railway line was relocated to Geumseong-ro 372 beon-gil and Geumseong-ro 368 beon-gil in today's Seonggeon-dong. A large-scale railway housing complex (1935, 150 m × 150 m) was established to southeast of the railway station in Hwango-dong. The complex consisted of 17 to 20 detached houses, connected to the city center through the underground road in Hwango-ri.

With the construction of Seo-Gyeongju Station in 1992, the Geumseong-ro railway line passing through the center of Gyeongju was relocated again to the current line passing by Hwangseong-dong beyond Bugcheon in the north. Thus, the Gyeongju railway line in Hwangseong-dong was located beyond Bugcheon north of Gyeongju Eupseong, avoiding Gyeongju Historic Areas designated as a UNESCO World Heritage Site and National Cultural Heritages around Namsan, south of Gyeongju.

The relocation of the Geumseong-ro railway line to Hwangseong-dong led to a decline in the commercial areas of the old city center in Hwango-dong, which had grown from the Korean War to the early 1990s with Seongdong Market near Gyeongju Station, and Hwarang-ro, a west–east road. From the 1970s, construction in Hwango-dong was restricted under the Cultural Heritage Protection Act, which resulted in a decline in Hwango-dong, one of the richest neighborhoods in Gyeongju. As Gyeongju City Hall was relocated from the old city center to Dongcheon-dong in the north, the commercial areas in the city center shrank further. In addition, as the passenger transport function in Gyeongju was relocated to Sin-Gyeongju Station in 2010, the decline in the surrounding commercial areas in the old city center further accelerated.

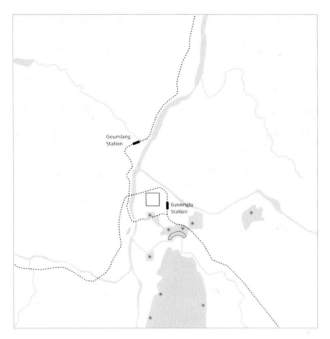

그림 6. 경주의 철도체계(1945) / Railroad System in Gyeungju, 1945

그림 7. 경주의 철도체계(1985) / Railroad System in Gyeungju, 1985

Cultural Heritage Preservation Law and Bomun Tourism Complex

Gyeongju became the first city subject to the 'Cultural Heritage Protection Law (문화재보호법, 1962), legislated by the Park Chung-hee administration (1961–1979) to preserve and use the country's cultural heritage as national resources. Accordingly, a part (19.6 km²) of the original city center of Gyeongju was designated as a 'Cultural Heritage Protection Area (문화재보호구역)' in 1962, which regulate development within its radius of 500 m. The designated Cultural Heritage Protection Area was further divided into the Hanok Aesthetics District (3.3 ha) and Building Height Restriction Area (5.94 ha), based on Gyeongju City's Zoning Ordinance (1972), which were established to implement the Cultural Heritage Protection Law.

Furthermore, to preserve cultural heritage and establish a tourism hub in Gyeongju from 1968 to 1974, the Park Chung-hee administration designated mountains and hills in 8 areas across Gyeongju as Gyeongju National Park (137 km²), Consequently, most of the old city center of Gyeongju has been designated under the Cultural Heritage Preservation Act since the 1970s, which has restricted development and construction by the private sector, and banned development in Gyeongju National Park and its surroundings.

Meanwhile, the Gyeongju Comprehensive Tourism Development Project (1971–1981) was conducted by central government around Bomun Lake in the northeast suburbs of Gyeongju, which was intended to be developed as a large-scaled tourism complex and transport infrastructure. Until that time, Bomun Lake was a reservoir built upstream of Bugcheon to prevent frequent floods. The purpose of the Gyeongju Comprehensive Tourism Development Project was to establish the city as Silla Capital Gyeongju, and develop an international tourism complex. Around Bomun Lake, 5 km away northeast of the center of Gyeongju, the Bomun Tourism Complex (1974–1979, 6.9 ha) and

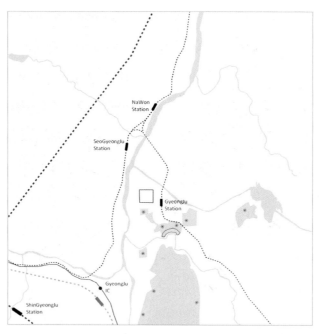

그림 8. 경주의 철도체계, 2010(출처: 한광야, 곽혜빈, 2018) / Railroad System in Gyeungju, 2010

이후 태종로의 경주역은 1936년 포항-경주-부산을 연결하는 북남방향의 동해남부선이 지나는 황오동의 경주역 부지로 이전되어 신축되었고, 이와 함께 태종로 철도선이 1943년 현재 성건동의 금성로 372번길과 금성로 368번길로 이선되었다. 그리고 경주의 황오동 철도역의 동남쪽에는 대규모의 철도관사(1935, 현재 원효로, 원효로 163번길, 원효로 177번길로 경계된 규모 150x150m 부지)가 조성되었다. 당시 철도관사는 약 17~20개의 단독주택으로 구성된 단지로서 현재 황오리 지하차로를 통해서 도심부와 연결되었다.

이후 경주 도시중심부를 관통하는 금성로 철도선은 1992년 서경주역의 조성과 함께 북쪽으로 북천을 넘어 황성동을 지나는 현재 노선으로 다시 이선되었다. 당시 황성동으로 이선된 경주의 철도선은 경주의 남쪽 남산을 중심으로 지정된 유네스코 세계문화유산과 국가문화유산 구역들을 피하며 경주읍성의 북쪽으로 북천 넘어 위치하게 되었다.

이러한 금성로 철도선의 황성동으로의 이전은 한국전쟁 이후 1990년대 초까지 경주역과 인접한 성동시장 그리고 서동방향의 중심도로인 화랑로를 중심으로 성장한 황오동의 구도심

transport infrastructure in Gyeongju were built with funding from the country's budget ($ 76 million) and the International Bank for Reconstruction and Development's aid ($ 25 million) to attract tourists within Korea, and from abroad.

New Development of the Seo-Gyeongju Station Area and Hwangseong-dong

Along with the Bomun Tourism Complex northeast of the city, university campuses and large-scale residential complexes were developed in Seokjang-dong and Seonggeon-dong with Seo-Gyeongju Station (1992) newly opened outside the Hyeongsan River/Seocheon. Moreover, with the relocation of the Wolseong-gun Office (1985), the Dongcheon-dong development (1970–2000), Hwangseong-dong development (1980–2000), and Yonggang Industrial Complex and Yonghwang District development (2007–2013) took place suburban areas outside Bugcheon.

Moreover, urban expansion in the west of Gyeongju since the 1970s was influenced by the Seoul Metropolitan Area Population Relocation Master Plan (1977–1986), prepared by by the Park Chung-hee administration as a national plan. The implementation of the Seoul Metropolitan Area Population Relocation Master Plan led universities located in Seoul to build campuses in other regions and guided national universities in other regions than Seoul to grow into local hubs with investment from central government. Dongguk University Gyeongju Campus (1978) was built in Seokjang-dong, northwest of the Hyeongsan River/Seocheon, Gyeongju, and the campuses of Sorabol College (1981) and Gyeongju University (1981) served as a hub of development west of Gyeongju.

University campuses took their shapes west of the Hyeongsan River/Seocheon hand in hand with the establishment of Geumjang Station for the Jungang line. in 1985. Since the Jungang line in Gyeongju was relocated to today's location in Hwangseong-dong, 2

상권의 와해를 초래했다. 특히 1970년대부터 문화재보호법에 묶여 건축행위가 제한되면서 당시까지 경주의 대표적인 부자 동네로 성장했던 황오동은 쇠퇴하기 시작했고, 경주시청이 구도심에서 북쪽의 동천동으로 이전해 나가면서 도심상권이 더욱 위축되었다. 여기에 주변 구도심 상권의 쇠퇴는 경주역의 여객운송기능이 2010년 신경주역으로 이전해 나가면서 가속화되었다.

원·구도심의 문화재보호법과 보문관광단지

경주는 박정희 정부(1961~1979)가 국가의 문화재를 보존하고 국가 자원으로 활용하기 위해 입법한 '문화재보호법(Cultural Heritage Preservation Law, 1962)'을 근거로 가장 먼저 적용대상 도시가 되었다. 이에 따라 경주 원도심의 일부(19.6km²)가 1962년 '문화재보호구역'으로 지정되어 지정구역으로부터 500m 내의 개발이 규제되었다. 이렇게 지정된 문화재보호구역은 이후 문화재보호법의 실행을 위해 제정된 경주시 조닝법규(Zoning Ordinance, 1972)에 근거해 한옥미관지구(3.3ha) 그리고 고도제한지구(5.94ha)로 다시 나뉘어 지정되어 규제되어 왔다.

또한 박정희 정부는 1968~1974년 경주의 문화재를 보존하고 관광거점을 조성하기 위해 경주시 일대 총 8곳에 산과 구릉지를 중심으로 사적형의 경주국립공원(Gyeongju Nat'l Park Area, 총면적: 137km²)이 지정되었다. 이에 따라 경주의 원·구도심의 상당부분은 1970년대부터 문화재보호법으로 지정되어 민간의 개발 및 건축행위가 규제되기 시작했고, 경주국립공원과 그 주변지의 개발도 금지되었다.

한편 경주의 동북쪽 외곽에는 중앙정부의 주도하에 보문호수를 중심으로 대규모의 관광 콤플렉스와 교통 인프라 개발인 경주관광종합개발사업(GyeongJu Tourism Development Project, 1971~1981)이 진행되었다. 물론 보문호수는 당시까지 주기적인 범람으로 인한 홍수피해를 막기 위해 북천의 상류에 조성된 저수지였다. 당시 경주관광종합개발사업의 목적은 경주를 '신라수도 경주'로 조성하고 이를 통해 국제적인 관광단지로 개발하려는 것으로, 경주 중심부로부터 북동쪽으로 약 5km 떨어진 보문호수를 중심으로 보문관광단지(1974~1979,

km north of its original location in 1992, Geumjang Station was newly established into Seo-Gyeongju Station connecting with Nawon Station. Furthermore, the establishment of Seo-Gyeongju Station led to the development of a large-scale residential complex in Seonggeon-dong, near Seokjang-dong.

It was also in the 1970s that Gyeongju expanded into Dongcheon-dong beyond Bugcheon. Around this time, the Gyeongju Comprehensive Development Plan (1971–1979) was implemented, and development in Dongcheon-dong (1970–2000) and Hwangseong-dong (1980–2000) has taken place upon the relocation of Wolseong-gun Office to north of Bugcheon. The Yonggang Industrial Complex and the Yonghwang District Land Development Project (2007–2013) proceeded along with National Route 7 and Sindang Intersection in Yonggang-dong in the west.

UNESCO World Heritage Site and Gyeongju Old Capital Preservation Project

Until Gyeongju was designated as a UNESCO World Heritage Site in 2000, the central government made various efforts to have Gyeongju listed as a World Heritage Site from the late 1970s and the 1988 Olympics. In particular, the Kim Dae-jung administration (1998–2003) and Gyeongju City hosted Gyeongju Culture Expo in 1998, and in that year, the Gyeongju National Research Institute of Cultural Heritage discovered the site of a temple from the Unified Silla period around Namsan, Gyeongju, and increased the city's possibility of being selected as a UNESCO World Heritage Site.

Thanks to these efforts, five historic areas in Gyeongju were listed as a UNESCO World Heritage Site in 2000 thanks to the historic value of areas including masterpiece

부지면적 6,9ha)와 경주의 교통 인프라를 국내외 관광객의 유치를 목적으로 당시 국비($76 million)와 국제재건개발은행(Int'l Bank for Reconstruction and Development)의 지원비($ 25 million)로 건설되었다.

외곽의 거점개발, 서경주역 개통과 황성동 개발

이러한 상황에서 경주는 흥미롭게도 경주 북동부의 보문관광단지와 함께, 형산강/서천 밖에 새롭게 개통된 서경주역(1992)을 중심으로 석장동과 성건동에서 대학캠퍼스 개발과 대단지 주거지개발이 진행되었다. 또한 북천 밖의 교외지에는 월성군청의 이전(1985)과 함께 동천동 개발(1970~2000)과 황성동 개발(1980~2000)과 함께 용강산업단지와 용황지구 택지개발사업(2007~2013)이 진행되었다.

먼저 1970년대 말부터 시작된 경주의 서쪽 도시확장은 박정희 정부가 국가기본계획으로 수립한 '수도권 인구재배치 기본계획(1977~1986)'의 영향을 받으며 진행되었다. 당시 '수도권 인구재배치 기본계획'의 시행은 서울 소재 주요 대학교들의 지방분교 설립과 중앙정부의 투자로 진행된 지방 국립대학교가 지역거점화를 유도했다. 이 시기에 경주의 형산강/서천의 서쪽부에 조성되기 시작한 석장동의 동국대 경주캠퍼스(1978)가 조성되었고, 이어서 서라벌대(1981)와 경주대(1981)의 대학캠퍼스는 경주의 서쪽개발의 거점으로 작용했다.

이러한 형산강/서천 서쪽부의 대학캠퍼스 조성은 1985년 중앙철도선의 금장역 신호소의 조성과도 연계되어 진행되었다. 이후 경주의 중앙철도선 구간이 1992년 기존의 철도선 위치에서 북쪽 약 2km 황성동의 현재 철도선 위치로 이선되면서, 기존 금장역은 나원역과 연결되는 서경주역(1992)으로 새롭게 조성되었다. 또한 서경주역의 조성은 석장동과 인접한 성건동에서 대규모 주거단지 개발로 이어졌다.

경주가 북천을 넘어 동천동으로 확장한 시점도 1970년대이다. 이 시기는 경주종합개발계획(1971~1979)이 실행되며, 북천의 북쪽으로 월성군청의 이전(1985)과 함께 동천동 개발(1970~2000)과 황성동 개발(1980~2000)이 진행되었다. 뒤 이어 서쪽으로 국도 7과 용강동 신당교차로를 따라 용강산업단지와 용황지구 택지개발사업(2007~2013)이 진행되었다.

Korean Buddhist artworks, such as sculptures, reliefs, stupas, temples, and royal palace ruins from the 7th to 10th centuries, the heyday of the Silla dynasty.

Gyeongju Historic Areas designated as a UNESCO World Heritage Site consisted of 3 areas + 2 areas with the total area of 2,880 ha, including 52 State-Designated Cultural Heritages, and a buffer zone (350 ha) around their surrounds. To preserve Gyeongju Historic Areas, the UNESCO designated a 7-25 m height restriction in the buffer zone in Seonggeon-dong, Hwango-dong, Seongnae-dong, and Jungang-dong.

Gyeongju Historic Areas designated as a UNESCO World Heritage Site consisted of the Mount Namsan Belt, which has Korea's unique Buddhism heritage from the 7th to 10th centuries; the Wolseong Belt, which used to be a palace; the Tumuli Park Belt, which includes many tombs; the Hwangnyongsa Belt, which is a Buddhism temple; and the Sanseong Belt, which has a defensive fortress. The level of authenticity is considered high, as these 5 areas have different types of heritage in their original locations, and their architecture, sculptures, pagodas, royal tombs, and mountain fortresses have remained mostly intact.

Meanwhile, Gyeongju became the first city subject to the 'Old Capital Preservation Law, (고도보존에 관한 특별법, 2005, 2011)', legislated by the central government in 2005. The purpose of the the Law on the Preservation and Promotion of Ancient Cities is to preserve historic heritage and the property rights of local residents in ancient cities, such as Buyeo, Gongju, and Iksan, which are cities from the Baekje dynasty that, along with Gyeongju, are listed as UNESCO World Heritage Sites.

To implement the the Law on the Preservation and Promotion of Ancient Cities, Gyeongju designated the 'Old Capital Preservation Area (고도보존 지구)', and established the 'Old Capital Preservation Plan (고도보존계획, 2011)' in Gyeongju, which was supplemented by the 'Gyeongju Old Capital Preservation and Empowerment Plan (경주

유네스코 경주 역사구역과 경주 고도보존사업

경주가 2000년 유네스코 세계문화유산으로 지정되기까지에는 중앙정부는 이미 1970년대 말부터 1988년 올림픽 개최를 전후로 이후 약 20년 간 경주를 세계문화유산 등재를 위해 오랫동안 노력을 진행했다. 특히 김대중 정부(1998~2003)와 경주시는 1998년부터 경주문화엑스포를 개최했고, 국립 경주문화재연구소는 1998년 경주 남산 일대에 통일신라시대의 사찰터를 발굴해 유네스코 문화유적으로 선정되기 위한 가능성을 높였다.

이러한 노력의 결과로 마침내 경주의 5개 구역이 2000년 유네스코 세계문화유산인 경주역사구역(Gyeongju Historic Areas)으로 등재 되었다. 유네스코가 이 구역을 경주역사구역으로 지정한 이유는, 신라왕조의 절정기인 7~10세기 한국 불교예술의 뛰어난 전형들이 조각,

그림 9. 경주 원구도심의 건물고도제한과 유네스코세계문화유산, 2010(출처: 한광야, 곽혜빈, 2018)
Building Height Regulation in the Old City and UNESCO World Heritage Sites, Gyeungju

고도보존과 활성화 계획, 2016)'.

During this process, Gyeongju's effort to designate the districts first and implement the plan later was met with huge opposition from residents, as it was understood as a violation of private properties, just like the 'Cultural Heritage Protection Law (문화재 보호법, 1962)'. Furthermore, the compensation the government paid to buy residents' land or buildings was very lower than what residents expected, and the construction of cultural facilities for local residents was mostly restricted for the sake of protecting cultural heritage.

Ancient City Preservation Districts in Gyeongju were designated 6 years later in 2011 when Old Capital Preservation Law was amended. At that time, 9 Ancient City Preservation Districts in Gyeongju were designated, including Daereungwon, Gyochon, Wolseong, and Hwangnyongsa as 'Special Preservation Districts (특별보존 지구)' maintained under the 'Historic and Cultural Context Area (역사문화환경지구)' and Hwangnam-dong Hanok District, a historic cultural environment protection district as secondary Special Preservation Districts. Thus, the designation (2011) of Ancient City Preservation Districts in Gyeongju aims for win-win ancient city management, which improves living conditions from the perspective of residents who own land and buildings, which is unlike the Cultural Heritage Protection Law and previous Old Capital Preservation Law.

부조, 탑파(塔婆, stupa), 사찰, 왕궁 유적 등의 형태로 집중되어 모여있기 때문이다.

유네스코 세계문화유산의 경주 역사구역은 3+2개 구역으로 구성되며, 이 구역의 전체 면적은 2,880ha으로 국가지정문화재 52개를 갖고 있으며, 그 주변에 완충지역(350ha)을 두고 있다. 유네스코는 이러한 경주역사유적구역의 보존을 목적으로, 특히 성건동, 황오동, 성내동, 중앙동) 일대의 완충지역에 7~25m의 고도제한을 지정했다.

유네스코 세계문화유산의 경주 역사구역은 7~10세기의 한국의 특별한 불교유적을 갖고 있는 남산 벨트(Mt. Namsan Belt, possessing various Buddhism remains), 옛 왕궁 터였던 월성 벨트(WalSung Belt, used to be old palace site), 다수의 고분들이 모여있는 대릉원 벨트(DaeReungWon Belt, has old tombs), 그리고 불교사찰 유적지인 황룡사 벨트(HwangRyong Temple Belt, the site of Buddhism Temple remains), 방어용 산성이 위치한 산성 벨트(SanSeong Fortress Belt, a site where defensive facilities were located)로 구성되어 있다. 이 5개 구역은 "각기 다른 종류의 유산이 원래의 위치에 남아있어 진정성이 높으며, 건축, 조각, 탑파, 왕릉, 산성은 모두 그 원형을 상당 부분 유지"해왔다. 한편 "사지 또는 궁궐지의 경우, 그 터만 갖고 있으며 건물의 원래 배치 형태를 보존하기 위해 기존의 상태를 그대로 유지" 하도록 하고 있다.

한편 경주는 2005년 중앙정부가 '고도보존에 관한 특별법(법률 제9213호 Old Capital Preservation Law, 2005, 2011)'을 입법하면서, 그 첫 번째 대상도시가 되었다. 당시 '고도보존에 관한 특별법'의 목적은 경주와 함께 백제왕조의 도시이며 역시 유네스코 세계문화유산으로 등재된 부여, 공주, 익산을 비롯한 고도(古都) 지역의 역사문화유산 보존과 지역 주민의 재산권을 보호하기 위해서 제정되었다.

이에 따라 경주시는 '고도보존에 관한 특별법'의 시행을 위해 '고도보존지구(Old Capital Preservation Area)'를 지정하고 '고도보존계획(Old Capital Preservation Plan, 2011)'을 수립했다. 이후 경주의 고도보존계획은 '경주 고도보존과 활성화계획(Gyeongju Old Capital Preservation and Empowerment Plan, 2016)'으로 보완되었다.

Construction of the High-Speed Railway and
Sin-Gyeongju Station

The latest suburban expansion of Gyeongju has been driven by the construction of the Gyeongju section of the Gyeongbu High-Speed Railway and Sin-Gyeongju Station (2010) in Geoncheon-eup at the rear of Gyeongju IC, 9.5 km southwest of the old city center of Gyeongju. The Gyeongbu High-Speed Railway Project, which runs at 300 km/h between Seoul and Busan (426 km, Seoul–Daejeon–Daegu–Gyeongju–Busan), began in 1990 under the Roh Tae-woo administration, proceeded in different stages: the construction of the Seoul-Daegu railway in Stage 1 (1998–2004) and the construction of the Daegu–Gyeongju–Busan railway and Gyeongju Station in Stage 2 (2004–2010).

Meanwhile, the first Gyeongju line proposal for Gyeongbu High-Speed Railway named 'Hyeongsan River line (1992, Ministry of Construction and Transportation)' was suggested with Bugnyeokdeul (now Yool-dong) as a railway station. The Ministry of Construction and Transportation highlighted that the Hyeongsan River line was an outcome that reflected the results of public hearings and cultural heritage field surveys and could minimize damage to the cultural heritage of Gyeongju, although the line would pass through the city center of Gyeongju. The Ministry evaluated that it was an optimal option, considering that relocating the Jungang line and the Donghae Nambu line would cause great damage to the cultural heritage of Gyeongju.

However, religious and academic communities opposed the Hyeongsan River line, citing damage to the historic heritage and landscapes.

The second Gyeongju line proposal suggested the 'Ijo-ri line (1995, Ministry of Construction and Transportation, and Korail)' in Ijo-ri, 5 km south of Bugnyeokdeul.

이 과정에서 경주시의 '선 지구지정, 후 보존계획수립' 추진은 지역 주민들에게 과거 문화재보호법(1962)과 유사한 사유재산의 침해로 인식되어 큰 반대를 받았다. 또한 주민 소유의 토지나 건물의 정부매입액수가 주민의 기대액수와 큰 차이로 예산 범위 내에서 매입 지원되기 않게 되어 문제가 발생했으며, 지역 주민을 위한 문화시설의 조성이 문화재 보호를 위해 대부분 제한되었다.

결국 경주의 고도보존지구가 지정된 시점은 6년이 지나 '고도보존에 관한 특별법'이 개정된 2011년이었다. 당시 경주 고도보존지구는, 문화재보호법에 의해서 관리되는 '특별보존지구(Special Preservation Area)'인 대릉원, 교촌, 월성, 황룡사지, 그리고 특별보존지구의 배후지로서 '역사문화환경지구(Historical and Cultural Context Area)'인 황남동 한옥지구 등 9곳으로 차별화하여 지정되었다. 이러한 경주 고도보존지구의 지정(2011)은, 기존의 '문화재보호법(1962)'과 '고도보존에 관한 특별법(2005)'과는 다르게 토지 및 건물의 소유자인 주민의 관점에서 주민생활여건을 개선하는 '상생적인 고도관리'를 지향하고 있다는 점에서 차이를 찾을 수 있다.

고속철도선의 건설과 신경주역

경주의 최근 교외지 확장의 큰 계기는 단연 경부고속철도의 경주구간 철도선과 그 고속철도 정차역으로 경주 원구도심으로부터 남서쪽 9.5km에 경주 IC 배후에 자리잡은 건천흡 신경주역(2010)의 건설이다. 원래 서울과 부산을 연결하는 최고 시속 300km의 경부고속철도(총길이 426km, 서울-대전-대구-경주-부산) 사업은 노태우 정부의 주도로 1990년에 착수되었다. 이후 경부고속철도는 1단계의 서울-대구 노선 건설(1998~2004)과 2단계의 대구-경주-부산 노선 건설 및 경주역사 건설(2004~2010)로 진행되었다.

한편 경부고속철도의 첫 번째 경주노선 계획안인 '형산강 노선(1992, 건설교통부)'의 철도선과 북녘들(현 율동)의 철도역으로 제안되었다. 건설교통부는 형산강 노선은 경주 도심을 지

This proposal accepted the opinion of the Buddhist community and the Ministry of Culture, Sports and Tourism that prevent the potential damage to cultural heritage that would arise from having the High-Speed Railway pass through the old city center of Gyeongju. At the same time, it was also considered to accept the opinion of the Ministry of Construction and Transportation that the station should be located in the city center of Gyeongju for passengers for Pohang and Ulsan to transfer with prospective double tracked Donghae Nambu line.

Meanwhile, UNESCO responded that if the High-Speed Railway passed through Gyeongju, it would be difficult to designate Gyeongju as a World Heritage Site. the second proposal was also rejected. due to the opinion of the academic community that Ijo-ri had many historic sites from the Silla period, and the opposition of residents in Gyeongju to the location of the station, which would be 10 km south of the old city center.

After a joint investigation (1996) by the central government and a survey (1996) for the new line, the third Gyeongju line proposal, which was the 'Hwacheon-ri line (1997, Ministry of Construction and Transportation, and Korail)', was finally accepted, resulting in the construction of the current line and Sin-Gyeongju Station.

The Hwacheon-ri line is 58.3 km long passing south of the Hyeongsan River, and 10 km shorter than the 68 km of the previous line, which would pass through the city center of Gyeongju. As a result, construction cost and time would be reduced. Furthermore, Hwacheon-ri, which houses Sin-Gyeongju Station, is surrounded by mountains, and does not stand out from the surrounding landscapes in Silla Wanggyeong, Seondosan, and Mangsan.

In addition, the new station has provided convenient transfer connection for residents in neighboring cities such as Pohang and Ulsan with direct connection with the existing

나지만 공청회, 문화재 지표조사 등에서 나타난 것을 반영하여 경주지역의 문화재 훼손을 최소한으로 하는 입지 선정임을 강조했다. 특히 경주지역 문화재를 크게 훼손하고 있는 중앙선과 동해남부선 이설을 고려한, 오히려 문화재를 보호할 수 있는 최적대안이라고 했다.

그러나 형산강 노선은 유물·유적 피해 및 경관훼손 등의 문제점이 거론되며 종교계 및 학계를 중심으로 반대되었다. 당시 건설교통부는 역사를 도심 외곽으로 이전하고 지하화 구간을 연장하는 대안을 검토하였으나, 결국 관계장관회의(1995)에서 형산강 노선 재검토를 결정하게 되었다.

이를 보완한 두 번째 경주노선 계획안인 '이조리 노선(1995, 건설교통부와 고속철도건설공단)'은 철도역을 기존 북녘들에서 남쪽 5km 떨어진 이조리에 제안되었다. 이는 고속철도가 경주 원구도심을 통과할 때 우려되는 문화재 훼손을 막자는 불교계와 문화체육관광부의 주장을 받아들인 것이다. 동시에 동해남부선 복선전철화 등과 연계하여 포항·울산 지역의 수송인구를 위해 경주 도심에 정차역을 두어야 한다는 건설교통부의 주장도 받아들여진 것으로 볼 수 있다.

하지만 유네스코 측에서는 경주 지역의 고속철도 관통 시 경주의 문화유산지역 지정이 어렵다고 대응했다. 또한 이조리에 신라시대 유적이 많이 분포한다는 학계의 여론과 철도역이 기존 도시 중심부로부터 남쪽으로 약 10km에 입지하여 이에 따른 경주 주민의 반발로 결국 반대되었다.

이후 중앙정부의 합동조사단 조사(1996) 및 신노선 용역조사(1996)을 거쳐 세 번째 경주노선인 '화천리 노선(1997, 건설교통부와 고속철도건설공단)'이 마침내 받아들여지며 현재의 노선과 신경주역(2010)이 건설되었다. 이 과정에서 경주 화천리, 방내리, 안심리, 덕천리를 두고 네 가지의 노선안들을 놓고 평가작업이 전행되었으며, 이들 네 가지 노선 중 문화재가 10여 개로 가장 적고 건설에 따른 사업비도 당시 2조1천억원으로 가장 적게 드는 화천리 노선이 최종안으로 확정되었다.

화천리 노선은 형산강 아래쪽을 지나 58.3km에 이르는 노선으로 경주 도시 중심부를 통과하게 돼 있던 당초 계획노선 68km에 비해 10km 가량 짧아 공사비가 경감되고 공사기간

road and railway network. Furthermore, in contrast to those of the Bugnyeokdeul area (825 ha) or the Ijo-ri area (495 ha), a large tract of land (1,650 ha) in Geoncheon is considered to be advantageous for urban expansion in the future.

In the end, the Gyeongju line in the Gyeongbu High-Speed Railway was arranged in parallel to the Gyeongbu Expressway with Dongdaegu Station and Ulsan Station. Currently, Sin-Gyeongju Station is a central station for the Gyeongbu High-Speed Railway, the Jungang line, and the Donghae line, absorbs the function of the existing Gyeongju Station in the old city, and encourages development around the station. In addition, the relocation of the Jungang line and the Donghae line which were previously located around Gyeongju Station, would effectively contribute to the preservation of the cultural heritage of Gyeongju.

Closing Remarks

Gyeongju is an interesting deposit of four historical stages of the capital of the Shilla kingdom, regional administrative center of the Goryeo–Joseon dynasties, modernized city center of administration and business establishments anchored by the railroad station and inter-city road during the Japanese occupation of Korea, and contemporary city of both cultural heritage preservation in the old district and new development in the outskirts.

Yet, contemporary Gyeongju is experiencing the spatial separation of cultural heritage preservation in the old district, and large-scale development of residential, leisure, and industrial complexes in suburban areas. In the spatial separation process, physical distance also generates a socio-economic gap between the old districts

도 단축되었다. 또한 신경주 철도역이 입지한 화천리 일대는 산으로 둘러싸여 있어 환경적 측면에서도 신라 왕경, 선도산, 망산지역의 경관으로부터 두드러지게 보여지지 않고 있다.

더불어 정차장 직선구간 3km 구간을 확보하는 데 다소 어려움이 있지만 기존의 도로망이나 철도망과 연계가 용이해 포항과 울산 등 인근 도시의 주민들에게도 편리하다는 장점이 있다. 또한 건천지역의 500만평 규모의 개발가능 지역은, 북녘들(250만평)이나 이조리(150만평)와 비교하여 미래의 도시 확장이 상대적으로 가능하다고 평가되고 있다.

결국 경부고속철도 경주 노선은 동대구역과 울산역으로 경부고속도로와 기본적으로 평행하게 배치되었다. 현재 신경주역은 경부고속철도선, 중앙철도선, 동해철도선이 모두 만나는 중심 철도역으로 기존 경주역 기능을 흡수하며 역세권 개발을 유도하고 있다. 또한 기존 경주역을 중심으로 배치되었던 중앙철도선과 동해철도선이 이선되면서, 경주의 문화재 발굴과 보존에 효과적일 듯하다. 그럼에도 신경주역이 경주의 새로운 신개발거점으로 등장하면서, 기존 경주 원·구도심의 쇠퇴가 가속될 것으로 우려되고 있다.

맺으며

이상에서 고찰한 것처럼, 경주는 신라왕조의 수도, 고려-조선의 지방행정거점, 일제강점기의 모던 경주, 그리고 현대 경주의 역사유적의 보전과 관광지의 개발이 함께 진행되고 축적되어, 역사유적의 보전과 새로운 개발이 첨예하게 충돌해온 도시이다.

이러한 경주의 도시변화는 결국 오래된 원·구도심의 유적보존과 이에 따른 외곽지의 대규모 주거·레저·산업단지 개발간의 상호 충돌과 격리의 관계로 진행되어 왔다. 이 과정에서 원·구도심과 외곽의 신도심 사이의 물리적인 거리와 사회·경제적 격차는 지속적으로 늘어나고 있으며, 원·구도심의 인구는 지속적으로 축소되어 왔다. 이러한 상황에 대응해 고도 경주가 앞으로 대응해야 하는 도전은 다음 세 가지이다.

첫째, 주변 도시와의 보다 큰 지역권 내에서의 협력과 미래의 구상이다. 고대 및 중세 경주

and the suburban areas, which continues to grow with the influx of public facilities relocated from the old districts.

Under such a situation, the ancient city of Gyeongju faces the following three challenges: Firstly, cooperation for the future with neighboring cities in a regional setting is required. If classical and medieval Gyeongju functioned with a coastal–inland system with Pohang, modern Gyeongju was connected with Daegu, and contemporary Gyeongju is under the influence of the Pohang–Ulsan industrial complexes. Such tendency can be proven by the concentration of manufacturing corporations near National Routes 7 and 35 on the northern and southern borders of Gyeongju, respectively.

An active response to such regional condition would be industrial development through the universities. Gyeongju can be the core of government research facilities and the R&D centers of private companies through the accumulated knowledge and technologies of already existing universities. Inflow of young research human resources must be secured, and outflow of such human resources should be prevented. To do so, social and education facilities of the universities and the research infrastructure of private corporations should be aggressively encouraged to move into the old districts.

Secondly, the city must decide on the function of Sin-Gyeongju Station of the GyeongBu railway line as an anchor of development, as well as substitutional use of the old railways and the stations in the old district. Gyeongju moved from a river- and stream-based city of the classical and medieval eras to a railroad and highway-based modern city. The greatest change of the century in Gyeongju would be the transition from ancient city to high-speed railway city.

Such change can be compared to that of Kyoto. An interesting fact is that UNESCO

가 포항과의 내륙-해안의 연계 체계하에서 기능했다면, 근대기 경주는 대구, 그리고 현대기의 경주는 오히려 포항과 울산의 생산활동권 내에 속해 있다. 이러한 경향은 경주의 북쪽과 남쪽의 외곽 경계지에 조성되는 상업단지에 국도 7과 국도 35를 중심으로 입주하는 기업들의 특성에서도 확인된다. 그리고 대규모 아파트 주거단지들이 산업단지들과 연계되어 개발되고 있다.

지역권 내에서 경주의 보다 능동적인 대응은 대학교들을 이용한 도시의 산업적 진화이다. 이미 경주에 입지하는 대학캠퍼스들의 지식 및 테크놀로지의 생산활동을 통해 공공연구시설과 민간기업의 R&D 연구시설들의 유치와 개발이 핵심이 될 수 있다. 이를 통해 젊은 지식 인구의 유입을 지속적으로 확보하며, 물리적 환경의 질적 향상을 도모하여 인재인구의 유출을 막아야 한다. 이 과정에서 대학의 지역사회시설과 교육시설 및 민간기업의 연구인프라들을 구도시 안으로 적극 유입시켜야 한다.

둘째, 경부고속철도의 신경주역(2010)이 외곽으로 확장되고 있는 경주에서 어떠한 새로운 개발거점의 역할을 할 것이며, 이에 따른 경주 구도심의 경주역의 대안적 이용과 기존 철도선의 이선에 관한 효과적인 대응이다. 경주는 고대와 중세의 하천 중심의 도시에서 근대 이후 육상교통체계 중심의 고속도로와 철도 중심의 도시로 진화해 왔다.

한편 최근 한 세대기 동안 경주에서 가장 큰 변화는 무엇보다 역사도시로서 고속철도선을 갖춘 도시로의 변화이며, 이는 교토의 경험과 매우 비교될 수 있다. 흥미로운 점은 경주의 문화재가 경부고속철도선 경주구간의 계획과정에서, 철도선과 철도역의 입지를 세 번 변경시키며 마침내 2010년을 전후로 20년의 계획과정이 종료되었다. 신경주역은 이미 옛 철도역의 기능을 흡수했으며, 이에 옛 경주역은 철도운송의 기능을 잃고 건축적으로 보전되었다. 문제는 옛 철도역과 연계되어 성장해온 100년 된 구도심의 상권과 시장의 와해가 진행되고 있다는 점이다. 또한 최근까지 경주를 관통했던 기존의 중앙철도선과 동해철도선의 이선이 과연 경주의 문화재 발굴 이외에 어떠한 긍정적 영향을 끌어낼 것인가에 대한 대응전략이 필요하다.

셋째, 경주는 1000년 전의 문화유산의 박물관이 아닌, 이 시대의 시민과 주민의 생활환경과 밀착된 역사문화도시를 지향하는 문화재 보존과 생활환경 개선이 시급하다. 경주는 1000

World Cultural Heritage sites three times caused changes in the alignment of railway and the location of station sites. This planning process expensively extended the planning period, which took 20 years to complete. Yet, it appears that it is a good job worth the change. Meanwhile Sin-Gyeongju Station has already absorbed the function of the old railroad station, which is currently undergoing demolition. The demotion of the old station, which underwent building preservation while losing its transportation function. Such change is causing breakdown of the markets and business center in the old district.

Thirdly, the ancient city of Gyeongju is more a museum of thousand year-old relics. Rather, it is a real city that combines the efforts to preserve the cultural heritage and improve the everyday environment of its local population. Registration of major historical sites in Gyeongju on the UNESCO World Cultural Heritage list gave Gyeongju the potential to become a global tourist site. The records show that from 2012, Gyeongju has welcomed twelve million tourists per annum for five years in a row. Even with negative factors, such as the SeWolHo ferry accident, the MERS epidemic, and earthquakes, the number of visitors to Gyeongju rivals that of Jeju-do Island (15 million). However, the number of visitors is far less than that of Kyoto, which is estimated to be around 54 million. Naturally, the question arises: which policies and projects must be undertaken to improve the number of visitors and quality of experience in Gyeongju?

Many cities in Korea had been forming large-scale tourist areas and housing on suburban areas, rather than the old core districts. However, such a trend should be reversed, to instead focus on the old districts. Fortunately, the national government has issued the Old Capital Preservation Law, and Gyeongju City is matching pace with the national government by undertaking the Gyeongju Old Capital Preservation Project,

년이 지난 현재 신라의 문화유산에 근거한 관광도시로 기능해 왔다. 특히 유네스코 세계문화유산(2000)의 도시로 경주의 주요 역사구역들이 지정됨으로써 경주를 한국학생의 수학여행지를 넘어, 국제관광지로서의 성장의 계기가 되었다. 실제로 경주는 2012년 이후 5년 동안 매년 방문객 1,200만명을 가진 도시이다. 특히 이 기간에 발생한 세월호, 메르스, 지진 등의 대형사고에도, 경주 방문객의 실적은 한국의 대표 관광지인 제주도의 방문객(1,475만명, 2017)에 비교될 만하며, 그럼에도 교토의 방문객(5,360만명, 2017)에는 1/4에도 못 미치는 수준이다. 그렇다면 앞으로 한 단계 질적으로 성장한 경주의 방문과 체험을 위해, 공공과 민간이 어떤 정책과 사업을 수립하고 추진해야 하는가에 관한 전략과 실행이 필요하다.

국내 도시들의 관광정책은 최근까지도 원·구도심이 아닌 도시 외곽에 숙박과 엔터테인먼트를 위한 독립된 대규모 관광단지와 숙소를 조성해왔다. 그런데 이러한 과거의 추세가 점차 소규모 중심의 원·구도심 중심의 생활환경 방문과 체험으로 변화되고 있다. 다행히 중앙정부도 '고도보존에 관한 특별법(2005)'을 제정하고, 경주시는 이에 맞추어 그 동안 문화재보존 정책으로 인해 낙후되어온 생활환경의 개선과 문화복지시설의 조성사업을 담은 생활밀착형 경주 고도보존사업(2011, 2016)이 추진 중이다. 이러한 시대요구와 공공정책에 부응하며, 지난 도농통합도시(1995)의 설립에 따른 원·구도심 내의 공공행정기능의 도시외곽으로의 이전은 지양되어야 한다. 오히려 외곽에 입지한 대학의 문화복지 및 교육시설들이 현재 비워져 있는 원·구도심의 건물과 필지로 회귀해 유입되어야 한다.

which will improve the neglected living environment and establishment of welfare facilities in the old districts. To meet current demand and public policy, relocation of administrative functions from the old districts to suburban areas must be stopped.

참고문헌

1. 경주시. 경주시 관광실태조사 보고서, 2013

2. 국토개발연구원, 고속철도와 지역 균형개발에 관한 연구, 1995

3. 교통개발연구원, 경부고속전철 교통영향평가-경주권, 1992

4. 김성득·최양원, 경부고속철도 경주 - 울산구간과 일본 신간선의 비교, 대한교통학회지, 15(2), 131~150, 1997

5. 강갑생, [그래픽뉴스] 경부 고속철도 추진 과정, 중앙일보, 2001

6. 한광야. 도시에 서다: 한국도시의 형성과 진화. 상상출판, 2016

7. 유네스코한국위원회. 세계유산 새천년을 향한 도전, 2010

8. 국토교통부 홈페이지

9. 박유광. 조선일보. "〈경주고속철〉 현존노선이 훼손줄일 최적안.....", 1995

10. 서인석. 매일경제. "경부고속철도 경주노선 경주시 외곽 건설", 1996

11. 이선복, 경주:고속전철 왜 문제인가?, plus V106, 플러스문화사, 1996

12. ⓒ 매일경제. "[노선병경] 경주구간 화천리 통과 최종확정", 1997

13. KBS 뉴스. 경부고속철도노선 경주우회-문화재보호 우선. 1996.06.08 (21:00, 1996

14. https://news.kbs.co.kr/news/view.do?ncd=3762810

15. Han, GwnagYa and Haebin Gwak. 2018. Lessons from GyeongJu: Railroad in Historic City, unpublished paper. Chiang Mai UNESCO Symposium.

References

1. Gyeongju City. Gyeongju City Tourism Survey Report, 2013

2. Han, GwangYa and Haebin Gwak. "Lessons from GyeongJu: Railroad in Historic City", Chiang Mai UNESCO Symposium, 2018

3. Han, GwangYa. Reading the Cities: Formation and Evolution of Korean Cities. SangSang Publishing, 2016

4. KBS News. Gyeongbu High-Speed Rail Line Bypass Gyeongju-Priority on Cultural Heritage Protection. 1996.6.8, 1996.

5. Kim, Seong-deuk, Yang-won Choi. "Comparative Study of Gyeongbu High Speed Rail Gyeongju-Ulsan Section and Japanese Shinkansen", Journal of the Korean Transportation Association, vol.15, no.2, pp.131-150, 1997

6. Korea Research Institute for National Land Development. Research on High-Speed Rail and Balanced Regional Development, 1995

7. Korea Transportation Development Institute. Gyeongbu High Speed Rail Traffic Impact Assessment-Gyeongju Region, 1992

8. Lee, Seon-bok. "Gyeongju: Why is the High-Speed Train a Problem?", Plus vol.106. Plus Munhwasa, 1996
Park, Yoo-gwang. "The Best Way to Reduce Damage for Existing Routes of Gyeongju High-Speed Rail", Chosun Ilbo, 1995.11.18, 1995

9. Republic of Korea Ministry of Land, Infrastructure and Transport Website

10. http://www.molit.go.kr/portal.do

11. Seo, In-seok. "Construction of Gyeongbu High-Speed Rail Gyeongju Route outside Gyeongju", Maeil Economic Daily, 1996.06.05, 1996

12. UNESCO Korean Committee. World Heritage Challenges for the Millennium, 2010
Final onfirmation of Reassignment of Hwacheon-ri in Gyeongju Section", Maeil Economic Daily, 1997.3.31, 1997

Chapter 5

일상의 문화공간

Cultural Space in Everyday Environment

21세기는 '문화의 시대'라고 불린다. 창조적 문화도시는 많은 지자체가 지향하는 개념이며, 미술관과 콘서트홀, 문화예술회관과 같이 공공이 지원하는 문화시설의 거점 공간화는 도시공간의 구조와도 연결된다. 최근에는 전통적으로 구분된 문화시설보다는 가변성이 있는 복합문화공간이 많이 만들어지고 있다. 더불어 상업공간의 문화화는 문화시설과 상업시설의 경계를 모호하게 만들고 있다. 궁극적으로 '일상 속 문화, 일상의 문화화'와 같은 화두가 실현되기 위해서는 다층의 조직적인 문화공간과 지원 프로그램이 필요하다. 오래된 유휴공간의 문화공간화는 여러 도시에서 동시다발적으로 일어나고 있으며, 도시에 새로운 거점지역를 만들거나 도시공간을 재구조화할 때 문화시설은 그 시대의 아이콘으로 자리매김하고 있다.

The 21st century is called the 'age of culture'. A creative cultural city is a concept pursued by many local governments, and the spatialization of publicly supported cultural facilities such as art museums, concert halls, and culture and arts centers is also connected with the structure of urban spaces. Recently, many complex cultural spaces with variableness are being created rather than traditionally separated cultural facilities. In addition, the culturalization of commercial spaces is blurring the boundaries between cultural and commercial facilities. Ultimately, in order to realize such topics as 'culture in daily life, culturalization of daily life', a multi-layered organizational cultural space and support program are needed. Cultural spatialization of old idle spaces is occurring simultaneously in several cities, and cultural facilities are establishing themselves as icons of the times when creating new base areas in cities or restructuring urban spaces.

인천 아트플랫폼,
유휴공간의 예술활동 거점

김환용 (한양대학교 에리카 건축학부 교수)

들어가며 : 인천아트플랫폼의 시작

인천아트플랫폼은 2002년에 인천 중구 해양동 일대의 '미술문화공간' 사업계획으로 조성된 예술문화공간이다. 인천아트플랫폼은 총 연면적 5,593m²에 지상 4층, 지하 1층 규모로 조성되어, 인천 개항장 내부 근대건축물의 문맥을 보전해온 의미있는 공간이다.

인천아트플랫폼은 2003년에 시작되어, 1930~40년대 건축물을 리모델링하여 창작스튜디오, 전시장, 공연장, 생활문화센터 등 총 13개 동의 규모로 조성되었다. 인천아트플랫폼은 과거의 흔적을 보유하고 있는 건축물 집단이 도시에서 그 기억의 보존을 통해 현대 도시에 공간적 의미를 재해석할 수 있게 만들어준다는 관점에서 의미있는 시도라 할 수 있다.

역사의 긴 시간 속에서 장소만들기의 작업으로 시작된 인천아트플랫폼 프로젝트는 창작, 전시, 교육, 유통 등의 기능을 충족시키기 위해 전체 계획방향을 보존 건축물과 기능 보완을 위한 최소한의 신축건축물로 분류하고, 개항기에 형성된 가로 구획과 역사적 경관을 유지하도록 하였다.

인천아트플랫폼의 전체적인 디자인 컨셉은 '비움과 채움', '기억과 향유', '소통'으로 잡혔으며 기능적 효율성을 위해 오픈스페이스를 확보하고 장소의 기억을 유지하기 위해 투명한 유

Incheon Art Platform, from Void Space into Regional Art Hub

Hwanyong Kim (Professor, Hanyang University ERICA, Dept of Architecture)

그림 1. 인천아트플랫폼 도입부 / Incheon Art Platform Facade

Introduction : The beginning of Incheon Art Platform

Incheon Art Platform is an art and culture space that was created in 2002 according to the 'Art and Culture Space' masterplan in Haean-dong, Jung-gu, Incheon. The

리를 덧대어 과거를 투영시켰다. 회랑과 오버브리지를 설치하여 건물 사이를 연결함으로서 기존 개항장의 도시 질서를 유지하고자 전체 단지를 열려있는 공간으로 조성하였다. 인천아트플랫폼은 개항장에 존재하는 도시공간의 일부로 열려 있으되 자기 정체성을 갖고 있는 고유한 문화공간이다.

인천 제물포항은 최초로 서구문물이 들어온 중심 항구로서 여러 나라의 문화가 공존하는 국제도시로 번성하였다. 외국인들의 유입은 조계지의 탄생을 이루었고 각국의 조계지가 모인 이곳이 인천 개항장의 시작이 되었다. 현재 삼국지 벽화거리의 끝자락에 구축된 조계지 계단을 중심으로 오른쪽은 개항장 문화지구로 왼쪽은 차이나타운으로 구분되고 있다. 한국전쟁을 통해 많은 근대건물들이 소실되었으나, 아직까지 근대건축물의 의미와 가치를 목격할 수 있는 공간으로 남아있다.

이러한 근대문화의 중심에 위치한 인천아트플랫폼은 복합문화시설로서 문화와 예술을 통해 예술가와 시민들이 소통할 수 있는 교류의 장을 제공한다. 예술가는 스튜디오, 공동 작업 공간, 미디어랩실 등 물리적인 창작공간과 협업 프로젝트, 오픈 스튜디오, 공연기획 등 행정적 지원도 받을 수 있으며, 시민들은 각종 공연과 교육 프로그램에 참여할 수 있다.

유동적 공간구성

아트플랫폼은 크게 8개동으로 나뉘어져, 개별 동마다 고유의 역사와 프로그램이 존재한다.

A동은 옛 군회조점(조운업을 하는 회사)의 사무실로 1902년 벽돌로 건축한 2층 건물로 20세기 초 사무공간 인테리어 특징을 보여준다. 2009년 인천아트플랫폼의 교육공간과 전시장 '크리스탈 큐브'로 사용하다가 2016년부터 인천생활문화센터로 운영되고 있다.

B동은 1948년 대한통운 창고로 사용하던 건물을 개조한 곳으로 아트플랫폼이 착공되던 때까지 계속하여 창고로 사용되었다. 현재 인천아트플랫의 주 전시장으로 입주작가들을 위한 창작무대이자 지역의 다양한 예술활동을 위한 공간으로 활용되고 있다.

building was originally built in the 1930s and 40s, and then remodeled in 2003 to create a total of 13 buildings that include creative studios, exhibition halls, performance halls, and living culture centers. This project, which started as the recreation of historical heritage, consists of preserved buildings and minimal new buildings to supplement functions to provide exhibition, education, and distribution, while maintaining the horizontal and historical landscape.

The design concepts of the art platform are 'Void and Occupancy', 'Memory and Enjoyment', and 'Communication'. For functional efficiency, transparent glass was added to secure open space and maintain history. Corridors and over-bridges were installed to connect buildings for fluent circulation, and the entire complex was created as an open space to harmonize with the pre-existing urban order around the port. The Incheon Art Platform is unique in its openness to pre-existing urban space, while maintaining its initial identity.

Incheon Art Platform provides a rendezvous focus for artists and citizens to communicate through culture and art. Artists are provided with creative spaces, such as studios, joint workspaces, and media lab rooms, as well as administrative support, such as collaboration projects, open studios, and performance planning, and citizens can take part in various performances and educational programs.

Flexible Space Composition

The Incheon Art Platform consists of eight large individual buildings, each with its own story and use. Building A is a two-story brick building that was built in 1902 as an office for a military branch, and represents the characteristics of office space in the

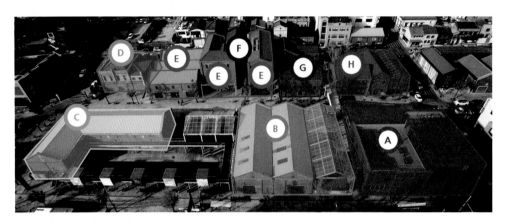

그림 2. 인천아트플랫폼 공간배치 / Spatial arrangement of Incheon Art Platform

　C동은 B동과 마찬가지로 대한통운 창고로 사용되어왔다. 현재 가변적으로 활용 가능한 블랙박스 형태의 공연장으로 운영되며 다양한 장르의 공연을 시민들에게 제공하고 있다.

　D동은 운영사무실로서 1888년 지어진 일본우선주식회사의 인천지점 건물로 2009년 아트플랫폼 개관 당시 아카이브관으로 활용되다 2016년부터 운영사무실로 사용되고 있다.

　E, F, G동은 1933년에 지어진 창고건물을 지역예술가들이 '피카소작업실'로 사용했던 곳이다. 리모델링 후 현재 인천아트플랫폼의 스튜디오와 게스트하우스로 사용되고 있다.

　H동은 1943년에 건축된 일본식 점포건물로 양쪽 벽을 벽돌로 쌓은 목조건물이었으나 현재 유리로 상층부를 리모델링하여 커뮤니티관과 인천생활문화센터로 사용되고 있다.

　인천아트플랫폼은 기존의 도시적 문맥을 최대한 보존하고 그 의미에 현대적 해석을 입혀 새로운 공간적 활용으로 탄생시키기 위해 노력한 결과이다. 플랫폼의 공간으로 진화하기 위해 각각의 공간은 규정되어 있지만 시간의 흐름에 맞추어 개연성 있게 변화하며 거주자, 사용자, 관람자의 입장에서 새로이 규정되고 진화한다.

early 20th century. It was used as an educational space and the exhibition hall 'Crystal Cube' for the Incheon Art Platform in 2009, and since 2016, has been operating as the Incheon Living Culture Center. In 1948, Building B was renovated as a warehouse for Korea Express, and continued its use until the art platform project began. Currently, it is the main exhibition hall of the Incheon Art Platform, and is used as the main creative stage for tenant artists and a space for various local art activities. Building C was used as a warehouse for Korea Express as well. Currently, it is operated as a black box-type concert hall that can be used flexibly, and provides various performances to citizens. Building D was the Incheon branch building of Japan Priority Co., Ltd., built in 1888; later, when the art platform opened in 2009, it was used as an archive hall, and has been used as an operating office since 2016. Buildings E, F, and G were warehouse buildings from 1933 that were used by local artists as 'Picasso Studio'. After remodeling, they are currently being used as studios and guest houses for the Incheon Art Platform. Building H was a Japanese-style wooden store building with brick wall reinforcement from 1943, but is currently remodeled with glass. It is now a community hall and the Incheon Living Culture Center.

The Incheon Art Platform is the result of efforts to preserve the existing urban context as much as possible, and to create new spatial utilization by applying a modern interpretation. For space to evolve into platform, each space needs to be defined, but vary by the flow of time, and built through the point-of-view of the tenant artists, users, and visitors.

계획과 연계를 통한 지역상생전략

인천아트플랫폼은 지역연계사업으로 미술문화공간 기본계획 및 지구단위계획과 함께 만들어졌다.

인천발전연구원이 준비한 인천 구도심의 문화적 재생과 미술문화공간의 효율적 운영방안과 미술문화공간 기본계획을 통해 인천 중구 해안동 일원의 문화적 재생과 미술문화공간으로 근대건축물을 보존하면서 시민들이 함께 이용할 수 있는 문화거점으로 인천 구도심을 활성화시키는 방법을 제안하였다. 이를 위해 사업대상지를 세 가지(개항기 근대건축물, 기존 건축물 활용, 철거 건축물) 타입으로 구분하여 도시경관적 측면, 건축적 측면, 미술문화적 측면을 고려하여 분석하였고 외관에 따른 기준설정, 공간규모, 동선계획을 지정하여 미술문화공간 중심의 지역 활성화를 유도하였다.

또한 인천광역시가 준비한 근대건축물밀집지역 지구단위계획을 통해 중구지역 중 역사적

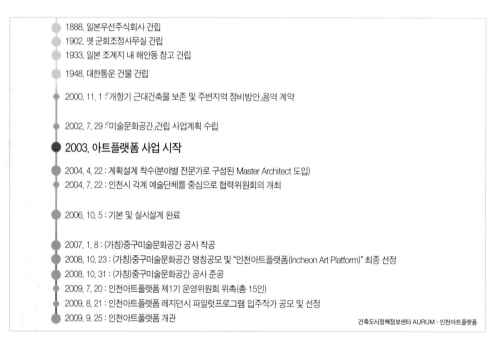

그림 3. 인천아트플랫폼 추진경과 / Progress of Inchein Art Platform

그림 4. 건물과 건물의 연결: 각 건물들은 독립된 구조로 활용되고 있으나 전체적인 맥락을 유사하게 유지하고 있다
Relationship between exisiting buildings

A win-win Strategy through Synergy with Local Assets

The Incheon Art Platform was part of a linkage project within the region, planned along with the basic art and culture space plan, and district unit plan. The basic art and culture space plan was proposed to revitalize the old urban center in Incheon as a cultural base for citizens to use while preserving modern building heritage as an art and culture space. The target site was divided into three types (modern buildings during the harbor-opening period, existing buildings, and demolished buildings), and analyzed through consideration of the urban landscape, architectural, and art & cultural aspects. Standard settings with exterior, size of the space, and circulation plan were designated to induce local revitalization.

가치를 갖는 근대건축물의 노후 및 훼손 방지, 무계획적인 개발로 인한 도심경관 저해를 방지하였다. 기존 지구단위계획에서 근대건축물 밀집지역의 대표 건축물을 기준으로 외관계획, 경관계획을 조성하고 특화가로구역 및 조성계획을 구축하였다.

아트플랫폼은 일반적인 도시사업의 기준으로 볼 때 상당히 긴 호흡으로 이루어진 프로젝트이다. 본 프로젝트의 MA(Master Architect)인 황순우 건축가의 표현대로 인천아트플랫폼의 성공적인 정착은 10년 간의 장기적인 안목으로 건물이 지닌 역사성을 보존하고 지역 활성화와 예술의 결합을 통해 도시를 지속 가능하게 재생시킬 수 있는 '문화 인큐베이터'로서 역할을 만들고자 한 노력이 있기에 가능했다.

근대건축물 활용을 통한 지역재생

인천 원도심에서도 개항장 문화지구는 정체성이 매우 강한 지역이다. 차이나타운, 동화마을 등으로 대표되는 이 지역은 인천 내에서도 근대건축물의 흔적이 곳곳에 남아있는 인천의 이미지를 상징하는 장소이다.

1930~1940년대 인천은 국내 최대의 쌀 집산지이자 미곡 유출의 중심지였다. 자연스럽게 배가 드나들던 해안동 일대에는 물건을 보관할 창고들과 큰 규모의 건물들이 들어섰다. 이렇게 건축된 거대 구조물들은 사실상 원도심의 도시공간적 성격을 규정하기에 이르렀으며, 2003년에 인천시가 본격적인 인천아트플랫폼사업을 시작하기 전에는 낙후된 건물이 밀집된 공간으로 인식되었다.

또한 인천시가 당시 처한 상황을 볼 때 시가 주도적으로 추진한 도시재생사업에 발맞춰 문화인프라 확보 문제는 지속적으로 제기되어왔다. 이에 따라 인천의 낙후된 문화시설에 대한 예술계의 지적을 해결하기 위해, 100년 이상 버려진 공장건물을 리모델링하여 작가들의 창작활동을 지원하는 인천아트플랫폼 개발계획이 진행되었다.

인천일보 문화재생, 시민의 삶을 디자인하다. 2. 예술로 지역을 변화시킨 '인천아트플랫

In addition, the committee prevented the deterioration and damage of modern buildings with historical value in Jung-gu, and prevented the urban landscape from being disturbed by unplanned development through the district unit plan in the dense area of modern buildings. In the existing district unit plan, exterior plan and landscape plan were created based on representative buildings in the modern building-intensive area, and a specialized road zone and a construction plan were established. The art platform is a long-term project for general urban projects. Architect Hwang Soon-woo, the Master Architect (MA) of this project, stated that the success of the Incheon Art Platform was possible because of efforts to preserve the historic buildings with a long-term perspective for 10 years, and to create a 'cultural incubator' that can sustainably regenerate the city through regional revitalization, and in combination with art.

Regeneration of the Region through the Use of Modern Buildings

The cultural district around the port had a very strong identity, even within the original city center. This area, represented by Chinatown and Donghwa Village, is a place that symbolizes the image of Incheon, and modern architecture remains everywhere. In the 1930s and 1940s, Incheon was the largest rice collection site in Korea, and the center of rice outflow. Large-scale buildings like warehouses were built in Haean-dong, where ships visit. The large structures naturally defined the urban spatial characteristics of the original city center. Before the city initiated the Incheon Art Platform project in 2003, the area was recognized as a dense space full of underdeveloped buildings.

폼'에 의하면, 인천 아트플랫폼은 '근대건축물의 활용'을 도시재생의 주요 수단으로 단기간 회복이 아닌 장기간 회복을 목표하였다. 이에 조성사업이 2000년부터 시작되어 2009년에 오픈하고 자리잡기까지 10년, 총 20년을 계획한 프로젝트이다. 인천 아트플랫폼은 1900년대 인천 개항장 인근이라는 도시의 역사성과 장소적 특성을 살려 문화적으로 재활용하자는 시민들의 뜻을 기반으로 공공이 개입하여 탄생하게 된 근대 문화유산을 재생한 문화예술 창작공간이다.

장소만들기로서의 플랫폼 전략

공간적 변화를 만들고자 황순우 건축가가 선택한 키워드는 '예술'이다. 다만, 예술의 공간적 적용과 장기적 활용성을 담보하고자 다른 예술공간계획과는 차별화될 수 있도록 아트플랫폼 내부에 앵커시설로서 거주와 창작이 결합된 레지던시를 구축하는 것이었다. 물리적으로 창작활동에 필요한 공간을 제공하고 이에 필요한 행정적 지원을 포함하여 아트플랫폼이 단순히 일회성 도시공간으로 자리잡기보단 공간의 변화를 통해 원도심 내부에 새로운 장소성과 지역성을 부여할 수 있는 플랫폼이 될 수 있도록 만들어간 것이다.

이제 개항장은 국내외 예술가들이 개성과 목소리를 표출하는 창작공간이자, 시민들이 일상에서 문화예술을 향유하고, 예술가들과 소통하며, 평생교육의 가능성을 열어준 열린 도시공간의 멀티플랫폼으로 변모하였다.

인천아트플랫폼에 고정된 것은 움직일 수 없는 건물뿐이다. 사무실, 전시장, 공연장, 자료관, 스튜디오, 게스트룸 등 각각의 동마다 부여된 기능은 있지만 그 기능에 국한되지 않고 공간은 변화하며 사용된다. 각 공간에서 일어나는 행위는 공간의 기획자와 사용자간의 암묵적 합의를 통해 새로이 태어나고 규정된다. 이렇게 자유롭고 유동적인 분위기가 인천아트플랫폼 자체를 하나의 유기적인 예술공간으로 만든다. 아트플랫폼은 그 이름에서 보이듯 문화예술공간이라는 기능적 접근보다는 플랫폼으로서의 활용이 더욱 돋보이는 하나의 장소이자 의미이다.

그림 5. 도입부 광장 / Open space at the beginning of Incheon Art Platform

Additionally, the issue of securing cultural infrastructure has been continuously raised due to the situation at the time, along with the city-led 'Urban regeneration project'. To solve the criticism concerning Incheon's underdeveloped cultural facilities, the 'Incheon Art Platform project' was developed. The project supports the creative activities of artists in abandoned factory buildings. The Incheon Art Platform aims for long-term restoration by utilizing modern buildings as a major means of urban regeneration. It is a cultural and artistic creation space that was planned with public intervention based on the will of citizens to utilize the historical and spatial characteristics of the city near Incheon port in the 1900s.

그림 6. 건축물 사이 오픈스페이스 / Open space in between buildings

인천아트플랫폼 입주작가는 개인 스튜디오를 개방하여 시민이 찾아와 작업과정을 볼 수 있게 하는 오픈 스튜디오를 해야 한다. 리서치 투어도 입주작가의 의무다. 인천 곳곳을 돌아다니며 인천을 보고 듣는다. 그러면서 작가는 자연스럽게 인천을 작품에 담는다. 인천을 담는 행위로서의 아트플랫폼은 공간 내부에서 시민들과의 소통을 통해 경험의 공유가 이루어지는 공간으로도 활용된다. 시민의 경험, 작가의 경험, 역사의 경험 등이 어우러져 예술가와 지역주민과의 자연스러운 교류를 통해 예술의 전달, 지역에 대한 기억의 전달이 이루어지는 하나의 교육적 공간으로도 사용된다.

Platform Strategy to Create a Place

'Art' is the keyword chosen by architect Hwang Soon-woo to make spatial changes. To ensure the spatial application and long-term utilization of art, a residence for artists with housing and workshop was established as an anchor facility to distinguish the art platform from the others. Rather than physically providing the space for creative activities and administrative support, the art platform can become a platform as landmark, and through spatial changes provide locality within the original city center. The port has now become a creative space for domestic and foreign artists to express their individuality, and an open urban multi-platform where citizens can enjoy culture and the arts, communicate with artists, and engage in lifelong education.

The only things that are fixed to the Incheon Art Platform are the buildings. Although specific programs are assigned to each building, such as offices, exhibition halls, performance halls, archives, studios, and guest rooms, they are not limited to those functions. Actions occurring in each space are reborn and defined through implicit agreement between the planner and the user of the space. This free and fluid atmosphere makes the Incheon Art Platform an organic art space. As the name suggests, the art platform is a place of various uses, rather than just art and cultural space. The artist who moves into the Incheon Art Platform needs a public studio for citizens to come and see their work. Research though local tours is also the duty of tenant writers. They tour Incheon to see, and to listen. In the meantime, these artists naturally capture Incheon in their work. The art platform as an act of containing Incheon is also used as a space where experiences are shared through communication with citizens inside the platform. It is also used as an educational space where the art

맺으며 : 또 다른 10년을 위한 기대

　지난 수년간 인천아트플랫폼은 안정적인 내부 시스템과 입주작가 중심의 콘텐츠 개발에 주력해왔다. 개관 초기부터 플랫폼으로의 공간적 활용에 대한 기획, 행정, 시스템적 노하우가 부족하였고 그에 따른 콘텐츠 개발 및 실행이 쉽지 않았다. 20년이 지난 지금 인천아트플랫폼은 시민과 거주 예술인들에게서 더 나아가 지역 전반에 기여할 수 있는 공간으로 변화하고자 한다.

　근대건축물의 보존을 통한 지역 정체성의 확보, 레지던시 프로그램을 통한 예술가들의 창

그림 7. 작업실 외부공간 / Working space and its outside

그림 8. 오버브리지와 회랑 / Overbridge and its corridor

and memories of the region are transmitted through exchanges between artists and local residents, with the combination of local residents, writers, and historical heritage.

Closing Remarks : Expectations for the Next 10 Years

Over the past few years, the Incheon Art Platform has focused on developing stable internal systems and contents centered on tenant artists. From the beginning, there has been a lack of planning, administration and systematic know-how for spatial use as a platform, and it has not been easy to develop and implement the contents accordingly. But now, after 20 years, the Incheon Art Platform is trying to change from

의성 보장, 가변적 프로그램의 적용과 공간의 유동적 활용, 참여와 공유를 통한 교육적 경험의 확장 등 인천아트플랫폼은 건축물군의 집합으로 이루어진 도시공간적 의미를 뛰어넘어 하나의 플랫폼으로 개항장 일대에 기여할 수 있는 장소를 만들었다는 점에서 충분히 훌륭한 시도로 볼 수 있다. 다가오는 또 다른 10년이 기대되는 이유는 이렇듯 유동적인 공간 활용을 통해 지역성 구축을 할 수 있는 인천아트플랫폼 고유의 가능성에 있을 것이다.

참고문헌

1. [문화재생, 시민의 삶을 디자인하다] 2. 예술로 지역을 변화시킨 '인천아트플랫폼', 인천일보, 2018.7

2. 인천 구도심의 문화적 재생과 미술문화공간의 효율적 운영 방안, 이현식, 서동희, 인천발전연구원, 2004

3. 개항기 근대건축물 밀집지역 지구단위계획, 인천광역시, 2021.10

4. 도약과 변화: 인천아트플랫폼 10주년, SPACE, 2019.8

5. 인천아트플랫폼에 나타난 도시재생의 특성연구, 김지수, 윤재은, 김민정, 한국공간디자인학회 논문집, 2017.4

6. 인천아트플랫폼 홈페이지: http://inartplatform.kr/sub/place.php?mn=info&fn=place&bn=hana_board_03&zest_bn=hana_board_02

citizens and tenant artists to a space that can contribute to the entire region.

Incheon Art Platform resides in the portside region as an urban landmark, securing regional identity through the preservation of modern buildings, ensuring the creativity of artists through residential programs, flexible use of space, and expansion of educational experiences through participation and sharing. The unique potential of the flexible space of Inchon Art Platform is another reason for high expectation in the next decade.

References

1. [Culture Regeneration, Design the life of citizens] 2. 'Incheon Art Platform' that transformed the area with art, Incheonilbo, 2018.7

2. Cultural regeneration of the old city center of Incheon and efficient operation of art and culture spaces, Lee Hyun-sik, Seo Dong-hee, Incheon Development Institute, 2004

3. District unit plan for contemporary buildings dense area during the harbor-opening Period, Incheon Metropolitan City, 2021.10

4. Takeoff and change: The 10th Anniversary of Incheon Art Platform, SPACE, 2019.8

5. Study on characteristics of urban regeneration in Incheon art platform, Jisoo Kim, Jeaeun Yoon, Minjung Kim, Journal of Korea Intitute of Spatial Design, 2017.4

6. Incheon Art Platform homepage: http://inartplatform.kr/sub/place.php?mn=info&fn=place&bn=hana_board_03&zest_bn=hana_board_02

광주 아시아 문화의 전당

임동원 (도화엔지니어링 상무)

그림 1. 남동쪽에서 바라본 광주 아시아문화의 전당 / Southeast birdview of Gwangju Asia Culture Center

들어가며

국립아시아문화전당(Asia Culture Center)은 아시아 문화에 대한 교류·교육·연구 등을 통한 국가의 문화적 역량 강화를 위해 만들어진 국가급 문화시설이다. 연면적 156,817m²로

Gwangju Asia Culture Center

Dongwon Lim (Senior Director, Dohwa Engineering)

Introduction

The Gwangju Asia Culture Center (hereinafter ACC) is a national-level cultural facility created to strengthen the country's cultural capabilities through exchanges, education, and research on Asian cultures. With a total floor area of 156,817m², it is a large-scale facility, and includes the site of the former Jeonnam Provincial Government Complex, a sacred site of the 5.18 Democratization Movement, which is a symbol of Korean democracy.

In this case, in the history of urban development in Korea, regarding that 1) the use of the relocated site (old Jeonnam Provincial Government Office), 2) site selection process for cultural facilities as a national policy, 3) the alternative selection process and approach, the paper would like to highlight 1) how historical and political meanings were embodied architecturally and 2) how publicity such as accessibility is achieved in this urban public facility.

서 지상 4층, 지하 4층규모의 대규모시설로서 한국민주화의 상징과도 같은 5.18 민주화운동의 성지인 옛 전남도청사부지를 포함한다.

이 사례는 한국 도시개발사에서 1) 이전적지의 활용(전남도청사) 2) 국가정책으로서의 문화시설의 입지 선정과정, 3) 대안의 선정과정과 접근법 등을 통해 1) 기념비적인 역사의 현장을 문화시설로 바꾸면서 어떻게 역사적, 정치적 의미를 건축적으로 구체화시켰는지, 그리고 2) 이러한 도심공공시설에 있어 접근성 등의 공공성을 어떻게 이루어냈는지에 대해 주로 살펴보고자 한다.

대상지 위치

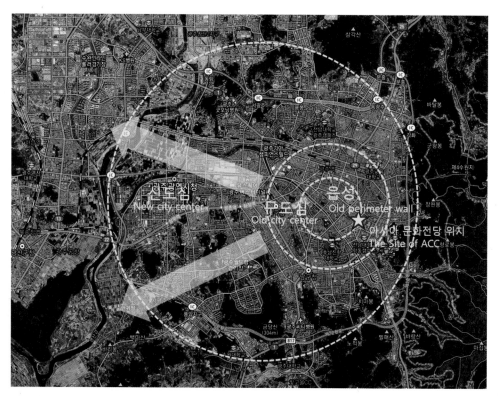

그림 2. 대상지 위치 / Site location in Gwangju

Site Location

Gwangju was originally a hub city in Jeon-nam province. The city also had 'Ep-seung' structure as the perimeter wall to demarcate the old city boundary. Most of major old cities in Korea has its own structure. The east side of the city was limited by Mu-deung mountain as the tributary of the So-baek mountain range. Owing to this natural limit of the east, the city expanded to the west. As time went on, the urban area has been expanded. Finally, horizontal distance of urban realm reaches around 15 km long. (refer to Seoul as 20km), while the population did not increase from 1.5 million units. The site is located on the western side of the old city center, and assuming from the old townships picture, it is located in the very site of a old regional government building. With regards of such historic meaning, the rehabilitation of old perimeter wall was partially realized in the project.

Relocation of Jeonnam Provincial Office

This was the place where the Jeollanam-do Provincial Office had been located since 1910, during the Japanese colonial period. After that, as Gwangju became a metropolitan city, it became independent from the jurisdiction of Jeollanam-do, and the issue of the relocation of the Jeonnam Provincial Office emerged. Thereafter it was relocated to Mokpo Namak New Town in 2005, the issue of utilizing the old provincial government building as a relocation site was emerged. Moreover, since this place was the sacred site of the 5.18 Democratization Movement in the 1980s, its utilization was a big task for the government at the time.

광주는 본래 읍성을 가진 전남내륙의 거점도시였다. 동쪽으로는 소백산맥의 한 지류인 무등산의 산세에 막혀 서쪽으로 도시의 팽창을 이루었다. 시간이 흐르면서 시세가 확장되어 시계는 가로축으로 약 15km에 달하는 시계 면적(서울이 개략 20km)을 가지는 반면 인구는 150만대에서 크게 증가하지 않았다. 대상지는 구도심의 서쪽에 입지하며, 읍치의 그림들을 분석한 것으로 추정하면 무등산을 진산으로 하는 관아자리에 입지하고 있으니, 그 역사적 의미는 꽤 오래된 지역이며, 현재 건축물 상부에 읍성의 일부가 복원되어 있다.

전남도청의 이전과 현 부지(이전적지)의 활용

이 곳은 전남도청이 일제강점기 시대인 1910년부터 자리하고 있었던 장소였다. 이후 광주가 광역시가 되면서 전라남도의 관할에서 독립하게 되었고, 전남도청의 이전문제가 대두되어 이를 목포 남악신도시로 확정하여 준공, 이전(2005)되면서 이전적지로 남은 옛 도청사의 활용문제가 대두되었다. 더구나 이곳은 80년대의 5.18 민주화운동의 성지이기에, 이에 대한 활용은 당시 정부의 큰 과제였다.

5.18 민주화운동과 대상지의 의미

구 전남도청사는 5.18 민주화운동의 상징과도 같은 장소이다. 당시의 군사독재정권에 반기를 둔 시민군들이 저항하기 시작했고, 이를 막는 과정에서 정부군은 무고한 많은 민간인의 희생을 요구하였다. 구 전남도청사는 이 과정에서 마지막 시민군 최후의 저항지였고, 당시 수십명의 시민군이 정부군에 희생당한 한국 민주화운동의 가장 뼈아픈 장소가 되었다.

KWANG-JU(1946. 6) .

그림 3. 1940년대의 전남도청사 / Junnam Province office (1940)

The 5.18 Democratization Movement and the Meaning of the Site

The former Jeonnam Provincial Government Complex is a symbol of the 5.18 Democratization Movement. The citizen militia against the military dictatorship at the time began to resist, and in the process of suppressing, the government forces demanded the sacrifice of many innocent civilians. This former Jeollanam-do government building was the last place of resistance for the militia in the process, and it became the most painful place for the Korean democracy movement, where dozens of militias at that time were sacrificed by the government forces.

그림 4. 5.18 민주화운동 당시의 전남도청과 도청앞 광장 / 5.18 Province office and square at 5.18 movement

초기 개발방향과 위치선정

이러한 5.18 민주화운동의 승화적 차원에서 광주에 아시아 문화의 전당을 만들겠다는 이 사업은 전 노무현 대통령의 공약사업이 되었다. 구체적 비전은 수도권의 집중된 문화자원을 분산시킴과 동시에 문화예술 교류의 장으로서 아시아 문화중심도시를 만들겠다는 것이었다. 집행예산이 약 7천억이었으니 그 규모 또한 상당한 사업이었다. 건립백서(아시아 문화중심도 시 백서, 2009)에 따르면 건립후보지는 현 부지인 A 전남도청, 그리고 후보지로서 거론된 지 역은 B 송원학원, C 광주 국군병원, D 중앙공원 및 E 서창지역으로 다음 그림으로 표기하고 있다.

부지 선정상의 특이점은 이 사업이 5년 내에 실현되어야 한다는 조건을 실현시키기 위해 대부분의 부지가 대지확보가 현 대상지역보다 용이한 이전적지가 많았다. 그러나 시 및 정부

Initial Development Direction and Site Selection

With the aim at the sublimation of 5.18 Movement, this project was promoted by former President Roh Moo-hyun. ACC was suggested to be a representative venue for cultural and artistic exchange and developed to make Gwangju city as a cultural hub city, not only in Korea but also in Asia. It also aimed at dispersing the concentrated cultural resources of the metropolitan area of Seoul. The executive budget was about 700 billion won (6 billion USD), so the scale was also considerable. According to the ACC white paper, the candidate sites are Jeon-nam Provincial Office A as current site, and B Song-won Academy, C Gwangju Armed Forces Hospital, D Central Park, and E Seo-chang area on the following figure.

A considerable fact is that most of the candidate sites were 'relocated' areas where it was easy to secure the site in order to realize the condition that the implementation

그림 5. 동쪽에서 바라본 광주 아시아문화의 전당 부지 (2005) / East birdview of the Site of Gwangju Asia Culture Center (2005)

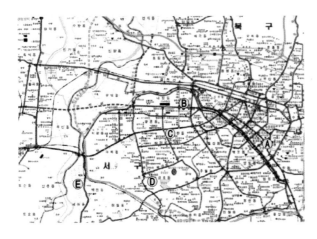

그림 6. 후보지역의 위치도
Candidate site location

는 입지의 상징성, 교통 접근성, 지역경제효과 등을 최우선으로 고려한 결과 구 전남도청부지,
현 대상지를 선택하였다. 그러나 이는 많은 양의 멸실과 보상을 필요로 하는 사업이 되었다.
백서에 따르면 총 368개의 필지에 대한 토지보상과 지장물 4,544개에 대한 보상이 2005년
부터 2008년까지 지속되었다고 기록하고 있다.

국제설계경기와 당선작 선정

이후 본격적으로 사업이 추진되었다. 2005년에 국립아시아 문화전당 국제설계경기가 이
루어졌고, 200여 개의 예선안 중 5개의 수상작이 선정되어, 당선안으로는 5.18 민주화운동
의 주요 건물인 도청사를 포용하고, 과감한 지하공간과 상부의 공원의 활용을 제시한 우규승
건축가의 '빛의 숲(Forest of light)'이 선정되었다. 수상작을 포함한 대부분의 수상안이 5.18
의 주요 건물만을 보존한 반면, 2위였던 이로재건축의 안은 대상지내부의 많은 기존 건축물들을
남기면서 새로운 문화공간의 프로그램을 오버랩시킨 매우 다른 접근법을 보여주기도 하였다.

당선안이 결정되면서 사업은 본 궤도에 올랐으나, 초창기 많은 디자인적 불만 등이 제기되
었던 것으로 보인다. 특히 전당 건립지역인 동구청에서는 빌바오의 구겐하임이나 시드니 오
페라하우스 같은 상징건축물을 원했고, 더욱이 광주광역시장 또한 랜드마크 기능에 대한 부

of the project shall be realized within 5 years. However, the city and the government chose the former Jeon-nam provincial office site as a result of considering the symbolism of the location, transportation accessibility, and regional economic effect as the top priority. However, the site required a large amount of housing loss and subsequent land compensation.

International Design Competition and Selection of Winners

Thereafter, the project was moved forward. An international design competition for the ACC was held in 2005, and 5 winning works were selected out of 200 preliminaries. 'Forest of light' by architect Woo Gyu-seung was selected as the winning proposal. The plan embraced the old provincial government building, the main building of the 5.18 democratization movement, and proposed the bold use of the underground space and the upper park. While most of the award-winning projects, including the award-winners, only preserved the main buildings of 5.18, the second-placed Iroje Architects showed a very different approach that overlapped the program of a new cultural space on many existing buildings inside the site.

Although finalizing the design selection, it seems that many design complaints were raised in the early days. In particular, the Dong-gu Office, where the hall was built, wanted symbolic structures such as the Guggenheim in Bilbao and the Sydney Opera House, and moreover, the mayor of Gwangju also expressed a negative opinion about the landmark function. Since 2007, 'the Degisn Committee' has been held under the government, reflecting the needs of local experts, etc., and after 7 rounds of expert discussions, revisions, changes, and supplements have been made to prepare a basic

그림 7. 당선안(우규승 안), 2위안 (이로재건축 안) / Winner (left) and 2ⁿᵈ prize proposal (right)

정적 견해를 표출하였다. 이후 2007년부터 정부 주제로 갈등조정위를 개최, 지역전문가등의 요구사항을 반영하며, 7차례의 전문가 토론 등을 거처 수정·변경·보완하여 기본설계안을 마련하였다.

비전과 스페이스 프로그램의 설정

먼저 왜 이 사업의 이름이 "아시아 문화의 전당인가?"부터 살펴보아야 한다. 현재 해당 홈페이지에 게시된 소개에는 "세계를 향한 아시아 문화의 창"이라는 이름으로 "5,18 민주화운동의 인권과 평화의 의미를 예술적으로 승화한다는 배경에서 출발한다"는 구절이 있다. 이는 분명히 당시의 정치권은 건립 당시부터 5.18을 분명히 의식했으며, 이를 민주주의의 요람에 예술

design.

Vision & Space Program

First of all, it is worthy to consider the reason of ACC naming. why is this project called "Asian Culture Hall?" Currently posted on the website of ACC, a verse is noted that under the notion of "A window of Asian culture to the world," ACC seeks to sublimate the meaning of human rights and peace of the May 18th Democratization Movement.

It is noted that the politicians at the time were clearly conscious of 5.18 from the start of the project, and they tried to create a dual image by overlapping the image of Gwangju as the city of art and the image of 5.18 as the cradle of democracy. This can be interpreted to realize an image of harmony that projects the future into the history of tragedy, not an approach like a Jewish museum that revives the concept of dark tourism as it is. In addition, the intention of positioning in the world especially targeting Asia is also noted, it is beyond the concept of a domestic cultural center image.

After the goal and vision was set, the issue was "what kind of ACC as cultural complex will we create?" Firstly, it aims to create an ecosystem of creative art or industry where the production and creation of cultural contents, research and sales (performance and exhibition), etc. can be concentrated in ACC. Further, the archival function of the Asian Culture Research Institute, support of the cultural content industry, Asian cultural networking and clustering, and tourism attraction are seen as major program strategies. This was specifically leveled down as an architectural space program. Detail program is tabled as below.

의 도시 광주의 이미지를 오버랩시켜, 문화도시를 만들려고 했음이 읽혀진다. 이는 다크 투어리즘(Dark Tourism) 개념을 그대로 되살리는 유대인박물관 같은 접근이 아닌, 비극의 역사에 미래를 투영하는 화합의 이미지를 구현하고자 했던 것으로 해석할 수 있다. 또한 아시아를 대상으로 한 것은 다분히 국내문화센터 개념을 넘어 세계속으로 포지셔닝(Positioning) 하고자 하는 의도 또한 읽을 수 있다.

개략적인 과업의 목표와 비전이 설정된 이후에 과연 이곳에 '어떤 문화시설을 지향할 것인가?'의 방향설정이 필요하다. 이는 문화콘텐츠의 제작과 창작, 연구와 판매(공연 및 전시) 등이 한곳에 집중해서 일어날 수 있는 창작의 생태계를 지향하였다. 크게는 아시아 문화연구소의 아카이브 기능, 문화콘텐츠산업의 후원, 아시아 문화네트워킹 및 클러스터링, 관광명소화 등이 주요 프로그램 전략인 것이다. 이는 구체적으로 건축적 프로그램으로 레벨 다운(Level Down) 되었다.

일련의 프로그램 및 건축물들은 다음과 같다.

	성격	구체적 프로그램
민주평화교류원	방문자센터 및 기념관	방문자센터, 기념관 1~5관, 문화콘텐츠 창제작, 국제 레지던시 시설
어린이문화원	국내 최대의 어린이 문화시설	콘텐츠 연구개발실, 창작실험실, 유아놀이터, 카페, 어린이극장, 어린이도서관, 다목적홀, 어린이 체험관
문화정보원	전당 내 최대시설로 아시아 문화자원을 연구, 수집함	대강의실, 국제회의실, 아시아 문화연구소, 아시아 문화 아카데미, 극장, 라이브러리 파크, 자료 열람실, 아시아 문화자원센터, 수장고
문화창조원	복합 콘텐츠 기획 개발, 랩기반의 콘텐츠 창제작 공간과 전시실로 구성 복합관, ACT 스튜디오	
예술극장	극장 1, 2 및 야외무대	
아시아 문화광장		

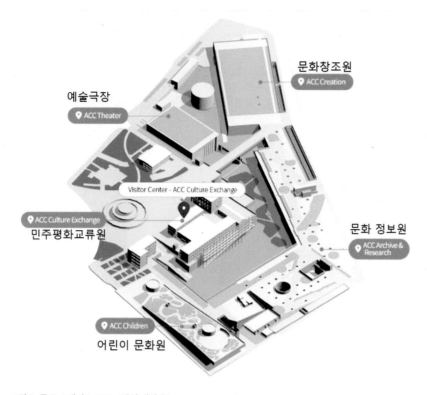

그림 8. 주요 스페이스 프로그램의 배치 / Space program layout

Main Architectural Characters

Physical Form

Centering on the former Jeon-nam Provincial Government Complex, each program group of ACC was placed along the roadside to enclose the entire site. Through this, it is read that it was intended to achieve harmony between the old and the new by drawing a smooth connection with the street and highlighting the old provincial government building as a central position.

주요 건축적 특징

개략적인 형태

구 전남도청사를 중심으로 문화의 전당의 각 프로그램군들이 도로변을 따라 배치되어 대지 전체를 에워싸는 형상을 취했다. 이를 통해 비교적 용이하게 가로와의 연계를 이끌어내고, 도청사를 중심건물로 부각시켜, 신구의 조화를 이루고자 한 것으로 읽혀진다.

주변에서의 접근성

대상지는 가로 300m, 세로 500m가 넘는 대규모 도시 조직(Urban Tissue) 를 가지고 있다. 대규모 도심개발에서 가장 문제가 되는 것이 어떤 시설일지를 막론하고 주변을 향해 얼마나 열려 있는가(Permeability)가 가장 중요한 도시건축의 공공성의 척도이다.

먼저 대지 북서 쪽에서 오는 3개의 큰 흐름이 있다. 이는 충장로로 대표되는 도심상업거리의 축과 교통대로인 금남로, 그리고, 광주예술의 거리로 대표되는 화방가로에서부터 오는 보행자의 축이다. 이는 모두 구 도청사 앞의 5.18 민주광장으로 수렴되며, 시각적 초점자리에 구 도청건물이 있어 강한 시각적 모뉴멘텔리티를 부여한다. 또한 가장 중요하게 다루어진 접근성 측면은 동서를 횡단하는 서석로와 제봉로의 보행축의 연결이다. 이는 518 민주광장을 지나 지하광장을 공중으로 횡단하는 극적인 공간감을 제공하며, 새롭게 조성된 오픈스페이스 사이를 연계하는 주 보행축으로 기능한다.

다양한 오픈스페이스

먼저 이 프로젝트의 가장 큰 특색은 대부분의 건물을 지하로 입지시켜, 중앙광장을 만들고, 건물의 상부는 시민을 위한 오픈스페이스를 만들고자 했던 것이다. 가뜩이나 녹지 및 오픈공간이 부족한 광주 구도심의 현황을 고려한다면 매우 고무적인 접근이 아닐 수 없다. 건물상부를 바로 이용할 수 있는 녹지는 문화생태공원이라 명명하여 지상에서 레벨차 없이 바로 연계

Permeability

The site has a large-scale urban fabric that exceeds 300m in width and 500m in length. Regardless of the urban function and architectural program, the most problematic issue is acquiring 'permeability' as the most important issue of publicity in large urban architecture.

Firstly, main pedestrian flows come from the northwest side of the site. Those are the axis of downtown commercial street, represented by Chungjang-ro, Geumnam-ro, and Hwabang-ro, represented by Gwangju Art Street. All of these converge to the 5.18 Democracy Plaza in front of the old provincial government building as the visual focal point, provides a strong visual monumentality. Also, the most important aspect of accessibility is the connection of the pedestrian axis of Seoseok-ro and Jebong-ro that crosses east and west. It also provides a dramatic sense of space that passes through the 5.18 Democracy Plaza and crosses the underground plaza in the above, and functions as a main axis to link the newly created open spaces.

Diverse Open Spaces

The biggest characteristic of this project is that most of the buildings are located underground to create a central plaza, and the upper part of the building is intended to create an open space for citizens. Considering the current situation of downtown Gwangju, which already lacks green and open space, it is an encouraging approach. The green area as the upper part of the building is named 'Cultural Ecological Park', and it is connected directly from the ground level, and in the underground level, a passage to the underground plaza and a corridor are created, so it provides a variety of spatial changes. In addition, this building has two 'sky parks' (the northern sky garden

되며, 지하광장으로 가는 통로 및 회랑동선을 만들어 다양한 공간적 변화감을 선사한다.

또한 이 건물은 2개의 하늘공원(북쪽 하늘마당과 어린이 문화원 상부)이 있는데, 고답적인 옥상정원이 아닌, 지상에서 자연스럽게 경사로 연계되어 새로운 문화명소로 자리잡고 있다.

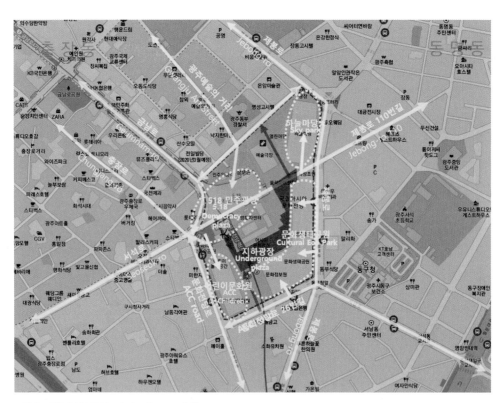

그림 9. 외부공간 분석 / Analysis of accessibility

지하광장의 모뉴멘텔리티

처음 이 건물을 볼 때면 으레 구 도청사의 기억과 장중한 지하광장을 꼽을 것이다. 도심지의 광대한 스케일의 지하광장은 일상에서 경험했던 지상의 소음, 번잡함, 움직임 등을 한번은 걸러진 비일상의 공간을 제공한다. 중앙의 광장을 지하로 내림으로써 도청의 뒷면이 새로운 정면으로 기능하고 있기도 하다. 그러나 역시 지하광장이 가지는 접근성의 한계가 있으며, 이에 따라 광장이 번잡하지 않은 점은 있다.

and the upper part of the Children's Cultural Center), and it is positioned as a new cultural attraction as it is naturally connected with a slope from the ground, rather than an isolated rooftop garden.

Monumentality of the Underground Plaza

In the first look of the ACC, most visitor may think of the memory of the old Provincial Government Building and the extraordinary scale of underground plaza. The vast underground plaza in the downtown provides an unusual space in which the noise, bustle, and busy movement of the daily life on the ground are filtered out. Further, by lowering the central square to the underground, the backside of the old provincial government office gains a new front facade. Nevertheless, it is noted that the limitations of accessibility of the underground plaza are unavoidable. The square is not crowded.

Closing Remarks

ACC was a very difficult task from the establishment of the program and vision setting. In the strong historic layer of unforgettable 5.18 Democratization Movement, the future of Gwangju, a city of art, was intended to sublimate into 'a cultural center that embraces Asia' as a consideration that architecture and society can provide. Therefore, this building is seen as a Korean example of ideological architecture implemented in a top-down manner, rather than a building based on a certain local administrative demand. Fortunately, if most of the ideological top-down architectures have monumentality that shows off, this building ACC does not. It would be a point

그림 10. 지하광장의 전경 / Overview of underground floor

맺으며

광주 아시아 문화의 전당은 프로그램과 비전수립부터 매우 어려운 작업이었다. 5.18 민주화운동이라는 무거운 역사의 레이어에 예향 도시 광주의 미래를 '아시아를 품는 문화의 전당'으로 승화시킨 것은 건축과 사회가 줄 수 있는 도시에 대한 배려라 할 수 있을 것이다. 그렇기에 이 건물은 다분히 어떤 지역적 행정수요에 기반한 건물이 아닌, 탑 다운(Top Down)방식으로 구현된 이데올로기적 건축의 한국적 사례로 보여진다. 다행인 것은 보통의 이데올로기적 탑다운 건축들이 대부분 자기를 과시하는 모뉴멘텔리티를 가지고 있다면 이 건물은 그렇지 않고, 역사를 마주보며, 부족한 도심의 오픈스페이스를 제공하고자 하는 공공성을 추구했다는 것은 평가할 만하다.

시민들의 평가에 대해서는 다양한 의견이 있다. 주로 콘텐츠 관련으로 보다 다양한 이벤트, 전시, 행사, 공연 등이 폭넓게 이루어져야 함을 지적하고 있다. 그럼에도 유동인구가 많은 광주의 도심과 연계되어 문화전당이라는 인지도를 높여가고 있다는 긍정적 측면이 많다. 그런 면에서 아시아 문화의 전당은 광주비엔날레 등의 문화인프라와 함께 양질의 아시아적 문화콘텐츠가 생산되어 전국적 아시아 스케일로 팔릴 수 있는 문화장터가 될 수 있을거라 믿는다.

worth evaluating.

Various opinions about the evaluation of citizens are collected. Mainly, more diverse events, exhibitions, events, performances should be made more widely in relation to contents. Nevertheless, many positive trends are captured that it is increasing the recognition as a cultural center in connection with the downtown area of Gwangju, which draws a large floating population. Moreover, it is noted that the ACC can become a marketplace where high-quality Asian cultural contents can be produced and sold on a national and Asian regional scale along with cultural infrastructure such as the Gwangju Biennale.

참고문헌

1. 아시아 문화중심도시 백서, 문화체육관광부, 2009
2. 2030 광주도시기본계획, 광주광역시, 2019
3. 국립아시아 문화전당 홈페이지 https://www.acc.go.kr/

References

1. Asia Cultural Center White Paper, Ministry of Culture, Sports and Tourism, 2009
2. 2030 Gwangju City Master Plan, Gwangju Metropolitan City, 2019
3. Asia Culture Center website https://www.acc.go.kr/

세종시 어반 아트리움

김영석 [건국대학교 건축학부 교수]

들어가며

새로 건설된 행정중심복합도시는 대한민국의 도시화와 경제 및 산업발전과정에서 야기된 수도권의 과도한 집중을 해소하고 지역 간의 균형있는 발전을 실현하기 위하여 조성되었다. 정

그림 1. 세종시 내 사이트 위치
Site Location in the Sejong Administrative City

Urban Atrium, Sejong City

Youngsuk Kim (Professor, Konkuk University, Dept of Architecture)

Introduction

To resolve excessive concentration in the metropolitan area caused by rapid urbanization and economic and industrial development, the newly constructed Sejong City aims to improve national competitiveness and balance regional development. In particular, District 2-4, where the Urban Atrium was built, is planned to play an important role in representing the value of Sejong city, nicknamed 'Happy City', as a center of commerce and culture.

Through the plan and implementation of the Urban Atrium, I examine the relationship between the site and the city, and compare the master plan of the Urban Atrium with the winning plans of the competition through physical form, function, and programs.

책적 목표는 전략적으로 국가균형발전 선도 및 국민통합의 구심적 역할을 통한 국가경쟁력 제고 및 도시 수준을 향상시켜 미래세대를 위한 지속 가능한 세계적인 모범도시의 구현이었다.

특히 어반 아트리움이 건설된 2-4생활권은 전체 생활권 가운데 상업 및 문화의 중심지로, 도시의 중심인 도시 상징광장 및 중심 상업지역이 위치하여 행복도시의 가치를 이끌어가는 중요한 역할을 하도록 계획되었다. 이 글에서는 어반 아트리움의 계획 및 실행을 통해 대상지와 도시와의 관계를 고찰하고, 어반 아트리움의 마스터플랜과 공모전을 통해 당선된 안들을 비교하여 대상지의 물리적인 형태와 기능, 프로그램을 정리하고, 우리나라 신도심계획의 사례로 정리하고자 한다.

사업 개요 및 배경

지리적 위치상, 대전, 청주, 천안, 공주 등 대도시와 인접하고 있는 세종특별자치시는 행복도시 건설지역과 기존 연기군의 읍면지역 등이 포함되어 형성된 도·농 통합형 도시로서, 중앙행정기관 이전에 따른 건설지역이 계획되었다. 행복도시 건설지역은 총 6개의 생활권으로 구성하여 도시계획을 진행하였다.

행복도시는 우수한 도시개념을 찾기 위한 국제공모로 진행되었다. 공동 당선된 5개의 안들 중 스위스의 뒤리그와 스페인의 페레아의 두 안을 중심으로 발전

그림 2. 세종시 도시개념 국제공모에 제출한 오르테가(위)와 뒤리그(아래)의 안 / International Urban Ideas Competition for the Multi-functional Administrative City, The City of the Thousand Cities by Andres Perea Ortega (above), An Orbital Road by Jean- Pierre Dürig (below)

Project Overview

Sejong City, which is adjacent to large cities such as Daejeon, Cheongju, Cheonan, and Gongju, has an integrated form of urban and rural fabric within the city. It is composed of six districts based on the phases of construction of the city. The city was planned to have an overall circular shape based on the international urban idea competition winners, Dürig and Perea's scheme with the main public transportation of Bus Rapid Transit (BRT) looping around the central open space.

Among these districts, the project site of the Urban Atrium is divided into areas P1 to P5, and has a series of lots stretching from north to south. It was important to establish a plan to link it with central commercial, business, and residential complexes, as well as central administrative agencies, the city square, and adjacent lots for future department stores and other interactive functions. The Urban Atrium is adjacent to the Event Road along the north—south axis. Directly crossing the site, along the east—west axis, is the scenery road and pedestrian road. The City Square connects the nearby residential district to the central open space (which includes Sejong Lake Park, Jungang Park, and the National Arboretum). BRT stops, branch bus stops, and taxi stops are located in the area containing the UEC and department stores next to the Urban Atrium. The atrium juxtaposes the open spaces to the east and west directions, and also connects the government complex and Sejong City Hall to the north and south.

시켜 중앙녹지공간을 비워두고 순환형 개발축을 따라 환상형 도시로 계획되었으며, 중심부에 BRT(Bus Rapid Transit)가 순환하도록 하여 도시교통체계를 대중교통 중심으로 구성하는 새로운 도시로 계획하였다.

이 중 어반 아트리움은 행복도시의 중심인 2-4생활권(중심상업업무기능) 중심을 관통하면서 도시 활동의 중심기능을 수행하는 입지적 특징을 가지며 중앙행정기관, 도시상징광장, 미래 백화점 부지 등을 비롯해서 중심상업, 업무, 주상복합 등 다양한 기능이 밀접해 있어서 이를 연계하는 계획을 수립하는 것이 가장 중요한 핵심과제이다. 어반 아트리움의 대상지는 P1~P5 구역으로 나누어져 남북으로 긴 형태의 대지모양을 가지고 있다. 또한 대상지는 남북 축을 따라 서쪽으로 도시의 이벤트가로가 접하여 있으며, 어반 아트리움을 가로지르는 동서 축의 경관 및 보행도로가 계획되어 있다. 도시 상징광장은 어반 아트리움을 중심으로 2생활권(주거지역)과 중앙 녹지공간(세종호수공원, 중앙공원, 국립수목원)을 연결해주는 공간이다. 어반 아트리움에 인접한 UEC·백화점 부지에는 BRT 정류장 및 지선버스정류장과 택시정류장이 위치하여 있다. 전제척으로 도시 중앙의 오픈스페이스로 연결되는 병치적인 위치이면서 남북으로는 정부청사와 세종시청을 잇는 연결점으로 작용한다.

어반 아트리움 마스터플랜

기본 컨셉

행복도시 2-4생활권의 중심이자 상업업무활동의 중심인 도시문화상업가로(UA)는 'Urban Atrium, 도시의 문화를 즐기기 위한 Cultural Landmark'라는 주제로 2-4생활권을 남북방향으로 가로질러 지상에 스트리트몰, 3층 레벨을 가로지르는 어반 프로미나드, 편안한 환경을 제공하는 캐노피 아래 옥상 정원을 이용하는 어반 클라우드 등의 주요 컨셉을 가지고 있으며 여기에 복합문화시설을 더해 1.4km 가량의 긴 띠모양으로 들어서는 복합상업문화거

Urban Atrium Master Plan

The Basic Concept

Urban Culture Commercial Street (UA), the center of the Happy City 2-4 living area and the center of commercial activities, has a complex concept, such as the Urban Atrium, a street mall across the 2-4 living area in the north—south direction, Urban Promenade across the third floor level, and Urban Cloud using a comfortable environment. The Urban Atrium's three goals are based on the concept of a hub for culture and commercial activity that links the surrounding space and functions of urban culture, walking streets where urban culture wants to live, and eco-friendly trails that draw emotions. The Urban Atrium was developed in conjunction with the urban symbol of square in the shape of a cross, creating the concept of the most energetic three-dimensional space in the city. In particular, to realize the central space of the carefully planned city, we introduced the Master Architect (MA) system, and tried to implement a three-dimensional city with detailed MA guidelines, rather than simply implementing a flat city with a district-level plan.

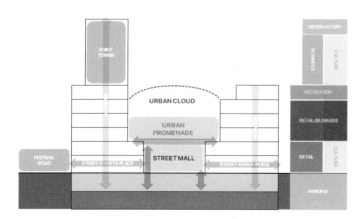

그림 3. 어반 아트리움 단면 컨셉 / Urban Atrium Concept in Section

리이다. 어반 아트리움의 3대 목표는 도시 문화가 살아 숨쉬는 걷고 싶은 거리, 감성을 끌어내는 도심 속 친환경 생태 산책로, 주변 공간과 기능을 연계하는 문화와 상업업무의 허브개념을 주축으로 한다. 어반 아트리움은 도시 상징광장과 십자가모양으로 연계 개발되어 도시 내에서 가장 활력이 넘치는 입체적 공간을 컨셉으로 조성되었다. 특히 세심하게 계획된 도시의 중심 공간을 실현하기 위해서 MA(Master Architect)제도를 도입하고 단순히 지구단위계획으로 평면적인 도시를 구현하기 보다는 자세한 MA 지침으로 입체적인 도시를 구현하기 위해 노력하였다.

어반 아트리움 마스터플랜의 기본틀

어반 아트리움의 마스터플랜은 다음과 같은 기본틀을 바탕으로 만들어졌다.

• 인접부지와의 연계

대중교통중심도로, 도시상징광장, 백화점/UEC, 중앙정부기관, 문화시설 등 다양한 주변 기능을 적절히 고려해 상호연계가 가능하도록 계획하는 것을 주축으로 도시 골격과 주변생활권의 공간축을 반영한 공간구상을 목표로 했다. 블록의 내외부가 가로를

그림 4. 어반 아트리움 마스터 플랜
Urban Atrium Master Plan

중심으로 커뮤니티 활동이 촉진되도록 새로운 개념의 가로대응형 구조로 계획하고 어반 아트리움과 도시 상징가로의 이벤트 등을 연계하여 주변 상권 및 보행 활성화를 유도하기 위해 공공시설물, 공공공간 등이 차별화된 축제가로로 계획했다.

• 중심보행축 활성화

어반 아트리움을 관통하는 공공보행통로를 통해 상업업무가로를 연속적으로 배치하고,

The Basic Framework of the Urban Atrium Master Plan

The master plan of the Urban Atrium was made based on the following basic framework.

• Linkage with neighboring sites

The goal was to plan a space that reflects the spatial axis of urban framework and surrounding living areas, focusing on planning to enable interconnection by properly considering various surrounding functions, such as public transportation-centered roads, urban symbol squares, department stores/UEC, central government agencies, and cultural facilities. Public facilities and public spaces were planned as differentiated festivals to revitalize the surrounding commercial districts and encourage walking by linking events on Urban Atrium and Urban Symbol Street to promote community activities inside and outside of the block.

• Activating the central walking axis

Commercial streets were continuously arranged through public walking passages through the Urban Atrium, and attractive walking axes were created through block-specific themes and stories to reinforce the fun and purpose of walking. In addition, for walking safety, convenience, and continuity, a design element of a public walkway that maximizes access and linkage between adjacent blocks, and a systematic movement plan with a plaza in the Urban Atrium, are considered necessary. In addition, an Urban Prominade, which penetrates the Urban Atrium at the third-floor level, is planned to strengthen the walking axis between the two Koreas.

보행의 재미와 목적성을 확보하기 위해 블록별 테마와 스토리를 통한 매력적인 보행축을 조성했다. 또한 보행의 안전과, 편의 및 연속을 위해 인접 블록간의 접근과 연계를 극대화하는 공공보행통로의 디자인적 요소와 어반 아트리움 내 광장과의 체계적인 동선계획이 필요하다고 생각된다. 이와 함께 3층 레벨에서 어반 아트리움을 관통하는 어반프로미나드가 계획되어 남북 보행축을 강화한다.

• 공간의 차별성과 경관의 조화로움

어반 아트리움의 외부공간과 주변의 계획적 연계성을 갖는 공간으로 다른 곳과 차별화된 매력적이고 활력있는 걷고 싶은 거리가 되어야 한다. 특히 인접한 수변공간은 특화기능을 도입하여 도시의 휴식공간과 문화시설이 상업·업무기능과 적절히 연계되어 다양한 매력을 제공하는 장소로 계획한다. 어반 아트리움은 바깥에서 봤을 때 행복도시의 디자인 언어를 잘 따르면서도 내부적으로 전혀 다른 공간적 분위기를 연출하는 데 주안점을 두었다. 즉, 도시 기본경관과의 통일감을 주면서 어반 아트리움만의 차별화된 경관을 만들어야 하였다. 축제가로변을 비롯해 주상복합용지, 주변 상업업무용지와 면한 곳 등 대상지 외곽은 1~2층의 저층부와 3층 이상의 중층부로 구분하여 기존 행복도시 경관지침을 반영해야 한다. 어반 아트리움 저층부, 중층부와 디자인 특화를 위해 도입된 포인트 타워를 적극적으로 활용해 수직적 경관을 조성하는 등의 입체적 계획으로 조화되면서도 차별화되는 특별한 경관을 만들고자 하였다.

어반 아트리움 마스터플랜의 블록별 테마

미래에 계획되는 어반 아트리움의 전체적 테마는 다음과 같으며 실제적으로 건설되는 것은 Urban Activator, Urban Crossing, Urban Kitchen의 테마를 갖는 세 블록이 대상지로 선정되었다.

• Differentiation of space and harmony of scenery

This is a space that has a planned connection with the outer space of the Urban Atrium, and should be an attractive and energetic walking concourse that is differentiated from other places. In particular, the adjacent waterside space is planned to be a place where urban rest areas and cultural facilities are properly linked to commercial and business functions to provide various charms by introducing specialized functions. The Urban Atrium focuses on creating a completely different spatial atmosphere internally, while following the design language of the Happy City externally. In other words, it was necessary to create a differentiated landscape that is unique to the Urban Atrium while giving a sense of unity with the basic urban landscape. The outskirts of the target site, including the festival roadside, residential and commercial complex, and the surrounding commercial business sites, should be divided into low-rise parts on the first and second floors, and reflect the existing Happy City landscape guidelines. It was intended to create a special landscape that is harmonized with three-dimensional plans, such as creating

그림 5. 어반아트리움 P1-P3 블록 전경 / View of Urban Atrium P1-P3 Blocks

- **자연환경 연계형 중심의 GREEN SQUARE**

  어번아트리움의 시작점으로서의 의미를 가지고 있으며, 자연에서 도시로 들어오는 관문
  이자 아웃도어로의 출발점의 개념을 가지고자 하였다.

- **대형 문화광장을 중심으로 하는 URBAN ACTIVATOR**

  세종시의 문화 중심점으로서 대형 백화점과 도시 상징광장을 연결시켜주는 역할을 하며
  다양한 전시가 중심이 되는 정적인 문화공간들이 1~2층 거리와 3~4층 어반프로미나드,
  5~6층 어반클라우드에 위치하게 하였다.

- **도시 상징광장을 가로지르는 URBAN CROSSING**

  도시 상징광장과 더불어 문화의 중심이 되는 어번크로싱과 도서관과 같은 문화중심공간
  이 어반프로미나드에 위치하게 하였다. 미디어를 중심으로 하는 문화공간을 조성하여 예
  술 지향 미디어파사드가 단순하지만 강렬한 문화 메시지를 전달할 수 있는 공간이 되고
  자 하였다.

- **상업업무지구를 관통하는 먹거리의 중심 URBAN KITCHEN**

  다양하면서도 식당들이 모여 있으며 특히 남쪽의 재래시장인 South Market이 위치하
  여 24시간 사람들이 활기차게 이용하는 공간으로 계획하였다.

- **수변환경을 최대한 즐길 수 있는 탁 트인 전망을 가진 광장과 건물의 조합 BLUE
  SQUARE**

  금강을 바라보는 친수공간으로서 어번아트리움의 여정을 마무리하는 공간으로 조성하
  였다.

마스터플랜 지침과 당선안들의 비교 및 조정

- **공공대응형 입면 : 외부입면은 주변 가로경관의 연속성 유지**

  외부입면은 포인트타워를 제외한 입면에 대해서 2층 이하의 저층부와 3층 이상의 중층

a vertical landscape by actively utilizing the point tower introduced for design specialization with the Urban Atrium.

The Urban Atrium Master Plan's Theme by Block

The overall theme of the Urban Atrium, which is planned in the future, is as follows, and three blocks with the themes of Urban Activator, Urban Crossing, and Urban Kitchen have been selected as targets.

• **Green square centered on natural environment linkage**

This has the meaning of the starting point of the Urban Atrium, and it is intended to have the concept of a gateway from nature to the city, and a starting point to the outdoors.

• **Urban activator centered on a large cultural square**

As a cultural center of Sejong City, this plays a role in connecting large department stores and urban symbol squares, and static cultural spaces centered on various exhibitions are located on streets on the 1st and 2nd floors, Urban Promenade on the 3rd and 4th floors, and Urban Cloud on the 5th and 6th floors.

• **Urban crossing across the urban symbolic square**

Along with the urban symbolic square, cultural centers, such as the Urban Crossing and libraries, which are the center of culture, are located in the Urban Prominade. By creating a cultural space centered on media, it is intended to become a space where art-oriented media facades can convey simple but intense cultural messages.

• **Urban kitchen, the center of food through the commercial business district**

There are various cafés and restaurants gathered, and in particular, the South

부를 나누어서 2단 구성을 준수하여 통일성 입면을 가지고자 하였다. 또한 주변 가로경관을 고려해서 가로변 입면이 단조롭게 연속되는 경우에는 적절한 간격으로 분절하여 입면 디자인을 반영하며, 저층부를 제외한 입면에 대해 일정부분 세로창을 설치하여 다조로운 입면계획을 탈피하고자 하였다.

그림 6. 공공대응형 입면
Specialized Street Elevation Design Realization

- 스트리트몰(Street Mall) : 누구에게나 24시간 개방되고 열려 있는 스트리트몰로 조성

공공보행통로는 24시간 개방해야 하며, 곡선형으로 최소 10m 이상의 폭으로 계획하게 하였다. 공공 보행통로를 제외한 추가동선은 다양한 형태로 조성 가능하며, 폭이 6m 이상인 보조통로를 추가 설치하고자 했으며 공공 보행통로와 다른 도선이 만드는 곳에 소규모 광장을 설치하여 걷고싶은 보행환경을 조성하고자 하였다. 진입부는 폭 20m 내외의 개방형으로 설계하며 어떠한 형태의 차단시설도 허용하지 않아 시선의 차단이 이루어지지 않고자 하였다. P1블록의

그림 7. P1블록의 스트리트몰 전경 / Street Mall of P1 Block

광장과 스트리트몰의 설계에서 지침이 잘 실현되었다.

Market, a traditional market in the south, is located there, so it is planned as a space that people use actively 24 hours a day.

- Blue square, a combination of squares and buildings with an open view where the water conversion view can be enjoyed as much as possible

As a water-friendly space overlooking the Geumgang River, it was created as a space to finish the journey of the Urban Atrium.

Comparison of Master Plan Guidelines and Winning Proposals

- Public-responsive elevation: maintaining the continuity of the surrounding street scenery

The outer elevation was intended to present a unified elevation by dividing the lower floors, ground floor and 2nd floors, and the 3rd and higher floors, excluding the point tower. In addition, in consideration of the surrounding street landscape, if the roadside elevation is monotonously continuous, such as P4 block, the elevation design should reflect the vertical articulation by subdivision at appropriate intervals, and a certain vertical window situated in the elevation to achieve a multifarious and vivid atmosphere.

- Street mall: building a street mall that is open to everyone 24 hours a day

The public walkway must be open 24 hours a day, and a width of at least 10 m is planned in a curved shape. Except for public pedestrian passages, additional routes

• 어반 프로미나드(Urban Promenade) : 중층부 연계를 통해 활용도 제고

어반 프로미나드는 전체 구간에서 폭 6m 이상 유지하여 연속적으로 조성되어야 하며 주로 3층에 조성하되 3~4층을 활용하여 계획하는 것을 가능하게 하여 문화시설과 연계되도록 조성하고자 하였다. 어반 프로미나드에서 아케이드, 필로티 등의 형태로 디자인하였고, 상부와 하부로 조망이 가능하고 연결브리지 설치를 통하여 소통 가능한 공간으로 연결성을 강조하고자 하였다.

그림 8. P2 에서 P3 블록으로 넘어가는 어반 프로미나드
Urban Promenade Connecting P2 to P3 Block

• 어반 클라우드(Urban Cloud) : 옥상정원을 활용한 녹지 확충 및 휴식과 힐링의 장소 창출

5층 상부에 옥상정원과 산책로를 조성하고, 6층은 옥상정원과 공중보행로의 활성화를 위해 적극적인 기능 도입을 통해 특색있는 공중거리로 조성하여 상부층의 산책로를 만들고자 하였다. 휴식과 힐링을 위한 카페거리, 컨벤션 기능과 결합한 옥외문화공간을 비롯한 다양한 휴식공간을 조성하여 산책로의 공간을 강조하면서도 상부층을 사람들이 이용하면서 상업공간으로의 기능이 상실되지 않도록 하였다.

그림 9. P3블록의 어반클라우드 / Urban Cloud at P3 Block

can be constructed in various forms, and auxiliary passages with a width of 6 m or more were intended to be installed, and a small plaza was installed where other routes meet to create a walking environment. The entrance is designed as an open type with a width of about 20 m, and no type of blocking facility is allowed, so the gaze is not blocked. The guidelines were well realized in the design of the square and street mall of P1 block.

• Urban promenade: improving the utilization of middle floors through Interlayer linkage

The Urban Prominade should be continuously created by maintaining a width of 6 m or more for the entire section, and it was intended to be built mainly on the third floor to be linked to cultural facilities, by enabling planning using the third to fourth floors. The Urban Prominade is designed in the form of arcades and piloti, and attempted to emphasize connectivity as a space that can be viewed from the top and bottom, and communicated through the installation of a connecting bridge.

• Urban cloud: expanding green areas using rooftop gardens and creating places for relaxation and healing.

The rooftop garden and promenade were built on the upper part of the fifth floor, and the sixth floor was intended to create a promenade on the upper floor by creating a distinctive public street through the introduction of active functions to revitalize the rooftop garden and public walkway. Various rest areas, including cafe streets for relaxation and healing, and outdoor cultural spaces combined with convention functions, were created to emphasize the space of the promenade, while preventing inhabitants from losing its perception as a function of commercial space.

• 포인트 타워 : 어반 아트리움의 경관형성 및 도시 전망장소 조성

도시의 랜드마크 형성을 위한 포인트 타워 디자인 특화계획을 제시하였으며, 공공개방이 가능한 포인트 타워 최상부에는 전망공간과 부대시설을 설치하여 공간의 흐름이 수직적으로 연결되고자 하였다 5~6층과 포인트 타워 하부의 공간이 상호 결합되면서 형성되는 다양한 공간을 적극 활용하여 어반 클라우드의 활용성을 극대화하고자 하였다.

그림 10. P2블록의 포인트타워 / Point Tower of P2 Block

• 친환경 공간 : 어반 아트리움에 걸맞는 다양한 친환경 공간 창출

동선이 교차/연계되는 곳에는 광장을 계획하여 이용자와 시민들의 휴식 공간을 제공하고, 옥상조경은 형식적 조경이 아닌 친환경 생태적인 옥상조경과 산책로로 조성하여 이질적인 공간이 아니고자 하였다. 벽면조경을 비롯해서 다양한 신기술을 도입하여 친환경 도시환경 도시 세종시의 디자인 계획을 반영하고자 하였으며, 층별로 구성되는 다양한 기능과 건축형태를 고려해서 친환경 공간계획을 수립하고자 하였다.

그림 11. P1과 P2블록 사이의 광장
Street Plaza between P1 and P2 Block

• Point tower: developing urban view point and cityscape of the UA

A point tower design specialization plan for the formation of urban landmarks was proposed, and an observation space and auxiliary facilities were installed at the top of the point tower where public opening is possible to connect the flow of the space vertically. It was intended to maximize the usability of the Urban Cloud by actively utilizing various spaces formed by combining the 5th to 6th floors and the lower space of the point tower.

• Eco-friendly space: creating a variety of eco-friendly spaces suitable for the UA

The square was planned to provide a resting place for users and citizens where the movement lines intersected/connected, and the rooftop landscape was created as an eco-friendly rooftop landscape and promenade rather than a formal landscape, so it was not a heterogeneous space. Various new technologies, including wall landscaping, were introduced to reflect the design plan of Sejong City, an eco-friendly urban environment city, and to establish an eco-friendly space plan in consideration of various functions and architectural forms composed of each floor. The building solar facility plan was not simply listed on the rooftop, but was intended to be creative and reasonable, such as integrating with the building through various searches in terms of form and technology, using walls (BIPV), and in combination with landscape facilities.

• Design-specialized streets: preparing an appropriate improvement plan for outdoor advertisements and public facilities

Throughout the Urban Atrium, it was intended to establish plans and install distinctive public facilities according to the designation of streets as specialized

건축물 태양광설비계획은 단순히 옥상에 나열하는 것이 아니라 형태나 기술적인 측면에서 다양한 모색을 통해 건물과 일체화하거나 벽면을 이용(BIPV), 조경시설과 결합하는 등 창의적이면서도 합리적인 계획을 하고자 하였다.

- **디자인 특화거리 : 옥외광고물/공공시설물 특화거리에 적절한 수준의 개선방안 마련**

어반 아트리움 전체에서 옥외광고물 특화거리 지정에 따른 계획수립 및 특색있는 공공시설물 설치, 공공예술 미디어파사드 설치 및 옥외광고물, 공공시설물 등 공공 디자인과 관련해서 현대기술과 공간의 콜라보를 보이고자 하였다. 또한 종합적인 디자인 협정을 통해 관리하는 방안을 제시하고자 하였다.

그림 12. P5 블록의 공공예술작품
Public Art in front of P5 Block

당선작 선정 및 조정과정

당선작은 건축설계와 재무로 이루어진 심사위원단에 사업계획, 설계 및 가격에 대한 평가로 선정되었는데, 특히 자세하게 주어진 실현지침에 얼마나 충실하였는가가 평가에 반영되었고, 많은 공모안에서 창의적인 해석을 보여주었다. 그 결과 P1~P5까지 앞서 제시된 그림과 같은 당선작들이 선정되었다.

이후 당선작의 조정은 두 가지 체계로 이루어졌다. 행복도시의 총괄자문회의는 공공 건축물에 대해서는 필수적으로 자문회의를 거쳐 조정하도록 되어 있었다. 각 전문위원들의 자문 결과 공통적으로 당선안을 최대한 유지하여 설계품질을 향상시키는 방향으로 진행되는 것으

in outdoor advertisements, and install public art media facades, outdoor advertisements, and public facilities. In addition, it was intended to suggest a plan for management through a comprehensive design agreement.

The Process of Selecting and Coordinating the Winning Scheme

The winning work was selected after evaluation of both the business and architectural plans by a panel of jurors, and in particular, of how faithfully the detailed guidelines were reflected in the scheme. Many of the entries showed creative solutions. Since then, the winning work has been coordinated in two systems. The General Advisory Council for Happy Cities was required to coordinate public buildings through advisory meetings. As a result of each expert committee member's advice, the design quality was improved by maintaining the winning plan as much as possible, and the design was developed in the direction of unifying the spatial atmosphere in all blocks through material and color. The second was to form an integrated consultative body with the Master Architects (MA) and the committee members of the competition, and derive an integrated proposal through five advisory meetings. Through this integrated consultative body, more specific matters were discussed, and advice and design plans were updated on the actual development of the design plan and various problems to be considered in the implementation design. As a result, the name of Urban Atrium was changed to the newly coined Urban 'Artrium', and began to be built with the goal of becoming a new center of culture and art.

로 되었으며, 재료 및 색채 등 전 블록에 공간적 분위기를 통일하는 방향으로 디자인이 발전되었다. 두 번째는 공모진행위원인 MA(Master Architect)들과 통합협의체를 구성하여 5차에 걸친 자문회의를 통해 통합된 안을 도출하였다. 이 통합협의체를 통해서 좀더 구체적인 사항들에 대한 토의가 이루어졌으며, 실질적인 설계안의 발전과 실시설계 시 고려해야 할 여러 문제점에 대해 자문과 설계안의 업데이트가 이루어졌다. 결과적으로 어반 아트리움은 Atrium이 아닌 Artrium이란 신조어로 바뀌면서 새로운 문화와 예술의 중심지를 목표로 건설되었다.

맺으며

2021년 10월까지 P4블록을 제외한 모든 대상지에 건축물이 완공되었다. 하지만 아직도 세종시 전반적으로 공실률이 높은 문제를 안고 있으며, 어반 아트리움도 예외는 아니다. 최근 어반 아트리움 인근 주상복합단지 입주가 시작되면서 세간의 주목을 받았으나, 여전히 상당수가 공실로 남아있는 상태이다. 특히 P1을 제외한 P2~5 건물들은 대부분 제기능을 하지 못하고 있는 실정이다. 주로 대형 식음료업장이나 영화관 등의 업종이므로 완공후 지난 2년간 코로나19에 따른 집합금지령과 방역지침에 영향을 받았을 거란 분석이 지배적이다.

그림 13. 완공된 P3 블록 / Completed P3 Block

어반 아트리움을 조성하면서 마스터플랜의 작성 및 공모 과정 전체를 아우를 수 있는 MA(Master Architect) 제도를 도입하고, 국제공모와 구체적이고 세부적인 MA 지침들을

Closing Remarks

By October 2021, all the buildings were completed, except for P4 block. However, Sejong City still has a high vacancy rate overall, and the UA is no exception. Recently, while the residential and commercial complexes near the UA have attracted public attention, many of them also remain vacant. In particular, important features of the UA are cultural and commercial functions, such as market, restaurant, and movie theater, which have been affected by the collective ban and quarantine guidelines under COVID-19 over the past two years.

The introduction of the Master Architect (MA) program, which can encompass the entire process of master planning, the competition, and winner coordination, is considered a better way to produce high-quality winning works and their realization, as well as detailed MA guidelines and specific requirements for the competitions. However, that the winning work was not completed within the given time is considered a problematic issue of this project. It is also considered to be an area to be improved that after the MA coordination process, there is no means to enforce the winning design and material being implemented without substantial change.

If the 'Living with Covid-19' era truly begins, we can expect to revitalize the Urban Atrium commercial district. If the city square crossing UA is actively promoted, with additional green features and seating areas, the Urban Atrium street mall and urban promenade will be greatly affected. Additionally, if the P4 block construction resumes, Urban Prominade, which is currently cut off, will be reconnected, and will help revitalize the shopping malls in P4 and P5 blocks by restoring the connectivity from the square.

포함한 자세한 공모지침을 마련한 점은 수준 높은 당선작을 뽑을 수 있었던 좋은 방식이라 생각된다. 다만 당선작들이 여러 가지 상황으로 주어진 시간 내에 완공되지 못한 것은 이 사업의 문제점으로 생각된다. 여기에 더하여 당선안 대로 설계되지 않거나 재료 및 색채에서 변경이 된 경우 추후 강제할 수 있는 수단이 존재하지 못하는 점 또한 개선해야 할 부분이라고 생각된다.

그림 14. P4 블록 공사현장
P4 Block Construction Site

위드코로나 시대가 본격적으로 시작된다면 어반 아트리움 상권 활성화에 대한 기대를 할 수 있을 것이다. 어반 아트리움의 허리를 관통하는 도시 상징광장이 적극적으로 홍보되고, 녹지와 휴식공간이 제공되어 도시상징광장이 활성화 된다면 자연스럽게 어반 아트리움 거리 및 상가도 큰 영향을 받을 것이라 생각된다. 또한 아직 완공되지 않은 P4블록이 완성된다면 현재 끊어져 있는 어반프로미나드가 광장으로부터 연결성을 회복해 주어 P4, P5 블록의 상가 활성화에 큰 도움이 될 것이라 예측된다.

참고문헌

1. '2030 세종도시기본계획', 세종특별자치시, 2014

2. '행정중심복합도시 2-4생활권 도시문화상업가로 사업제안공모 관리 최종보고서', LH 한국토지주택공사, 2015

3. '행정중심복합도시 2-4생활권(중심상업지구)특별계획구역2내 도시문화상업가로(어반 아트리움) 사업제안공모, LH 한국토지주택공사, 2015

4. '행정중심복합도시 2-4생활권(중심상업지구) 특별계획구역2내 도시문화상업가로(어반 아트리움) 사업제안공모 지침서', LH 한국토지주택공사, 2015

5. 조판기, 정원기, '행정중심복합도시 건설과정과 완성과제', 국토연구원, 2021

6. 육심무, '행복도시 도시상징광장 설계공모', 충청포스트, 2015

7. http://m.ccpost.kr/news/articleView.html?idxno=3291

References

1. 2030 Sejong City Comprehensive Plan , Sejong Special Self-Governing City, 2014

2. Final report for the multifunctional administrative city 2-4 district (Urban Atrium), Korea Land and Housing Corporation, 2015

3. Competition Guideline for the multifunctional administrative city 2-4 district (Urban Atrium), Korea Land and Housing Corporation, 2015

4. International design competition for the multifunctional administrative city, 2-4 district (Urban Atrium), Korea Land and Housing Corporation, 2015

5. Cho, Panki, Chung, Won-Gi, The process and issues of multifunctional administrative city, Korea Research Institute for Human Settlements (KRIHS), 2021

6. Yuk, Sim-Mu, Happy city urban symbolic plaza design competition, Chungchung Post, 2015

서울 창동·상계 문화산업단지

장항준 (전 시아플랜건축 디자인 총괄, 현 MDA건축사사무소 대표)
구자훈 (한양대학교 도시대학원 교수)

들어가며

창동·상계지역은 '2030서울플랜'의 동북부 광역중심지로서, 그 위상에 맞는 경제와 문화의 중심지로 도약하기 위해 '도시재생 활성화 지역'으로 지정 되었다. 이 사업은 도시재생 선도사업의 일환으로서 창동·상계지역에 창업육성 및 문화기능을 조성하기 위해, '동북권 창업센터'에서 육성한 창업기업을 지원하고, '서울아레나'와 연계한 '창동·상계 창업 및 문화산업단지'를 조성하는 것이 목적이다. 창동·상계 창업 및 문화산업단지 조성사업은 서울특별시 도봉구 창동 1-9일원에 조성되고 있으며, 대지면적 10,746m², 건축면적 5,697.46m², 연면적 156,270m², 지하 7층, 지상 45층의 프로젝트이다. 주용도는 문화집회 및 관람시설, 창업창작 레지던스(오피스텔) 및 공영주차장으로 서울시와 SH공사가 발주처이다. 국제설계공모를 통해서 시아플랜건축/오즈건축/한아도시 컨소시움의 제안이 당선되었으며 기술제안 실시설계는 DL이앤씨 컨소시움에 의해 진행되어 2023년 5월 준공 목표로 조성중이다.

Changdong-Sanggye Start-up and Culture Industry Complex, Seoul

Hangjoon Jang (Former Design General Manager, SIAPLAN Architecture
/CEO, MDA Architecture Office)
Jahoon Koo (Professor, Hanyang University Graduate School of Urbanism)

The success of urban regeneration projects that revitalize declined areas lies in attracting new urban functions and creating jobs. The Changdong Sanggye area is located in the center of the four districts at the end of northeastern Seoul and has a population of 3.2 million (1.7 million in Northeast 4 districts and 1.5 million in neighboring Gyeonggi Province), but is a typical residential town with no jobs created for large-scale housing supply in the 1980s. Since 2014, the Seoul Metropolitan Government has been pushing for an urban regeneration project with the aim of 'creating a new economic center for Changdong Sanggye.' First, the goal of this project is to build a culture and arts base by attracting 20,000 seats and expand living culture infrastructure.

Among them, the 'Seoul Arena' project, a commercial cultural complex including 20,000-seat performance halls, 1,500-seat and 500-seat small and medium-sized performance halls, which will serve as anchors for the cultural industry, is undergoing detailed design as a private proposal project. In order to create jobs by fostering specialized industries such as knowledge-type R&D industries and cultural industries, Changdong Aurene, which has start-up and employment support spaces for middle-

프로젝트 개념과 디자인 제안

Continuous Open Space

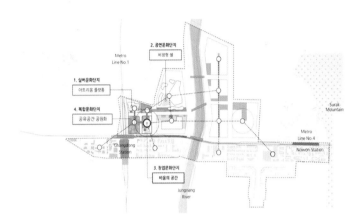

그림 1. Continuous Open Space

창동·상계 창업 및 문화산업단지 조성사업의 핵심은 '문화'이다. 따라서 사업구역 내 문화콘텐츠를 분류하여 문화공간을 다핵화하고 그 각각을 잇는 연속적인 오픈공간을 계획하여 단계적으로 도시를 연결한다. 각각의 문화 코어는 키워드를 가지는 건축공간이 되고 선큰과 오픈스페이스, 공원, 브리지 등의 입체적 연계방안을 통해 향후 도시와 건축의 변화에 유기적으로 대응할 수 있는 공간을 구축하고자 한다

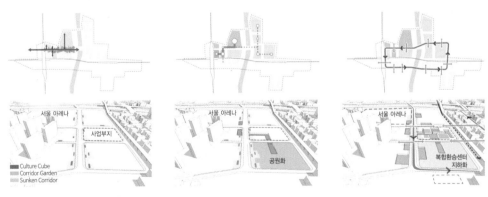

그림 2. ConversionPlatform

aged and senior generations, completed construction at the end of 2020, and a large-scale start-up cultural complex project is underway.

Large-scale start-ups.The Seed Cube Chang-dong project, a project to create offset start-ups and cultural industrial complexes, is being created in Chang-dong 1-9 area, Dobong-gu, 10,746m², 5,697.46m², total area 156,270m², 7 basement floors, and 45 ground floors, and various start-up support spaces and offices. Through an international design contest, the proposal of the SIAPlAN Architects /Oz Architects / Hana City Consortium was elected, and the technical proposal implementation design was carried out by the DL E&C Consortium and is being created with the goal of completion in May 2023.

Project Concept and Design Proposal

Continuous Open Space

The core of this project is 'culture'. Therefore, cultural contents in the business area are classified to multi-centralize the cultural space, and a continuous open space connecting each of them is planned to connect the cities step by step. Each cultural core becomes an architectural space with keywords, and through three-dimensional linkage plans such as sunken, open space, parks, and bridges, we intend to build a space that can organically respond to changes in cities and architecture in the future.

그림 3. Conversion Platform

VERTICAL OPEN SPACE

지속 가능하고 창의적인 환경의 중요한 요소는 자연친화적인 오픈스페이스 공간감에 있다. 부지 내 최대의 오픈스페이스를 확보하기 위해 수평적 공간의 한계에서 벗어나 수직적으로 확장된 오픈스페이스는 다면적으로 확장, 연계된다.

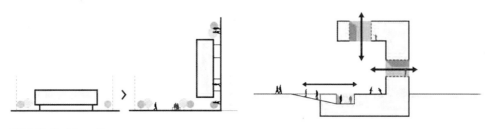

그림 4. Vertical Open Space

INTERACTIVE PLATFORM

다방향으로 Open된 Vertical Platform은 도시와 모든 방향에서 소통하며 접점을 만들어낸다. 다양한 문화와 시민이 마주치며 만들어 내는 Surrounded Activities는 새로운 문화

Vertical Open Space

An important element of a sustainable and creative environment lies in eco-friendly open space. In order to secure the largest open space on the site, vertically expanded open spaces link and expand in all directions beyond the limits of horizontal space.

Interactive Platform

The multi-directional Vertical Platform creates a contact point by communicating with the city in all directions. Also, surrounded Activities, created by various cultures and citizens form a new cultural start-up ecosystem.

Green Pieces

Human-centered architecture aims to give a comfortable natural experience close to the building. At least one side of each facility faces nature, and the green spaces run from the city to the inside of the building, forming a space for civic exchange and relaxation.

Open Space Network

The continuous open space connected around the four open spaces forms the axis of welfare from the complex transfer center to the Arena to Jungnangcheon Stream. Open space within each site becomes a trigger to play a sustainable and creative role.

The four Green Pieces formed through the Vertical Platform, which enhances the accessibility of the land, will become a resting place for residents and a attraction for the local community along with the Sunken corridor.

First phase – Open space proposes the integrated concept of creating a city and a step-by-step urban linkage plan.

창업 생태계를 형성한다.

GREEN PIECES

인간 중심의 건축물은 건물과 가까이 있는 편안한 자연의 경험을 목표로 한다.

각각의 시설은 최소한 한 면은 자연에 접하게 되며 녹의 공간들은 도시에서부터 건물 내부까지 이어지며 시민 교류와 휴식의 공간을 형성한다

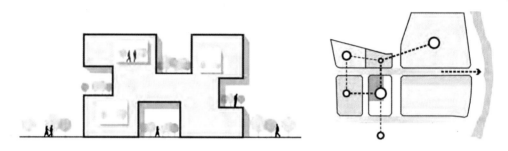

그림 5. Green Pieces

OPEN SPACE NETWORK

4개의 거점 오픈스페이스를 중심으로 연결된 연속적인 오픈공간은 복합환승센터-아레나-중랑천까지 복지의 축을 형성한다, 각 부지 내의 오픈스페이스는 지속 가능한 창조적 역할을 할 수 있는 매개장치(Trigger)가 된다

대지의 접근성을 높인 Vertical Platform을 통해 형성된 4개의 문화시설은 선큰과 함께 거주자의 휴식공간과 지역사회의 명소화가 될 것이다.

오픈스페이스 공간의 첫 번째 단계는 도시를 만들어 가는 통합 개념과 단계적 도시 연계 계획에 대하여 제안한다.

대규모 주택단지, 철로 등으로 파편화 된 도시조직 속 고립된 공간들을 연속적인 오픈스페이스로 연결한다.

The project connects isolated spaces in urban organizations fragmented by large-scale housing complexes and railroads through continuous open space.

Open space does not simply exist flat, but also exists in three-dimensional cross-sections and elevation. It also contains programs and connects cities through shared spaces.

Program Conversion

Reorganize Programs Based on Activities

It plans a new creative culture ecosystem by arranging programs connected to a lifestyle centered on movement, away from program arrangements sorted by existing fields. A music creation club room is located in front of the residency of cultural industry workers, and a sharing studio and creative practice room are located between

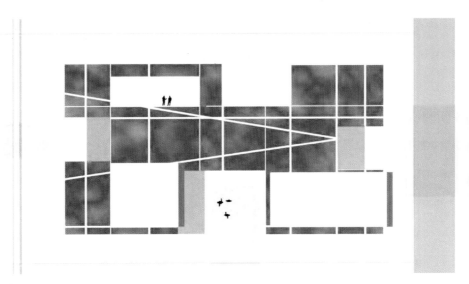

그림 6. Open Space Network

Master Plan
Continuous Open Space and Programmed Open Ground

그림 7. Master Plan

오픈스페이스는 단순히 평면적으로 존재하는 것이 아니라 3차원적인 단면, 입면에도 존재하며, 공유공간을 통해 프로그램을 담고 도시를 연결시킨다

Program Conversion

Reorganize Programs Based on Activities

기존 분야 별로 Sorting되어 있던 프로그램배치에서 벗어나 동선중심의 라이프 스타일과 연결된 프로그램 배치로 새로운 창작문화 생태계를 계획한다. 문화산업 종사자의 레지던스 앞에는 음악창작 동아리실이, 청년창업 오피스 사이는 공동작업실 및 창작실습실이 위치한다.

the youth start-up offices.

Vertically Lifted Platform

In order for cultural start-up facilities to be spatially and functionally linked and step-by-step development, people and information must be shared at each stage of growth through vertical space creation rather than horizontal space concentration. In addition, the movement is shortened vertically so that various three-dimensional shared spaces can be accessed, and a wide outer space can be created by integrating vertical facilities.

The idea of connecting urban spaces through continuous open spaces results in opening the ground floor. In order to overcome the limitations of the horizontal open space on the

그림 8. Conversion Platform

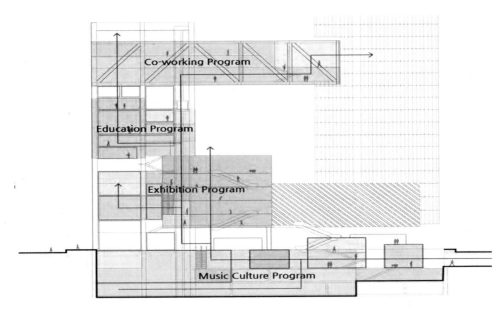

그림 9. Program Conversion

Vertically Lifted Platform

문화창업시설이 공간, 기능적으로 연계되고 성장 단계별 발전이 이루어지기 위해서는 수평적인 공간의 집약보다는 수직적 공간 조성을 통해 성장 단계별 인적, 정보 공유가 일어나도록 하는 구조로 계획했다. 또한 수직동선으로 동선이 단축되어 다양한 입체적 공유공간들을 접할수 있게 되고 수직적 시설 집적화로 넓은 외부공간 조성이 가능하다.

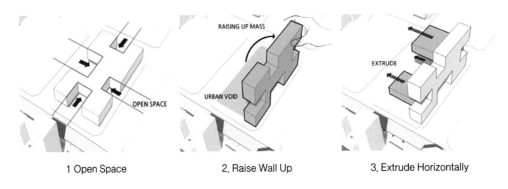

1 Open Space   2. Raise Wall Up   3. Extrude Horizontally

그림 10. Vertically Platform Design Process

Culture Palette

Cultural Programs Collaged in Sunken Level

In the underground sunken space, various cultures and behaviors are collaged in a pallet-shaped space. In addition, in the center of sunken, there is a large central walking space connecting the complex transfer center and the Arena. The central walking space collides with various cultures and reproduces new cultures through cultural exchanges. The underground space forms a second new ground layer and extends Open Space to the underground level.

The four Green Pieces, formed through the vertical platform, experience various spaces without boundaries between the ground and the basement with sunken corridor, creating a local community attraction where colorful cultural colors harmonize.

그림 11. Culture Palette

도시공간을 연속적인 Open 공간을 통해 연결하고자 하는 생각은 지상층을 여는 것으로 귀결된다 지상에 이루어지는 수평적인 Open 공간의 한계를 극복하기 위해서 건물의 축을 Vertical하게 설정하고 들어올려진 새로운 패러다임의 공간을 제시한다.

Culture Palette

Cultural Programs are Collaged in Sunken Level

지하 선큰공간은 다양한 문화와 행태가 팔레트 형태의 공간에 콜라주되어 있다. 선큰의 중심에는 복합환승센터와 아레나를 연계하는 큰 중심보행공간이 형성되어 있다. 중심보행공간은 콜라주화된 여러 가지 문화와 부딪히며 문화적인 교류를 통해 새로운 문화를 재생산한다. 지하공간은 두 번째 새로운 지상층을 형성하며 오픈스페이스를 지하 레벨까지 확장시킨다.

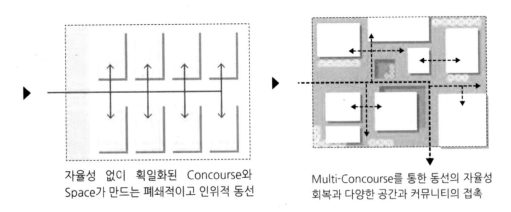

자율성 없이 획일화된 Concourse와
Space가 만드는 폐쇄적이고 인위적 동선

Multi-Concourse를 통한 동선의 자율성
회복과 다양한 공간과 커뮤니티의 접촉

그림 12. Sunken Corridor

Vertical Platform을 통해 형성된 4개의 Green Piece는 Sunken Corridor와 함께 지상과 지하의 경계없이 다양한 공간을 체험하며 그 안에서 다채로운 문화의 색감들이 어우러지는 지역사회의 명소를 만들어낸다.

Horizontal Cores Penetrate and Connect Vertical Platforms

The horizontal core inside the Open wall is the central axis and transition space of the Vertical Platform that connects rectangular spaces in connection with vertical movement. It is also a complex space that serves not only as a functional role of internal movement but also as a corridor where the community occurs and as a shade to control the amount of sunlight on the west side in consideration of an eco-friendly elevation system.

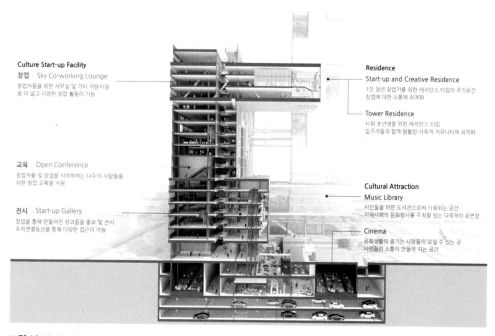

Culture Start-up Facility
창업 Sky Co-working Lounge
창업자들을 위한 사무실 및 기타 지원시설
좀 더 넓고 다양한 창업 활동이 가능

교육 Open Conference
창업자들 및 창업을 시작하려는 다수의 사람들을
위한 창업 교육을 지원

전시 Start-up Gallery
창업을 통해 만들어진 성과품을 홍보 및 전시
수직연결동선을 통해 다양한 접근이 가능

Residence
Start-up and Creative Residence
1인 청년 창업가를 위한 레지던스 타입의 주거공간
창업에 대한 소통에 최적화

Tower Residence
사회 초년생을 위한 레지던스 타입
입주자들과 함께 원활한 사회적 커뮤니티에 최적화

Cultural Attraction
Music Library
시민들을 위한 도서관으로써 사용되는 공간
지역사회의 문화행사를 주최할 있는 다목적의 공연장

Cinema
문화생활의 즐기는 사람들이 모일 수 있는 곳
시민들의 소통이 만들어 지는 공간

그림 13. Vertical platform section

View Point of Architectural Shape in The City

The image of the building seen in the Changdong and Sanggye areas becomes a building representing the rhythmical and new urban landscape image through its

Horizontal Cores Penetrate and Connect Vertical Platforms

열려 있는 내부의 수평적 코어는 수직동선과 연계하여 장방형에 배치된 공간들을 연결하는 Vertical Platform의 중심축이자 전이공간이다. 또한 내부 동선의 기능적인 역할뿐만 아니라 커뮤니티가 일어나는 복도의 역할을 하고 친환경 입면시스템을 고려해 서쪽의 일조량을 조절해주는 차양의 역할도 하는 복합적인 공간이다.

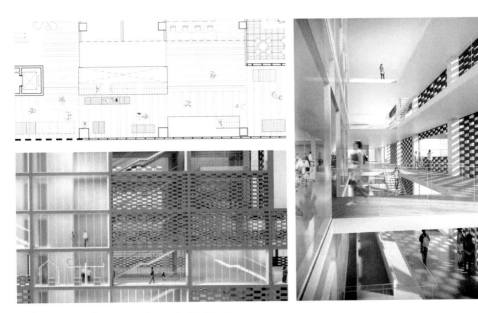

그림 14. Start- up Eco-system Connected by The Program

View Point of Architectural Shape in the City

창동·상계 권역에서 보이는 건축물의 이미지는 독특한 외관과 다양한 볼륨감을 통해 리드미컬하고 새로운 도시경관 이미지를 갖는 지역을 대표하는 건축물이 된다. 북쪽의 도로에서 마주하는 경관은 전면부의 켄틸레버 구조로 된 매스와 하부의 비움의 공간이 오픈된 공간감을 극대화한다. 남쪽 고가철로 방향에서는 타워와 건축부 사이를 열어 폐쇄적일 수 있는 공간에 개방감을 주고 고가철로와 단지가 연계된 외부공간을 조성한다.

unique appearance and various volumes. The scenery facing the road in the north maximizes the sense of open space through the mass of the cantilever structure in the front part of the open wall and the empty space in the lower part. In the south elevated rail direction, the space between the Visible Tower and the open wall is opened to give a sense of openness to a space that may be closed, and to create an external space connected with the elevated

International Competition Review

An international design contest for the Changdong-Sanggye Cultural Complex project was held through the second stage in 2018. The work of 7 participants (Siaplan Architects, Haenglim Architects, Nodo 17 Group, Toyo Ito & Associates, Cho Byeong Soo Architectural Institute, OCA, and Steven Holl Architects) selected through the first stage screening. Among them, Siaplan Architects was selected as the winner. The judges wanted to select a project that understood the characteristics of the land connecting the complex transfer center and Seoul Arena to be built in the future, and creative and high-quality project that could breathe new vitality into the northeastern part of Seoul.

The judges' evaluation of the winning works can be summarized in the following three categories.

First, the plaza was placed on the axis of the city connecting the transfer center and the Seoul Arena, and connected to the cultural and commercial spaces on the lower floors so that people could naturally enter and experience various things.

Second, in consideration of the complexity of projects led by SH but involving private

그림 15. New Landmark in the city

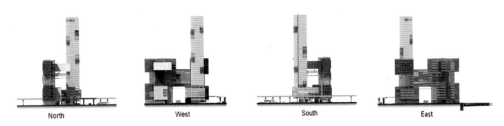

North   West   South   East

그림 16. Elevations

국제공모 심사평

창동상계 문화복합단지사업을 위한 국제설계공모가 2018년 2단계를 통해 개최되었다. 1단계 심사를 거쳐 선정된 시아플랜 건축사사무소, 행림종합건축사사무소, Nodo 17 Group, 과 지명초청 업체인 Toyo Ito & Associates, 조병수건축연구소, 건축사사무소 OCA, Steven Holl Architects, 총 7개 참여자의 제출 작품을 심사하여 당선안 으로 시아플랜 건축사사무소안이 선정되었다. 심사위원들은 향후 건설될 복합환승센터와 서울아레나를 잇는 대지의 특징과 여러 사업자가 참여하는 사업의 복합성을 이해하고, 서울 동북부지역에 새로운 활력을 불어넣을 수 있는 창의적이면서 완성도가 높은 안을 선정하고자 하였다.

REIT operators in accordance with Seoul's urban rehabilitation plan, the mass and movement of start-up spaces, residential spaces, and cultural commercial spaces were clearly separated horizontally and vertically so that they could flexibly respond to future conditions.

Third, a simple refined geometric type of building was expected to become a new urban architecture visually and spatially without overwhelming the surrounding urban landscape.

The judges selected the Conversion Platform, which has the highest publicity and feasibility, as the winning project, expecting economic, cultural, commercial, and residential attractions to be built in the northeastern part of Seoul in the era of urban regeneration.

Closing Remarks

SH Seoul Housing & Urban Corporation, the operator of the Changdong and Sanggye Startup Culture Industrial Complex projects, has established the Changdong Urban Regeneration REITs Project Group, an asset management company, to carry out the project. The Changdong Urban Regeneration REITs Project Group has a REITs project structure in which Seoul Urban Corporation participates with land shares, and HUG, which invests in the Housing and Urban Fund, and Seoul Asset Management Co., Ltd. participate as asset investors.

In the future, after the completion of the biomedical knowledge industry center cluster linked to the Hongneung Bio Complex in the Changdong vehicle base, and the Seoul Arena Complex and the Changdong Sanggye Startup Culture Industrial Complex

[1]심사위원이 당선작을 평가한 내용은 다음 세 가지로 요약할 수 있다.

첫째, 환승센터와 서울아레나를 연결하는 도시의 축선 상에 광장을 배치하고, 저층부의 문화, 상업공간과 연결하여 사람들이 자연스럽게 안으로 들어와 다양한 경험을 할 수 있도록 하였다.

둘째, 서울의 도시재생활성화계획에 따라 SH가 주도하는 공공사업이면서도 민간 리츠(REIT)사업자가 참여하는 사업의 복합성을 고려하여, 창업공간, 주거공간, 문화상업공간의 매스와 동선을 수평, 수직으로 명확히 분리하여 향후 여건에 유연하게 대응할 수 있도록 하였다.

셋째, 단순하면서도 정제된 기하학적 형태의 건축물은 주변 도시경관을 압도하지 않으면서 시각적 공간적으로 새로운 도시건축이 될 것으로 보았다.

심사위원은 도시재생의 시대에 서울 동북부에 경제, 문화, 상업, 주거의 명소가 들어설 것을 기대하면서 공공성과 실현 가능성이 가장 높은 Conversion Platform을 당선작으로 선정한 것이다.

맺으며

창동·상계 창업 및 문화산업단지 조성사업의 시행주체인 SH서울주택도시공사는 사업시행 및 관리감독을 위한 창동도시재생리츠사업단(자산관리회사)을 설립하였다.

또한 기술제안방식공모를 통해 실시설계와 책임준공사인 DL이앤씨 컨소시움을 선정하고 2019년 9월 착공, 2023년 5월 준공을 목표로 공사가 한참 진행 중이다.

향후 주변의 서울아레나 복합문화시설, 세대 융합형 복합시설 및 공공문화시설 등이 들어서면 서울 동북부 창업문화산업의 인프라가 구축되고, 창동역의 GTX 노선 광역연결망을 통해 문화와 스마트산업의 혁신성장의 거점으로 서울의 지역균형발전 및 선순환 산업생태계가 완성될 것으로 기대되어진다.

---

1 PROJECT SEOUL [2단계] 창동·상계 창업 및 문화산업단지 조성사업 국제설계공모 (seoul.go.kr)

will create 1,000 new businesses (including 500 start-up space utilization companies) and 80,000 new jobs. In this case, it is expected that the Changdong offset area will establish itself as a new center of new economy in the northeastern region of Seoul and Gyeonggi-do through the KTX and GTX regional networks at Changdong Station.

그림 17. A New cultural place "Conversion Platform"

구자훈

한양대학교 도시대학원 교수로 재직 중이며, 기획재정부 중장기전략위원회 위원, 국토부 도시재생 실무위원회 위원, 공공지원 민간임대단지 통합 심위위원장, 창동·상계도시재생 총괄CD를 담당하고 있다.

김영석

현재 건국대학교 건축대학 건축학부 교수로 재직하고 있으며, 서울대학교 및 하버드대학교에서 도시설계 및 도시계획을 전공하였다. 뉴욕의 컨설팅회사인 Sam Schwartz PLLC에서 뉴욕 월드트레이드센터의 보행환경, 미국 44대 오바마 대통령 취임식 보행자 계획 등을 책임졌었고 건축 및 도시설계 컨설팅을 주로 하는 다양한 실무를 경험하였다. 2011년부터 건국대학교에 재직하면서 건축학부와 건축전문대학원에서 졸업 설계를 담당하고 있으며, 도시와 건축의 접점에서 이론과 실무의 간극을 줄이기 위해 교육, 연구 및 실무분야에서 노력하고 있다. 현재 대한건축학회 교육원에서 초/중학생 건축교육을, 한국도시설계학회 국제교류위원회 위원장을 담당하고 있다.

김형일

단국대 건축학과를 졸업하고 동대학원에서 석사학위 취득 후 미국 일리노이 공대(Illinois Institute of Technology)에서 건축학 박사학위를 취득하였다. 2005년부터 일리노이 공대 건축학부에서 교수로 재직하였으며, 2012년부터 2018년까지 싱가포르 국립대 건축학과에서 교수로 재직하면서 동서양의 건축문화를 체험하고 건축도시를 연구하였다. CTBUH(Council on Tall Building and Urban Habitat) Singapore Chapter 창립멤버였으며 Singapore BIM 혁신연구소 센터장을 역임했다. 2018년 귀국하여 삼육대 건축학과에서 외래교수를 역임하고 있으며, MA 건축사 사무소에서 건축설계 및 건축설계공모 관리 전문인력으로 LH공사, SH공사, 서울시 등 공공기관들과 협력하며 공공주거를 비롯한 문화시설과 도시공간을 대상으로 이론과 실무를 병행하고 있다.

Jahoon Koo

He is a professor at Hanyang University's Graduate School of Urbanism, a member of the Mid- to Long-Term Strategy Committee of the Ministry of Strategy and Finance, a member of the Ministry of Land, Infrastructure and Transport's Urban Regeneration Working Committee, chairman of the Integrated Committee for Public-Supported Private Rental Complex, and Changdong-Sanggye Urban Regeneration CD

Youngsuk Kim

He studied urban design and planning at Seoul National University and Harvard University, and has worked and researched on the relationship between architecture, the site, and the city. While teaching both in the college of architecture and the graduate school of architecture at Konkuk University, he has been a leading figure in various international workshops as well as the architectural education program for young students at the Architecture Institute of Korea. Currently, he is a director of the Urban Design Institute of Korea and the chairman of the International Exchange Committee.

Hyeongil Kim

After graduating from Dankook University's Department of Architecture, he obtained a master's degree from the same graduate school, and then obtained a PhD in architecture from the Illinois Institute of Technology, USA. He served as a professor in the College of Architecture at the Illinois Institute of Technology from 2005 to 2009 and as a professor at the Department of Architecture at the National University of Singapore from 2012 to 2018, experiencing the architectural culture of the East and the West and researching architecture and cities. He was a founding member of the CTBUH (Council on Tall Building and Urban Habitat) Singapore Chapter and served as the director of the Singapore Center of Excellence (COE) in BIM Integration. After returning to Korea in 2018, he has served as an adjunct professor at the Department of Architecture at Sahmyook University. As a specialist in architectural design and design competition management at the MA Architects' Office, he works in cooperation with public institutions such as SH Corporation, LH, and Seoul City Hall, and combines theory and practice with public housing, cultural facilities, and urban spaces.

김호정

김호정은 현재 단국대학교 건축학부에 재직 중이며 건축 설계 스튜디오 및 인간행태론, 건축실무 관련 교과목을 가르치고 있다. 한양대학교에서 학부를 마치고 정림건축에서 3년간의 실무경험 이후 미국으로 건너가 M.I.T에서 건축학 석사학위(M. Arch)를 마쳤다. 이후 Gwathmey Siegel Architects (New York), Goody, Clancy & Associates (Boston)에서 건축가로 일하면서 종교시설, 교육시설 및 도서관 프로젝트를 진행하였다. 2004년 귀국 후 강원대학교에 재직하면서 홍천 광덕마을 공동체 팬션, 홍천 허브밸리, 웰니스 타운 계획과 같은 건축 설계 작업을 병행하였고, 2009년 단국대학교로 전직하였다. 건축과 도시를 만드는 중간영역에 대한 폭넓은 관심과 더불어 친환경 설계방법론, 성능 기반 디지털 디자인방법론에 대한 연구를 계속해 오고 있다.

김환용

중앙대학교 건축학과를 졸업 후 한양대학교 건축학 석사, 미국 텍사스주 소재 University of Texas at Austin에서 도시 및 지역계획 석사, Texas A&M 대학교에서 도시 및 지역계획 박사학위를 취득하였다. 학위를 마친 후 Texas A&M 대학교 도시계획조경학과 전임강사로 2년간 재직하였으며, 2015년 한국으로 귀국하여 인천대학교 도시건축학부를 거쳐 현재 한양대학교 ERICA 건축학부에서 건축학 전공 주임교수를 맡고 있다. 2020년 1월부턴 태국 방콕은행에서 후원하는 Thammasat 대학교 도시설계국제프로그램 리서치 펠로우도 겸직 중에 있다. 지리정보분석, 빅데이터 분석, 스마트시티, 디지털트윈 등 실증분석에 기반한 도시계획과 도시설계를 연구하고 있으며, GTX-B노선 환승센터 총괄계획단 건축MP, 군포/안산/의왕 3기 신도시 Urban Concept Planner로도 활동 중에 있다. 한국도시설계학회 이사, 한국 BIM학회 이사로 참여하며 한국연구재단, 국토교통과학기술진흥원, 국토지리원, 한국토지주택공사 등 다양한 기관과 함께 연구개발 프로젝트를 수행 중에 있다.

오 다니엘

미국 버클리대에서 조경설계를 공부하고, 하버드대에서 조경설계 석사(MLA), 도시계획 석사(MUP) 학위 취득 후 Skidmore, Owings and Merrill(SOM) Hong Kong, London, Bahrain, 그리고 AECOM New York에서 도시설계 실무를 수행하면서 중국, 동남아, 유럽, 북미까지 다양한 지역의 도시설계 특히 중심업무지구, 산업단지, 관광단지, 국가전략개발계획까지 다양한 규모의 도시설계 프로젝트를 전담하였다. 현재 고려대학교 건축학과에서 도시계획과 도시설계 관련 수업 및 연구를 수행하고 있으며, 고려대학교 스마트도시학과 주임교수와 고려대학교 도시재생 교육센터장을 겸직하고 있다. 도시재생, 스마트도시, 탄소중립도시 등 다양한 미래도시 패러다임 변화와 함께 변화하는 공공공간의 역할과 스마트화, 포용성과 타분야와의 융합방안에 대해 중점으로 연구하고 있다.

Hojeong Kim

Ho-Jeong Kim is a faculty member of Architecture at Dankook University, teaching architectural design studios, human behavior theory, and professional practice. After completing her undergraduate degree at Hanyang University, she gained three years of practical experience at Jeonglim Architecture. Afterwards, she moved to the United States and completed her Master's Degree in Architecture (M. Arch) at M.I.T. Afterwards, she worked as an architect for Gwathmey Siegel Architects (New York), Goody, Clancy & Associates (Boston), where she worked on religious facilities, educational facilities and public library projects. After returning to South Korea in 2004, while working at Kangwon National University, she worked on architectural design projects such as Hongcheon Gwangdeok Village Community Center and Village Hotel, Hongcheon Herb Valley, and Wellness Town. She transferred to Dankook University in 2009. Along with broad interest in the middle area that creates architecture and cities, she has continued to research environment-friendly design methodologies and performance-based digital design methodologies.

Hwanyong Kim

Dr. Hwanyong Kim is an associate professor in the School of Architecture and Architectural Engineering at Hanyang University ERICA. He joined ERICA in 2020 fall semester. Prior to ERICA, Dr. Kim served as a professor at Incheon National University and a visiting assistant professor in the Department of Landscape Architecture & Urban Planning at Texas A&M University in the U.S., where he had finished his Ph.D. in Urban and Regional Sciences. Dr. Kim studied architecture as his undergraduate degree at Chung-Ang University, S. Korea and finished his masterdegree in community & regional planning at the University of Texas at Austin, U.S. He holds another masterdegree in architecture design from Hanyang University, S. Korea. His research interests include geographic information systems (GIS), building information modeling (BIM), smart city development, AR/VR and drone applications, land use planning, and sustainable urban simulations.

Daniel Oh

Upon receiving a Master of Landscape Architecture (MLA) and Master of Urban Planning (MUP) from Harvard University, he gained professional experience as an urban designer at Skidmore, Owings, and Merrill (SOM) Hong Kong, London, Bahrain, and AECOM New York. His urban design experience ranges in scale and program from central business districts to national strategic development plans and in locations like China, Southeast Asia, Europe, and North America. In 2010, he started his academic career in South Korea, and currently, he's an Associate Professor in the Department of Architecture at Korea University. He is also serving as the Chair of the Department of Smart City and the Director of the Center for Urban Regeneration at Korea University. His research and professional focus have been on the valorization of public spaces in the rapidly evolving urban paradigms such as urban regeneration, smart city, inclusive city, and circular city

우신구

서울대학교 건축학과에서 학사, 석사, 박사과정을 밟았다. 현재 부산대학교 건축학과에서 도시건축, 건축과 사회, 설계과목을 가르치면서 도시건축연구실을 지도하고 있다. 대학이 소재하는 부산이라는 도시의 다양한 건축과 도시공간을 주요한 테마로 마을만들기, 공공공간, 도시경관 그리고 도시재생 분야의 연구와 실천을 진행하고 있다. 그동안 참여한 프로젝트로는 광복로일원 시범가로사업, '행복한 도시어촌 청사포만들기' 국토환경디자인 시범사업, 부산 서구 아미·초장 도시재생사업을 비롯하여 다양한 공공 프로젝트에서 총괄계획가와 총괄코디네이터로 참여하였다. 도시건축연구실을 중심으로 부산의 광복동과 서면 등의 공공공간, 서동과 반송동을 비롯한 정책이주지, 원도심의 초량동, 수정동, 영주동 등 산복도로 지역, 사하구 감천문화마을과 서구의 아미동 비석문화마을 등 도시마을에 대한 지역 리서치를 진행하여 아카이브를 구축하거나 단행본으로 출간하고 있다. 현재 한국도시재생학회 회장과 국무총리 직속 도시재생특별위원회 민간위원으로 봉사하고 있다.

이재민

이재민 교수는 울산대학교 건축학과에 재직중이며 도시의 형태와 공공공간에 대한 연구를 수행 중이다. 주요 연구 주제로 도시설계, 도시재생, 생태도시설계, 도시설계 정량분석 등이며 새로운 도시설계 수법에 관심을 가지고 있다. 이재민 교수는 중앙대학교 건축학사, 미시간대학교 도시설계 석사, 도시계획 석사, 유펜대학교 도시계획학 박사를 취득했고 Skidmore, Owings and Merrill LLP 시카고, 뉴욕사무소에서 도시설계 실무를 수행했다.

임동원

홍익대에서 건축을 공부하고, 네덜란드 델프트 공대에서 도시계획 석사학위를 받았다. 원도시건축, 삼성물산 등을 거쳐 현재 도화엔지니어링 상무로 재직 중이다. 신도림역 테크노마트 (2002), 중국 서안 Xian스마트시티(2014), 케냐 나이로비 중앙역 마스터플랜(2020) 등을 수행하였다.

Shinkoo Woo

He completed his bachelor's, master's and doctorate degrees in architecture at Seoul National University and is currently professor at the Department of Architecture at Pusan National University. Focusing on the various issues in Busan Metropolitan City where the university is located, he and his Lab. of Urban Space and Architecture have been conducting researches and practices in the fields of community design, public space, townscape and urban regeneration. He has been participated as master architect and general coordinator in various public projects including the Gwangbok Street Pilot Project, the "Making Happy Urban Fishing Village Cheongsapo" and the Ami-Chojang Urban Regeneration Project in Busan. Currently, he is the chairman of the Korea Urban Regeneration Association and a civilian member of the National Committee of Urban Regeneration led by the Prime Minister.

Jaemin Lee

Jae Min Lee is an assistant professor in the School of Architecture of the University of Ulsan. His research interests include non-conventional and underground public space, ecological modeling of cities and regions, and bridging the gap between quantitative and qualitative research traditions in urban design. He is an emerging scholar at the PENN Institute for Urban Research, and an external faculty collaborator at the Center for Environmental Building & Design, University of Pennsylvania. He has worked in a range of city building projects as an urban design associate in the Chicago and New York offices of Skidmore, Owings, and Merrill LLP

Dongwon Lim

After graduating from the Department of Architecture at Hongik University, Lim received his master's degrees of Urbanism MsC from the TU-Delft. Through the carieer of Wondoshi Architects, Samsung C&T, and Dohwa, Lim experienced Shindorim Technomart (Urban UEC, 880,000m^2, 2002), Xian Smart city (China, 120ha, 2012), Nairobi Railway city (175ha, Kenya, 2020) as main projects.

장항준

홍익대 건축학과에서 학사. 고려대 건축공학과에서 석사를 수학했다. 특히 인간의 감성과 자연, 기술의 융복합 디자인 분야에 관심을 갖고, 친환경부문의 작품수상과 다양하고 창의적인 작품을 수행하고 있으며. 시아플랜건축에서 디자인총괄로서 디자인 경쟁력을 이끌어왔으며, 현재는 MDA건축사사무소의 대표이사로 재직 중이다. 주요작품으로는 테스코 아시아 리더십 아카데미(2013 녹색건축대전 최우수상), 창동·상계 문화복합단지 국제공모(2018) 당선작 등이 있다.

최순섭

서울대학교 건축학과를 졸업하고 동대학원에서 석사 및 박사학위를 받았다. '건축 통합적 도시설계'라는 박사논문 주제를 통해 건축과 도시설계의 지속가능한 연계방식을 연구하였고, 2013년에는 캐나다 The University of British Columbia(UBC)의 방문연구원으로 초빙되어 참여형 커뮤니티 계획방식인 '디자인샤렛' 연구를 진행하였다. 이후 2014년부터 한국교통대학교 건축학부 교수로 근무하고 있으며 충주시 도시재생 사업 총괄 코디네이터 및 현장지원센터장(2015-2017)과 경기도 정주환경 개선사업의 총괄 코디네이터(2018-2019)를 역임하였다. 주요 연구로는 '사회실험 적용형 도시재생사업 체계', 'Heritage Density Transfer를 통한 건축자산 보존 방식', '지역재생을 위한 D.I.Y Spirit 개념과 실행 방식', '도시재생사업에서 유휴 국공유지 활용의 현장적 한계' 등이 있으며 저서로 '도시재생 후진지 되지 않기'가 있다. 현재는 지방중소도시의 스마트 축소를 위한 철도역과 역세권 적정 및 복합화계획 연구를 진행하고 있다.

Hangjoon Jang

Bachelor of Architecture at Hongik University. I studied master's degree in architectural engineering at Korea University. In particular, he is interested in the field of convergence design of human sensibility, nature, and technology, and is carrying out various and creative works by winning works in the eco-friendly sector. He has led design competitiveness as a design general manager in SIAPLAN Architecture, and is currently serving as the CEO of the MDA (MokYang Design Architects) Architecture Office. Major works include Tesco Asia Leadership Academy (2013 Green Architecture Competition Best Award) and Changdong Sanggye Cultural Complex International Competition (2018).

Soonsub Choi

After graduating from the Department of Architecture at Seoul National University, Choi received his master's and doctoral degrees from the same graduate school. Through the subject of his thesis, Integrated Urban Design', he explored the sustainable connection between architecture and urban design, and he was invited as a visiting researcher at The University of British Columbia (UBC) in 2013 to conduct cooperative researches on Charrette', a participatory community planning method. Since 2014, he has worked as an associate professor of The School of Architecture at Korea National University of Transportation, and joined the Urban Regeneration Project in Chungju (2015-2017) and the Settlement Environment Improvement Project in Gyeonggi-do (2018-2019) as a general coordinator. His major studies include Regeneration Project System applied with Social Experimental Projects', 'Preservation Method of Architectural Asset through Heritage Density Transfer', 'Concept and Implementation Method of D.I.Y Spirit for Local Regeneration', and 'Limits of Idle Public Land Utilization in Urban Regeneration Project Fields'. He also published a book titled from U.R. Field', which aims to change the mind of participants and stakeholders for the success of Urban Regeneration Project. Currently, Choi is conducting a research on optimization method of area and complex use planning for Smart Shrinkage centered on railway stations and station areas in local small and medium-sized cities.

최혜영

서울대 조경학과를 졸업하고 (미)펜실베니아대학교에서 조경학 석사학위, 다시 서울대학교에서 박사학위를 받았다. 서울시 공공미술위원회 위원, 서울시 교육청 공공건축심의위원회 위원으로 활동 중이다. 공저서로 용산공원 설계 국제공모 출품작 비평(2013), 공간닥터 프로젝트 vol.1(2020), 도시설계의 이해: 실무편(2020)이 있다. 최혜영은 뉴욕AECOM(전 EDAW), West8 뉴욕 사무실에서 전 세계 다양한 문화권의 유수 프로젝트를 담당했으며 대한민국 최초의 국가도시공원으로 조성되고 있는 용산공원 프로젝트의 국제공모전에서 팀의 당선을 이끌었다. 2012년부터 용산공원 기본설계 및 공원조성계획 수립 프로젝트를 포함해 용산공원 관련 다수의 프로젝트에 참여해왔다. 설계 과정의 경험을 토대로 용산 공원에 관련된 다양한 연구를 수행했거나 수행 중에 있으며 공원 접근성, 공원 아카이브 등으로 연구 주제를 확장하고 있다.

한광야

연세대에서 건축을 공부하고, 하버드대에서 도시설계 석사(MAUD), 펜실베이니아대에서 도시계획 박사(PH.D.) 학위를 받았다. 도시설계와 도시계획 분야의 학자이며 실무자로서 보스턴 Cecil and Rizvi 설계사무소와 필라델피아 Wallace Roberts and Todd 설계사무소에서 도심 블록설계부터 지역권 생태환경계획까지 프로젝트들을 수행했다. 현재 동국대 건축공학부 도시설계 전공 교수이며, 한반도의 물리적 국토계획, 동국대 캠퍼스-커뮤니티 계획, 서강대 남양주 캠퍼스 구상 등을 수립했고, 서울 잠실 도시재생 국제현상설계(2015)에서 당선했다. Global Universities and Urban Development, 도시설계, 대학과 도시, 동남아 도시들의 진화 등의 책을 쓰고 번역했다. 서울시 해방촌 도시재생과 신당5동 도시재생의 총괄계획가로 활동하고 있으며, 철도 메트로와 도시개발, 도시 네이버후드의 진화 등의 연구를 진행하고 있다.

Hyeyoung Choi

Hyeyoung Choi was in charge of emerging projects in various cultures worldwide at AECOM (formerly EDAW) New York and West8 New York office. In 2012, she led the team to win the international competition for the Yongsan Park Project, which is being built as Korea's first national urban park. Since then, she has been involved in a number of projects related to Yongsan Park, including the establishment of general park planning guidelines, setting up the park master plan, and the development of the schematic design. Based on her experience in the design process, she has conducted various research related to Yongsan Park, with topics of park accessibility, public engagement, and park archives.

Hyeyoung Choi received her master's degree in landscape architecture from the University of Pennsylvania (U.S.) and a doctorate from Seoul National University. She is a board member of the Korean Institute of Landscape Architecture and a member of the Seoul Public Art Committee. She wrote several books with co-authors, including Criticism of the Submissions from the International Competition for Yongsan Park Master Plan (2013), The Space Doctor Project vol.1 (2020), and Understanding Urban Design: A Practice (2020).

GwangYa Han

After studying architecture at Yonsei University, he received his M.A.U.D. from Harvard University and his Ph.D. in City and Regional Planning from the University of Pennsylvania. As a scholar and practitioner in urban design and planning field, he has worked on projects ranging from urban block design to region-scaled environment planning with Cecil and Rizvi Design in Boston and Wallace Roberts and Todd in Philadelphia. Currently, he is a professor in the Department of Architectural Engineering, Dongguk University, and has prepared a series of master plans including the physical plan of the Korean Peninsula, Dongguk University campus-community plan, and Sogang University Namyangju campus plan, and co-won the honorable award at the Seoul Jamsil Urban Regeneration Int'l Design Competition (2015). Professor Han has written and translated books, including Global Universities and Urban Development, Urban Design, Universities and the Cities, and the Evolution of Cities in Southeast Asia. He has worked with the communities as well as local governments and Seoul Metropolitan Government as Master Planner for urban regeneration projects in Haebang-chon and Sindang-dong in Seoul, and is conducting research on urban railroad-metro system and urban development, and the evolution of urban Neighborhood.

한국의 도시재생
Urban Regeneration in Korea
도시를 살리는 다섯 가지 해법

초판 1쇄 인쇄 2022년 04월 25일
초판 1쇄 발행 2022년 04월 29일

—

지은이 한국도시설계확회
펴낸이 김호석
펴낸곳 도서출판 대가
편집부 주옥경·곽유찬
디자인 전영진
마케팅 오중환
경영관리 박미경
영업관리 김경혜

—

주소 경기도 고양시 일산동구 장항동 776-1 로데오메탈릭타워 405호
전화 02) 305-0210
팩스 031) 905-0221
전자우편 dga1023@hanmail.net
홈페이지 www.bookdaega.com

—

ISBN 978-89-6285-356-8 (93530)